Parasites:
immunity and pathology

For

Malgosia, Kasia and Wojtek

Parasites: immunity and pathology

The consequences of parasitic infection in mammals

Edited by

Jerzy M. Behnke

University of Nottingham, UK

Taylor & Francis
London · New York · Philadelphia
1990

UK   Taylor & Francis Ltd, 4 John St, London WC1N 2ET

USA   Taylor & Francis Inc., 1900 Frost Rd, Suite 101, Bristol, PA 19007

British Library Cataloguing in Publication Data
Parasites.
    1. Mammals. Parasites
    I. Behnke, J. M.
    599.02

    ISBN 0–85066–499–3

Library of Congress Cataloging-in-Publication Data
Parasites, immunity and pathology: the consequences of parasitic infection in mammals / edited by Jerzy M. Behnke
      p.   cm.
    Includes bibliographical references.
    ISBN 0–85066–499–3: $55.00
    1. Parasitic diseases. 2. Parasitology. 3. Mammals—Parasites.
I. Behnke. Jerzy M.
    [DNLM: 1. Host–Parasite Relations. 2. Parasites—immunology.
3. Parasitic Diseases—immunology. 4. Parasitic Diseases—
pathology. WC 695 P2231]
    QR201.P27P367 1989
    599′.0233—dc20
    DNLM/DLC
    for Library of Congress   89–20403
                                               CIP

Printed in Great Britain by Burgess Science Press, Basingstoke
on paper which has a specified pH value on final paper
manufacture of not less than 7.5 and is therefore 'acid free'.

Contents

v

Preface

Parasites have been with us since the dawn of life, representing perhaps the most diverse group of organisms inhabiting any single, identifiable habitat and arguably reflecting the most popular life style adopted by living things. For obvious reasons the attention of parasitologists has focused on those species which are of medical, veterinary or economic importance, and aspects such as pathology and control of parasites have figured prominently among research priorities. Taxonomy and the elucidation of life cycles have also played an important, at times a dramatic, role in the history of parasitology. In more recent years the subject has been through apparent phases when trendy, often inspired, innovatory techniques have dominated the field, until superseded by the next generation.

Parasitology has blossomed as a result of the development of new technologies and as a result of collaborative studies across formally distinct academic disciplines. Nevertheless, much of modern day parasitology is still taught with a view to controlling those species which are a nuisance to man. Funds for research emphasize the need for practical relevance. Far less attention has been devoted to species which affect animals of little consequence to human life; many of the latter have evolved stable and non-pathogenic relationships with their hosts. It is not widely appreciated that the problems which we encounter through parasitic diseases are often of our own making, in the sense that the development of human civilization has proceeded at a pace considerably faster than that under which natural evolution operates. Analysis of host-parasite systems not affected by human interference, can illustrate how regulation of host and parasite populations evolved under the influences of natural selection. The study of parasite biology, without clear 'applied' objectives, has much to offer.

The first chapter in this book attempts to draw to the attention of classic parasitologists the world beyond the species of direct relevance to man and emphasizes that many of the pathological consequences of infection may be predictable if considered in the evolutionary context. In situations where there is little interference from man, parasites achieve a balance with their hosts which may be of benefit to both participants when viewed at the population level.

Parasitology, like all sciences, has grown considerably, and a number of distinct approaches are available for teaching and illustrating the scope of the subject. This book is primarily concerned with the interrelationship of parasites, immunity and the resultant pathology. To aid in this objective the text has been structured in an approach which does not follow the classic style of considering each phylogenetic group of parasites in turn. Rather, topics linking a variety of taxonomic groups have been identified and reviewed within each chapter. This approach, of course, poses problems peculiar to itself, not the least of which is the risk of repetition of life cycles throughout the book. In order to save on length, we have not presented detailed figures of all the parasite life cycles, and to avoid excessive overlap, life cycle illustrations accompany the first detailed consideration of the species in the text. Where necessary the contributors have given additional brief information relating to the biology of the topical species. Further, to simplify matters for the non-parasitologist, the index identifies sections in which the reader may find further details of particular life cycles, but on the whole each chapter has been written assuming reasonable familiarity with parasite biology and certainly with the organization and function of the immune system itself. A number of very useful books have been published on these subjects in recent years, and the reader is directed to relevant volumes for further details. Lani Stephenson's (*The Impact of Helminth Infections on Human Nutrition. Schistosomes and Soil-Transmitted Helminths*) and David Crompton's (*Human Helminths*) books, both in this series, give excellent life cycle summaries.

It was my intention when organizing the volume that we should produce a readable, up-to-date text within easy reach of students at the advanced undergraduate level. With this objective, the book could not have been totally comprehensive and the approach, therefore, is deliberately selective, but the degree of selectivity has varied among the contributors. It was never my intention that the entire subject should be covered at great depth and hence the better-studied systems have been discussed in more detail, but the balance has shifted from one chapter to the next depending on the topic and the interests of the contributor. It was my hope that the finished volume would stimulate interest and excitement for the subject among the student population and that it would serve to encourage new recruits to a research field for which all the contributors share a deep personal involvement.

Finally, I would like to express my very deep appreciation to my coauthors, particularly for tolerating my comments and responding to my suggestions for alterations. I am extremely grateful to all my colleagues in Nottingham who have been most patient with me during the period of preparation and who have helped me to review the contributions. I would particularly like to express my gratitude to Derek Wakelin and Tony Lammas for their constructive comments. My postgraduate students have been, over the years, a constant source of stimulation and inspiration to me and have reinforced my own interest in parasitology. Their understanding and active contribution to the research activities of the Department of Zoology is beyond measure.

Without their natural inquisitiveness and motivation, little would have been possible. My final words of thanks go to Professors P.N.R. Usherwood and D. Wakelin who have incessantly and unselfishly supported the development of parasitology as a major research interest of the department and to whom I am immeasurably indebted for the opportunity of participating in the process.

Jerzy M. Behnke,
Nottingham, August, 1989.

Contributors

C.J. Barnard, Behavioural Ecology Research Group, Department of Zoology, University of Nottingham, University Park, Nottingham NG7 2RD, UK.

J.M. Behnke, MRC Experimental Parasitology Research Group, Department of Zoology, University of Nottingham, University Park, Nottingham NG7 2RD, UK.

G.A. Castro, Department of Physiology and Cell Biology, Medical School, University of Texas Health Science Center, Houston, Texas 77225, USA.

I.A. Clark, Department of Zoology, Australian National University, Canberra ACT, Australia.

F.E.G. Cox, Division of Biomolecular Sciences, King's College London (KQC), University of London, 26–29 Drury Lane, London WC2B 5RL, UK.

K. Crook, MRC Experimental Parasitology Research Group, Department of Zoology, University of Nottingham, University Park, Nottingham NG7 2RD, UK.

R.K. Grencis, Immunology Group, Department of Cell and Structural Biology, Stopford Building, University of Manchester, Manchester M13 9PT, UK.

D.M. Haig, Department of Immunology, Moredun Research Institute, 408 Gilmerton Road, Edinburgh, EH17 7JH, UK.

M.J. Howell, Department of Zoology, Australian National University, Canberra ACT, Australia.

A.E. Keymer, Department of Zoology, University of Oxford, South Parks Road, Oxford OX1 3PS, UK.

A.J. MacDonald, Department of Allergy and Clinical Immunology, The Cardiothoracic Institute, Brompton Hospital, Fulham Road, London SW3 6LY, UK.

R.M. Maizels, Department of Pure and Applied Biology, Imperial College of Science, Technology and Medicine, Prince Consort Road, London SW7 2BB, UK.

D.J. McLaren, Division of Parasitology, National Institute for Medical Research, Mill Hill, London NW7 1AA, UK.

H.R.P. Miller, Department of Pathology, Moredun Research Institute, 408

Gilmerton Road, Edinburgh, EH17 7JH, UK.

R. Moqbel, Department of Allergy and Clinical Immunology, The Cardiothoracic Institute, Brompton Hospital, Fulham Road, London SW3 6LY, UK.

R.J. Quinnell, Department of Zoology, University of Oxford, South Parks Road, Oxford OX1 3PS, UK.

R.A. Wilson, Department of Biology, University of York, York YO1 5DD, UK.

1. Coevolution of parasites and hosts: host-parasite arms races and their consequences

J.M. Behnke and C.J. Barnard

1.1. INTRODUCTION

The relationship between a parasitic organism and its host's immune system is an extremely complex one. It represents but one aspect of the parasite's overall interaction with its host, and this like any other ecological relationship has many different facets all of which collectively constitute the characteristics of the parasite's niche (Kennedy, 1976). As experimental techniques have improved so our knowledge of the host response and the effector mechanisms which it unleashes has expanded, and this has been partially matched by an increasing awareness of the strategies used by parasites to evade host immunity. The conflict between a parasite and its host's immune system is in some respects a unique aspect of ecology; free-living animals do not contend with a concerted strategy by the environment to terminate their existence in a manner which is quite comparable with the highly specific mammalian response to micro- and macroorganisms (Wakelin, 1976).

1.2 HOST-PARASITE ARMS RACES

Hosts and their parasites can be thought of as being involved in an evolutionary arms race (see Dawkins and Krebs, 1979; Dawkins, 1982), a series of escalating mutual counter-adaptations by the two lineages to exploit or inhibit

1

exploitation by the other. Adaptations by hosts to avoid or resist infection by the parasite result in selection pressure on the parasite to circumvent them. However, while it seems clear that selection will produce such processes of response and counter-response, the rapidity with which each lineage should respond to the other on an evolutionary timescale and the extent to which it should respond are less obvious.

Host-parasite arms races are asymmetrical in the sense that the currency of adaptation in the two lineages is different (Dawkins and Krebs, 1979). Using a literal arms-race analogy, they are akin to a 'sharper swords by party A / thicker shields by party B' race rather than a symmetrical race along the lines of '100 megaton bombs by party A / 200 megaton bombs by party B' (Barnard, 1983). Since different adaptive currencies are used, asymmetric races are likely to have built-in differences in the pace and magnitude of change by the racing lineages simply because of differences in the metabolic pathways involved. However, even if differences in adaptive currency between lineages did not themselves lend an edge to one of them, there are other reasons why one lineage may have an advantage over the other. In the case of host-parasite arms races, the dice are likely to be loaded in favour of the parasite. Following Dawkins and Krebs (1979), we can identify at least three sources of imbalance.

Differences in evolutionary rates

In interspecific arms races, one lineage may have an inherently greater rate of evolution than another, perhaps due to differences in genetic diversity or developmental pathways. There is evidence that different taxonomic groups have evolved at different rates in the past (see Stanley, 1979). In a race between lineages with differing rates of evolution, the faster evolving lineage may be at an advantage since it would enjoy a longer period of grace before an adaptive step is countered. One factor that may bias relative evolutionary rates in favour of parasite lineages is differences in generation times between host and parasite. In those parasites which have shorter generation times than their hosts, the potential for rapid directional selection in the parasite lineage is clearly greater.

The 'Life-Dinner Principle'

Selection is unlikely to be acting equally strongly on both lineages within an arms race. There may be differences between lineages in the penalty of failure. In host-parasite races, individual parasites which fail to establish in or on a host are likely to die and their reproductive potential is zero. Hosts which fail to avoid or reject a parasite, however, may suffer only mild discomfort or temporary illness; their reproductive potential may be reduced but not by much. 'Failed' hosts may thus still contribute offspring to the next generation while 'failed' parasites almost certainly will not. As a result, selection is likely to act more strongly on parasites to resist rejection by hosts than on hosts to resist inroads by parasites. To that extent the parasite lineage has an advantage. By analogy with similar differences in predator-prey relationships, this asymmetry in selection pressure has been dubbed the 'Life-Dinner Principle' (Dawkins and Krebs, 1979).

Adaptation budgets

Continuing mutual counter-adaptation between host and parasite results in what parasitologists refer to as 'parallel evolution', an evolutionary tailoring of the host-parasite relationship so that associations develop within which lineages coevolve more or less in lock-step. One consequence of this process, and the high degree of specialization which may be needed to exploit a particular host, is narrow host specificity (Cameron, 1964; Croll, 1966, 1973). Platyhelminth parasites, such as the tetraphyllid and tetrarhynchid cestodes, provide good examples of narrow host specificity expected after generations of involvement in arms races.

This specificity may lend parasites another advantage in the race with their hosts. Adaptations are costly. They use up metabolic resources so that it is reasonable to assume that resources invested in one adaptation are not available, or are available in reduced quantities, for investment in other adaptations. Organisms can thus be thought of as having a limited adaptation 'budget' (Dawkins and Krebs, 1979) which is dispensed according to the various selection pressures calling upon it. The extent to which a lineage can invest in any one adaptation is therefore likely to be compromized by the strength of selection for other adaptations (Dawkins and Krebs, 1979; Barnard, 1984). This has important consequences for the degree to which each lineage can adapt to the other. A generalist parasite involved in arms races with several, phylogenetically distantly related hosts is unlikely to show fine-grained adaptation to any one of them because selection will be exerting conflicting demands on its resources and limiting its response in any particular direction. Exploitation of more closely related hosts, however, may mean that similar adaptations are selected for in each case, so that a greater proportion of the parasite's adaptation budget can be channelled into exploiting each host.

Furthermore, while each of the parasites exploiting a given host species is adapting to a single competing lineage and can thus specialize, the opposite may be true for hosts. Hosts are likely to be exploited by a range of parasite species (see below) and may thus be racing against diverse adaptations which compromise their response to any one of them. Host resources are, therefore, spread thinly across parasites while parasite resources are concentrated against hosts. Such asymmetry in resource allocation to a race is likely to put parasites at an advantage.

The consequences of arms races

Changes by parasites responding to changes in their hosts, must include adaptations which permit survival for long enough to reproduce and generate sufficient transmission stages to ensure that infection is passed on. An obvious route to take would be for the parasite to reduce the apparent antigenic difference between itself and its host. Sprent (1962) has thus argued that in the long term parasites may evolve to mimic closely their hosts' proteins. By reducing the antigenic disparity between themselves and their hosts, parasites would minimize their chances of being recognized and rejected (Chapter 13). Since infected hosts would be disadvantaged, selection pressure should in general favour strategies which increase the host's powers of discrimination. The end product on the host side appears to be a variety of distinct effector mechanisms controlled by a complex regulatory system, itself under the influence of highly polymorphic genes. On the parasites' side this is countered by a range of evasive mechanisms which enable some individuals to invade, survive and reproduce (Chapter 13). However, this should be seen as a continuing contest, with each side only achieving a temporary advantage. No host response is totally effective in all members of the species (Chapters 5 and 6), and no evasive strategy gives parasites unchallenged supremacy.

1.3. PARASITISM: ADVANTAGES AND DISADVANTAGES

The exploitation of animal tissues as an environment probably constitutes the most common mode of life style in the animal kingdom (Price, 1977). In nature almost all free-living animals carry parasites, many are infected with several species concurrently and a good proportion of the parasites will be totally specific to the host in question. Take, as an example, parasites of man: *Taenia solium*, *T. saginata*, *Ancylostoma duodenale*, *Necator americanus*, *Ascaris lumbricoides*, *Trichuris trichiura*, *Wuchereria bancrofti* and *Onchocerca volvulus* to name a few, are all specific to man as the definitive host. Some may cause incidental infections in other hosts, but essentially man is the important definitive host through which these organisms are propagated. In the Third World people may carry infections with several of these species concurrently (Croll and Ghadirian, 1981; Haswell-Elkins *et al.*, 1987) and may also be affected by one or more of the four *Plasmodia* and by intestinal protozoa (Crompton and Tulley, 1987).

The benefits which parasites derive from exploiting host tissues obviously vary depending on the exact nature of the relationship. Briefly, benefits may include a continuous supply of nutrients and relative constancy in the environment (if immunity is excluded). These aspects of the parasitic relationship are comprehensively reviewed by Kennedy (1976), Smyth (1976) and Whitfield (1979).

On the other side of the coin, exploitation of hosts presents particular problems at reproduction and transmission for the invading parasites. Hosts often represent a discontinuous environment and transmission strategies have to be well adjusted to ensure that parasite stages successfully bridge the gap between existence in one host and infection of the next. Thus reproductive strategies may include prodigious fecundity and asexual division; transmission strategies may include daily synchronization of the availability of transmission stages with vector feeding times and exploitation of host behaviour (see Hawking, 1975; Barnard and Behnke, 1990).

1.4. PATHOLOGY

Parasitic organisms have reproduction as their priority in their definitive hosts; sufficient progeny have to be produced to overcome the immense mortality at transmission in order to replace each reproductive individual in the next generation. Thus the host is deprived of vital resources, nutrients or tissues which the parasites sequester to support their reproductive efforts. The inevitable consequence is local or more extensive damage to host tissues (i.e. pathology), and if this becomes sufficiently severe to be perceivable externally causing noticeable deterioration of the infected host, the animal is referred to as suffering from the disease associated with the parasite (Figure. 1.1).

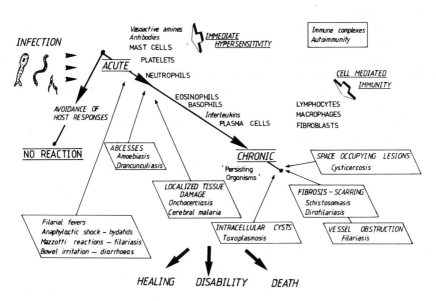

Figure 1.1. An outline of inflammation and healing in parasitic diseases (MacKenzie, 1986).

However, pathology may arise from a variety of unrelated causes and in case the term itself engenders confusion, the reader is referred to an article by MacKenzie (1986) in which a brief historical introduction to the subject is given and complemented by a discussion of the associated terminology, synonyms and ambiguities. In its broadest sense, pathology may be considered as the study of any abnormal body function (in comparison with a normal, healthy, uninfected host) and its associated causes. There are many possible reasons for the departure from normal body function in a particular host but in the context of this volume those changes which are directly or indirectly attributable to invasion by parasitic organisms will be the primary concern. Thus pathological changes in the host may be directly linked to micro- and macro-organisms feeding on cells or tissues and depriving the host of their normal contribution to general body function. Parasites may disrupt cellular activity by invading cells and usurping them for their own needs or they may secrete toxic products which cause local or more widespread damage.

Pathological changes may also ensue as a result of the normal host response to invasion by parasites or as a consequence of a malfunction in the homeostatic regulatory mechanisms whose primary function is to protect the host from infection. As will become evident later, many species of parasites have evolved ways of manipulating the host response and hence pathology, immunology and parasite biology are so highly interlinked that they cannot be easily divorced in a balanced consideration of the subject.

Indeed pathology is very much an important aspect of the parasitic relationship and for some species, it has been argued, pathology may be beneficial for transmission (Ewald, 1980; Anderson and May, 1982; Dawkins, 1982). A good example is found in rats infected with the lung parasitic nematode *Angiostrongylus*. A severe inflammatory response is required in the lungs to enable the eggs to pass out of the host (MacKenzie, 1986). When T-cell deprived mice are infected with *Schistosoma mansoni*, faecal egg counts are significantly depressed in relation to controls (Figure 1.2). Furthermore, when the granulomatous response of the liver to accumulating parasite eggs is experimentally suppressed, host mortality is significantly enhanced through the toxic products of the eggs, which are no longer contained and restricted to the sites of reaction. In *S. mansoni* infection a well-regulated host response appears to be beneficial to both host and parasite (Phillips and Lammie, 1986) the latter having evolved ways of exploiting the host's immune response and the accompanying immunopathological reactions to facilitate its own life cycle (Figure 1.3).

The importance to the parasite of the pathological changes it induces is what we might expect from a progressive arms race in which each counter-adaptive step provides the selective environment for the next step. In this sense it may be misleading to consider pathology as an advantage to the parasite. The reason parasites may now depend on or appear to benefit from the pathology they induce is that this provides the environment in which the parasite's current strategy was favoured. In some cases, this may mean the

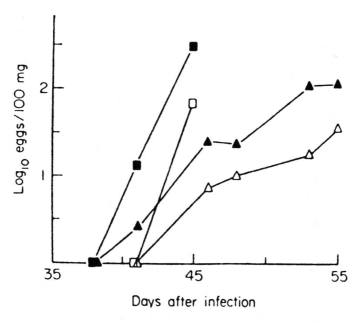

Figure 1.2. The mean number of eggs of *Schistosoma mansoni* detected in the faeces of normal (closed symbols) or T-cell deprived (open symbols) mice following infection with either 35 (triangles) or 175 (squares) cercariae. Each group comprized ten mice (Doenhoff *et al.*, 1978; see also Dunne *et al.*, 1983; Doenhoff *et al.*, 1986).

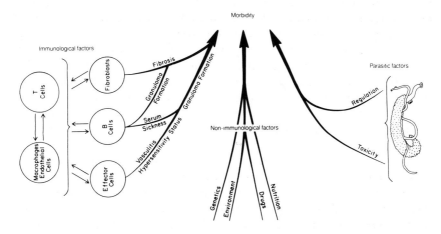

Figure 1.3. Some of the factors that contribute to morbidity in schistosomiasis (Phillips and Lammie, 1986).

parasite, through coevolution, is unable to survive or complete its life cycle in the absence of the pathological environment. However, were the pathological consequences of infection to be totally without benefit to either host or parasite, selection would presumably act to reduce pathology and lead progressively towards asymptomatic infection.

Short or even long-term advantages to one lineage are a predictable consequence of a continuously waged arms race, though the extent to which advantages are maintained in natural communities will depend on the processes regulating host vulnerability and parasite transmission. Where these are artificially increased, the potential natural advantage to the parasite may be increased considerably with the possibility of drastic consequences for both host and parasite populations. Familiar examples of increased pathology resulting from overinfection by parasites include weight loss and parasitic enteritis in lambs and calves raised on permanent pasture. Artificially high larval contamination of herbage is created when naturally roaming ungulates have restricted grazing space in fenced fields (Michel, 1976, 1985). *Fasciola hepatica* may decimate sheep flocks in warm, damp years; coccidiosis in domestic fowl represents another example (Long and Jeffers, 1986), but the list is long. Essentially in these situations the capacity for the counter-response evolved by the host prior to domestication is upset by human intervention. Losses to the parasite at transmission are reduced and hosts become overinfected. In these examples parasites can be seen to exert a significant and often quite drastic, regulatory effect on host population density, brought about by man-made restriction of movement and often exacerbated by atypical weather conditions (Thomas, 1974).

1.5. PARASITISM AS A FORCE IN EVOLUTION

Under natural conditions severe host mortality directly attributable to parasitic infections is difficult to establish and has seldom been reported (Gibson *et al.*, 1972). Nevertheless, evidence for host mortality is available (Wilson, 1971; Lester, 1977; Gordon and Rau, 1982; Kennedy, 1984) and is usually associated with species where the incapacitation of the intermediate host would be beneficial to the parasite in enhancing the host's susceptibility to predation by suitable definitive hosts (Brassard *et al.*, 1982). However, host mortality may occur in other instances. Cyclic fluctuations in red grouse populations have been attributed to the accumulation of heavy worm burdens of the parasitic nematode *Trichostrongylus tenuis* with resultant loss of condition in the birds, depression in breeding performance and in extreme cases, death (Wilson and Wilson, 1978; Potts *et al.*, 1984; Hudson *et al.*, 1985).

On the whole extensive host mortality of the type observed in viral and bacterial epidemics is seldom caused by parasitic organisms alone. However, there are exceptions, notably malaria which is still reputed to be responsible for over one million human deaths per annum worldwide. Rather, the effects of parasites on their hosts are generally insidious, undermining host health and increasing susceptibility to secondary acute infection with ensuing losses to the host population, particularly when compounded by additional stress such as overcrowding and/or malnutrition. Often parasites have very little effect on host fitness or survival and a minimal influence on reproduction, as has been well documented in the case of the red-spotted newts, *Notophthalmus viridescens* (Gill and Mock, 1985).

The relative importance of parasitism as a force reducing competition between the susceptible individuals in an affected population is still an issue of some debate, but host population growth and dynamics may be influenced by the detrimental effects of infection on individual health and reproduction (Anderson, 1978; Anderson and May, 1978; May and Anderson, 1978). Since reproductive success is likely to correlate positively with resistance to infection, parasites which elicit pathology may be an important force in evolution (Gillespie, 1975). Several lines of evidence support this idea. For instance mice infected with *Trichinella spiralis* lose behavioural dominance and assume a subordinate status (Rau, 1983a, 1983b, 1984) which is likely to affect their chances of mating (e.g. Wolff, 1985). A comprehensive comparative study of the reproductive performance of captive American kestrels (*Falco sparverius*) showed that *Trichinella pseudospiralis* exerted a marked detrimental influence, with infected birds producing fewer eggs and suffering greater losses from breakage and embryo mortality (Saumier *et al.*, 1986). Similarly, *Trichostrongylus tenuis* has been shown to reduce significantly the breeding success of red grouse on moors in the UK (Hudson, 1986).

If parasites have a detrimental effect on the reproductive performance of their hosts, it may pay individuals to choose as mates those members of the opposite sex which show evidence of resistance to infection. If there is a correlation between parasite load and general body condition and appearance, the outward appearance and vigour of potential mates could act as a useful choice criterion. Hamilton and Zuk (1982) have argued that mate choice based on apparent resistance may have provided the driving force for the evolution of exaggerated secondary sexual characters. Using birds as an example, Hamilton and Zuk suggest that since individuals infected with various parasites often have shabby or lacklustre plumage, the brightness and condition of plumage could act as a guide to resistant genotypes. Moreover, the coevolutionary arms race between hosts and parasites provides a mechanism for maintaining heritable fitness variation in disease resistance in the face of consistent preference for resistant mates by the choosing sex (Hamilton and Zuk, 1982).

The Hamilton-Zuk model has been tested in comparative analyses of male plumage brightness in different North American and European passerines. Comparisons controlling for the potentially confounding effects of mating systems and taxonomic grouping have shown the significant positive correlation between parasite load and brightness predicted by the model (Read, 1987,

1988; also Ward 1988 for similar results with fish but see Read and Harvey, 1989), but convincing evidence for mate choice based on resistance within species has not emerged (see Read, 1988). Although one study showed what appeared to be a preference by male mice for uninfected females (*versus* females infected with *T. spiralis*), the bias arose because infected females were more aggressive towards approaching males (Edwards and Barnard, 1987).

The effects of parasites on the evolution of sexual hosts may be more fundamental than just providing the impetus for sexually selected bodily appearance. By selecting for variability in host responses to infection, parasites may also be an important agent in the evolution and maintenance of sexual reproduction itself (Jaenike, 1978; Hamilton, 1982; Lively, 1987).

1.6. THE RELATIONSHIP BETWEEN IMMUNITY AND PATHOLOGY

The most sophisticated and advanced measures for dealing with invasion by parasites have evolved in the mammals, and the scope of this book will be limited to reviewing relationships involving mammalian hosts. However, exploitation by parasites is a potential hazard faced by all organisms and mechanisms of resistance have also evolved in lower groups (see Marchalonis, 1977; Lackie, 1986).

The complexity of the immune response and the extreme polymorphism of the controlling genes (Chapter 6) suggest that the system in mammals is indispensible to survival. In recent years the AIDS epidemic has emphasized precisely how important it is for a fully functional immune system to be maintained in operative condition; life expectancy is severely curtailed when the immune system is compromized by the HIV virus. The adaptive immune response has evolved with incredible complexity. It comprizes many component processes capable of identifying and regulating both the initiation of responses and the selection of the available effectors (Chapters 3, 4 and 5). Some of the effector mechanisms are extremely destructive not only to the target but also to the host (Chapter 7), and here again the close relationship between immunity and pathology is evident because some pathogens are best dealt with by mechanisms so toxic in their effects that the host itself may sustain severe damage [for example, tumour necrosis factor (cachectin), Beutler and Cerami, 1987; oxygen radicals, Clark, 1987; see Chapter 7].

Why then have effector mechanisms with such powerful host-damaging properties evolved? It may be that no other strategies were sufficiently protective to the host to maintain coexistence between host and parasite. In some cases, however, the deleterious effects may simply represent a fitness cost to the allele(s) coding for the mechanism in particular individual genotypes which is offset by fitness advantages in individuals of other genotypes who suffer only temporary inconvenience and eventually recover from infection.

Indeed such a spectrum of consequences is generally observed in the field in respect of species which regularly induce pathology. Onchocerciasis and Bancroftian filariasis are both examples of spectral diseases which affect communities comprizing some individuals who seem to be totally refractory to infection and disease, others that carry parasites but show no pathology, through to extremely sick persons showing extensive pathology but total resistance to the parasites (Figure 1.4, 1.5; Partono, 1985). Individuals with elephantiasis (Figure 1.6) do not have circulating microfilariae so they serve no useful purpose to the parasite in enabling transmission to occur and thus in perpetuating the causative organism.

It is also significant that immigrants from non-endemic areas, rarely develop microfilaraemia after settling in regions where *Wuchereria bancrofti* is prevalent, but elephantiasis and other pathologies associated with the species are more regularly encountered than among the indigenous population (Beaver, 1970). During World War II American servicemen stationed in the Pacific sustained significant casualties to filariasis and many were repatriated because

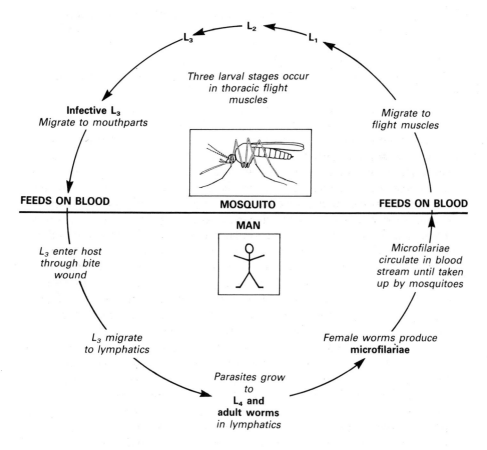

Figure 1.4. The life cycle of *Wuchereria bancrofti*.

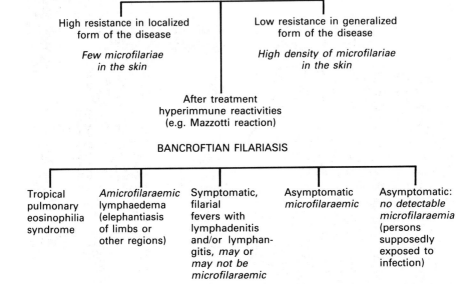

Figure 1.5. The spectrum of disease in onchocerciasis and Bancroftian filariasis (Haque *et al.*, 1983).

of the debilitating complications which ensued. Evidence of elephantiasis was still apparent up to 16 years after their return to the mainland, but throughout the period very few developed circulating microfilariae (Trent, 1963). The precise value to the host of elephantiasis and associated pathologies in filariasis is a subject for speculation although the symptoms almost certainly represent an exaggerated response against the adult worms. It is not certain whether immigrants are more susceptible to infection, as would be expected of hosts not involved in local coexistence with the parasite, and develop enhanced pathology because of more frequent invasion by infective larvae or whether they are no more susceptible but show greater pathology when they are infected because they do not respond to the strategies employed by filarial worms to tone down host responsiveness to parasite antigens; the latter being a mechanism better tuned to the genetic characteristics of local people and hence permitting a high frequency of microfilaraemia and lower pathology in indigenous populations. There is some evidence that animals exposed to microfilariae *in utero* become unresponsive to parasite antigens (Haque and Capron, 1982). People living in areas where filariasis is endemic would have been exposed to microfilarial antigens from conception onwards and unresponsiveness to parasite antigens, a phenomenon which is well established in human filariasis (see Chapter 13), may ensure minimal pathology in this group. Nevertheless, some cases of pathological reactions to filariasis do occur

Figure 1.6. Two individuals from India suffering from elephantiasis as a consequence of exposure to infection by *Wuchereria bancrofti* (photographs by courtesy of Dr D.A. Denham).

among local people, and the value of such responses to the hosts in question is still a major uncertainty. There are no similar, experimentally manipulable animal models which can be analysed to uncover the immediate cause (apart from *Brugia malayi* in cats), and the condition remains poorly understood.

Because of the potentially life-threatening consequences of unleashing effector mechanisms in the host, the mechanisms involved are intricately controlled by a variety of feedback, helper and suppressor systems (Klein, 1982; Green and Gershon, 1984). The pathways involved are kept finely tuned and a balance is maintained (Chapters 4 and 5). However, it is plainly apparent that many parasites do not succumb to potentially destructive responses. Thus one hallmark of parasitic diseases is their chronicity (Behnke, 1987). Affected animals may carry the organisms for years because the parasites have found ways of circumventing host immunity. A range of possible solutions adopted by parasites is reviewed in Chapter 13.

Earlier it was stated that some pathological consequences may actually be beneficial to parasites in aiding the completion of their life cycles. On a similar theme, Mitchison and Oliveira (1986) have argued that parasites which cause chronic infections may gain an advantage through generating pathology within the host because the host should respond in turn by limiting the

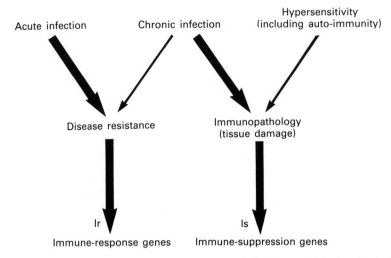

Figure 1.7. Selection for immune suppression (Is) genes (Mitchison and Oliveira, 1986). Acute infections (e.g. viral) select for host resistance since there is an obvious fitness advantage to be gained from the possession of responder genes (Ir). Chronic infections also contribute to the selective pressure for resistance. However, chronic infections are frequently associated with severe, long-lasting pathology, and processes which minimize such complications would also benefit the host. Therefore, immune-suppression genes (Is) exerting a down regulatory influence on immunopathological events would provide the host with a fitness advantage and would be subject to positive selection pressure in populations afflicted by chronic diseases. Few hypersensitivity reactions severely impair host fitness, but Is genes capable of preventing auto-immune disease would also provide a fitness advantage.

damaging processes through regulatory suppressor pathways (Figure 1.7). Indeed, it was suggested that the parasites' reproductive lifespan may be prolonged by invoking host immune-suppression genes (Is) to downregulate the immunopathological and tissue-damaging response to chronic infection (Figure 1.8). Immune-suppression genes, like the immune-response genes (Ir), are highly polymorphic and must therefore be subject to strong selection pressure on the individual alleles. Hosts which limit immunopathological responses but nevertheless restrict parasite development through non-damaging protective responses should be favoured by natural selection. The polymorphism of the Is genes may be maintained by the need to cope with a diversity of parasitic species from a range of taxonomic groups, each with its distinctive life cycle and host-parasite relationship. Many Is genes map to the MHC class II locus I–E in mice and DQ in man, in effect separating them from the Ir genes which in mice map to I–A and in man to DR. Thus there may also be an evolutionary advantage in keeping the immune-response genes dissociated from those for suppression.

Pathology may also be created in parasitic infections through malfunction in the host immunoregulatory systems. To give one example here, hyperreactive malarious splenomegaly is a life-threatening condition which is commonly

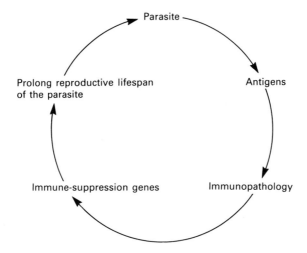

Figure 1.8. An evolutionary feedback loop selecting for immune suppression (Mitchison and Oliveira, 1986).

encountered in regions where malaria is widespread (Crane, 1986). It is believed to be caused by an imbalance in the T-lymphocyte control of antibody synthesis in response to infection (Figure 1.9). An exaggerated IgM response to parasite antigens is accompanied by polyclonal stimulation of B cells with the resultant overproduction of immunoglobulin, much of it to non-parasite-related antigens. The abnormally high concentration of circulating IgM results in the formation of immune complexes, leading to perturbed splenic function and in turn anaemia and leucopenia. Primary responses to heterologous antigens are depressed and together with neutropenia leave the host susceptible to a variety of secondary pathogens.

1.7. THE EPIDEMIOLOGY OF PARASITIC INFECTIONS

This interplay of host immunity, pathology and parasite strategies for evading such responses generates an extremely complicated epidemiological picture in affected communities (Chapter 12). It is more often than not impossible to unravel the reasons for the aggregated distribution of infection intensities among hosts, which is a feature so characteristic of parasitic diseases. Recent findings have confirmed a role for immunity, i.e. some individuals cannot mount effective protective responses even after chemotherapy and become heavily reinfected. A proportion of the population carrying few parasites before treatment do not reacquire intense infection afterwards. Others may be overwhelmed by intense challenge resulting from behaviourally determined

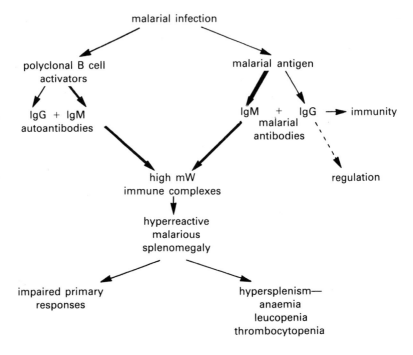

malarial infection

polyclonal B cell
activators

malarial antigen

IgG + IgM
autoantibodies

IgM + IgG
malarial
antibodies

immunity

high mW
immune complexes

regulation

hyperreactive
malarious
splenomegaly

impaired primary
responses

hypersplenism—
anaemia
leucopenia
thrombocytopenia

Figure 1.9. The pathogenesis of hyperreactive malarious splenomegaly (HMS). Mechanisms leading to the acquisition of acquired immunity to malaria appear intact, but an imbalance in T-lymphocyte control of humoral immune responses (possibly in association with particular HLA Class II antigens) leads to gross overproduction of antimalarial and other IgM antibodies. These are incorporated into circulating high molecular weight immune complexes, which are largely responsible for the perpetuation of splenomegaly, possibly by splenic reticuloendothelial blockade. Pancytopenia results from the splenomegaly itself. The leucopenic component of this hypersplenism is the major factor responsible for the high mortality from severe infections, to which impaired ability to mount primary immune responses to non-malarial antigens may also contribute (Crane, 1986).

heavy exposure to the infective stages (Hominick *et al.*, 1987). The problems which have been outlined are the subject of contemporary research and currently there are several epidemiological studies in various stages of completion which are trying to address the issues involved. Much depends on the outcome from such studies. Future strategies for control, including the investment of research funds for immunology and pathology of parasitic infections, will depend on the conclusions which are reached.

1.8.   PATHOLOGY AND VACCINATION

The importance of understanding comprehensively the relationship between disease and resistance to infection cannot be underestimated. If pathology is an indispensible component of host-protective resistance, should we even contemplate developing vaccines for such parasites? In vaccinating people affected by relevant species we would risk exacerbating and exaggerating the disease in individuals who may never have suffered pathological consequences if left alone. An equally important risk is the possibility that vaccines may enhance pathology without inducing protection. If, as Mitchison and Oliveira (1986) suggested, parasites have evolved antigens to elicit pathology as part of their own strategy for evading host immunity, then it is conceivable that such antigens may be selected inadvertently for inclusion into vaccines with resultant counter-productive consequences to the host.

Another danger may be the possibility of blocking host-protective immunity through vaccination. Experience with *Trypanosoma cruzi* has shown that such a possibility is realistic. Sera from chronically infected patients and from experimentally infected mice which controlled parasitaemia were found to contain antibodies (IgG2) which mediated complement-dependent lysis of blood trypomastigotes. Vaccination with *T. cruzi* antigens elicited an antibody response which was non-protective (Brener, 1986). Similarly, vaccination of mice against *Leishmania major*, using the membrane glycolipid antigen LPG (see Chapter 2), generated protective immunity, but when a small glycolipid subcomponent of the intact molecule was separated and used alone, the treated animals developed more intense pathology (Mitchell and Handman, 1986).

These are important issues to resolve before the implementation of vaccines in control programmes for human parasites in the tropics. Whereas standards for animal vaccines may be lower, an effective vaccine for human use would have to be free of all side effects, otherwise local people would have to be persuaded that the suffering they are being asked to endure is for the good in the long term. Indeed, well-meaning parasitologists may be driven out of communities which they are attempting to cure if they fail to anticipate and appreciate the possible reactions of local people to the side effects of the treatment. It is a fact that many people harbouring parasites, such as intestinal round worms, filarial worms and even low intensity schistosome infections, may not recognize that they are infected, so minor may be the symptoms of infection. Nevertheless, such persons serve as important sources of transmission stages which sustain the parasites' life cycles and ultimately generate the infective stages which invade people who do react and suffer more intense pathology.

1.9. THE VICIOUS CIRCLE

It should be apparent that parasitic infection, host immunity and pathology are intricately interwoven aspects of the host-parasite relationship. In human populations living in endemic regions, additional aspects include malnutrition, poverty and ignorance. Together they constitute a vicious circle of compounding factors from which indigenous people cannot break free without support from outside (Figure 1.10).

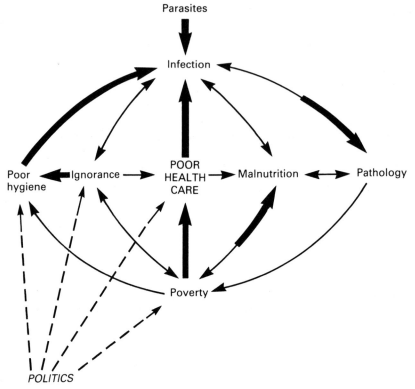

Figure 1.10. The vicious circle of parasitic diseases in Third World countries.

A complicating issue is the relationship between malnutrition and parasitic infection (see Crompton, 1985; Stephenson, 1987). It is now recognized that malnutrition has severe immunodepressive consequences to the host (Chapter 13), and this in turn may increase susceptibility to disease and enhance pathology (Keusch, 1985). The histopathological changes in the intestines of malnourished people are indistinguishable from those detected in chronically infected subjects. Villous atrophy, crypt of Lieberkuhn hyperplasia and reduced absorptive function are all commonly encountered in parasitic gastrointestinal infections and in people with tropical sprue and malnutrition.

If malnourished people are indeed weakened in their resistance to parasites, then the development of vaccination as a measure to aid in the control of parasites and in their ultimate eradication may be an impractical solution. It is a sobering thought that a vaccine for human parasites in the tropics may also have to contend with the additional problem of immunodepression in malnourished hosts, and this may be simply too much to expect in the foreseeable future.

REFERENCES

Anderson, R.M. (1978), The regulation of host population growth by parasitic species, *Parasitology*, **76**, 119–57.

Anderson, R.M. and May, R.M. (1978), Regulation and stability of host-parasite population interactions. 1. Regulatory processes. *Journal of Animal Ecology*, **47**, 219–47.

Anderson, R.M. and May, R.M. (1982), Coevolution of hosts and parasites, *Parasitology*, **85**, 411–26.

Anderson, R.M. and May, R.M. (1985), Helminth infections of humans: mathematical models, population dynamics and control, *Advances in Parasitology*, **24**, 1–101.

Barnard, C.J. (1983), *Animal Behaviour: Ecology and Evolution*, London: Croom Helm.

Barnard, C.J. (1984), The evolution of food-scrounging strategies within and between species, in Barnard, C.J. (Ed.), *Producers and Scroungers: Strategies of Exploitation and Parasitism*, London and Sydney: Croom Helm.

Barnard, C.J. and Behnke, J.M. (Eds.) (1990), *Parasitism and Host Behaviour*, London, UK: Taylor & Francis.

Beaver, P.C. (1970), Filariasis without microfilaremia, *The American Journal of Tropical Medicine and Hygiene*, **19**, 181–9.

Behnke, J.M. (1987), Evasion of immunity by nematode parasites causing chronic infections, *Advances in Parasitology*, **26**, 1–71.

Beutler, B. and Cerami, A. (1987), Cachectin-tumour necrosis factor: a cytokine that mediates injury initiated by invasive parasites, *Parasitology Today*, **3**, 345–6.

Brassard, P., Rau, M.E. and Curtis, M.A. (1982), Parasite-induced susceptibility to predation in diplostomiasis, *Parasitology*, **85**, 495–501.

Brener, Z. (1986), Why vaccines do not work in Chagas Disease, *Parasitology Today*, **2**, 196–7.

Cameron, T.W.M. (1964), Host specificity of helminthic parasites, *Advances in Parasitology*, **2**, 1–34.

Clark, I.A. (1987), Cell-mediated immunity in protection and pathology of malaria, *Parasitology Today*, **3**, 300–5.

Crane, G.G. (1986), Hyperreactive malarious splenomegaly (tropical splenomegaly syndrome), *Parasitology Today*, **2**, 4–9.

Croll, N.A. (1966), *Ecology of parasites*, London: Heinemann Educational Books Ltd.

Croll, N.A. (1973), *Parasitism and other Associations*. London: Pitman Medical.

Croll, N.A. and Ghadirian, E. (1981), Wormy persons: contributions to the nature and patterns of overdispersion with *Ascaris lumbricoides*, *Ancylostoma duodenale*, *Necator americanus* and *Trichuris trichiura*. *Tropical and Geographical Medicine*, **33**, 241–8.

Crompton, D.W.T. (1985), Chronic ascariasis and malnutrition, *Parasitology Today*, **1**, 47–52.

Crompton, D.W.T. and Tulley, J.J. (1987), How much ascariasis is there in Africa? *Parasitology Today*, **3**, 123–7.

Dawkins, R. (1982), *The Extended Phenotype*. Oxford: Freeman.

Dawkins, R. and Krebs, J.R. (1979), Arms races between and within species, *Proceedings of the Royal Society, London*, **B205**, 489–511.

Doenhoff, M., Musallam, R., Bain, J. and McGregor, A. (1978), Studies on the host-parasite relationship of *Schistosoma mansoni* infected mice: the immunological dependence of parasite egg excretion. *Immunology*, **35**, 771–8.

Doenhoff, M.J., Hassounah, O., Murare, H., Bain, J. and Lucas, S. (1986), The schistosome egg granuloma: immunopathology in the cause of host protection or parasite survival? *Transactions of the Royal Society of Tropical Medicine and Hygiene*, **80**, 503–14.

Dunne, D.W., Hassounah, O., Musallam, R., Lucas, S., Pepys, M.B., Baltz, M. and Doenhoff, M. (1983), Mechanisms of *Schistosoma mansoni* egg excretion: parasitological observations in immunosuppressed mice reconstituted with immune serum. *Parasite Immunology*, **5**, 47–60.

Edwards, J.C. and Barnard, C.J. (1987), The effects of *Trichinella* infection on intersexual interactions between mice. *Animal Behaviour*, **35**, 533–40.

Ewald, P. (1980), Evolutionary biology and the treatment of signs and symptoms of infectious diseases. *Journal of Theoretical Biology*, **86**, 169–71.

Gibson, G.G., Broughton, E. and Choquette, L.P.E. (1972), Waterfowl mortality caused by *Cyathocotyle bushiensis*, Khan, 1962 (Trematoda: Cyathocotylidae), St. Lawrence River, Quebec. *Canadian Journal of Zoology*, **50**, 1351–6.

Gill, D.E. and Mock, B.A. (1985), Ecological and evolutionary dynamics of parasites: the case of *Trypanosoma diemyctyli* in the red-spotted newt *Notophthalmus viridescens*, in Rollinson, D. and Anderson, R.M. (Eds.), *Ecology and Genetics of the Host-Parasite Interactions*, *Linnean Society Symposium Series* 11, pp. 157–83, London: Academic Press.

Gillespie, J.H. (1975), Natural selection for resistance to epidemics. *Ecology*, **56**, 493–5.

Gordon, D.M. and Rau, M.E. (1982), Possible evidence for mortality induced by the parasite *Apatemon gracilis* in a population of brook sticklebacks (*Culaea inconstans*). *Parasitology*, **84**, 41–7.

Green, D.R. and Gershon, R.K. (1984), Contrasuppression: the second law of thermodynamics revisited. *Advances in Cancer Research*, **42**, 277–335.

Hamilton, W.D. (1982), Pathogens as causes of genetic diversity in their host population. in Anderson, R.M. and May, R.M. (Eds.), *Population Biology of Infectious Disease*, Berlin: Springer-Verlag.

Hamilton, W.D. and Zuk, M. (1982), Heritable true fitness and bright birds: a role for parasites? *Science*, **218**, 384–7.

Haque, A. and Capron, A. (1982), Transplacental transfer of rodent microfilariae induces antigen-specific tolerance in rats, *Nature (London)*, **299**, 361–3.

Haque, A., Capron, A., Ouaissi, A., Kouemeni, L., Lejeune, J.P., Bonnel, B., and Pierce, R. (1983), Immune unresponsiveness and its possible relation to filarial disease, *Contributions to Microbiology and Immunology*, **7**, 9–21.

Haswell-Elkins, M.R., Elkins, D.B. and Anderson, R.M. (1987), Evidence for predisposition in humans to infection with *Ascaris*, hookworms, *Enterobius* and *Trichuris* in a South Indian fishing community. *Parasitology*, **95**, 323–37.

Hawking, F. (1975), Circadian and other rhythms of parasites, *Advances in Parasitology*, **13**, 123–82.

Hominick, W.M., Dean, C.G. and Schad, G.A. (1987), Population biology of hookworm in West Bengal: analysis of numbers of infective larvae recovered from damp pads applied to the soil surface at defaecation sites. *Transactions of the Royal Society of Tropical Medicine and Hygiene*, **81**, 978–86.

Hudson, P.J. (1986), The effect of a parasite nematode on the breeding production of red grouse, *Journal of Animal Ecology*, **55**, 85–92.

Hudson, P.J., Dobson, A.P. and Newborn, D. (1985), Cyclic and non-cyclic populations of red grouse: a role for parasitism? in Rollinson, D. and Anderson, R.M. (Eds.), *Ecology and Genetics of Host-Parasite Interactions*, *Linnean Society Symposium Series*, 11, pp. 77–89, London: Academic Press.

Jaenike, J. (1978), An hypothesis to account for the maintenance of sex within populations, *Evolutionary Theory*, **3**, 191–4.

Kennedy, C.R. (1976), *Ecological Aspects of Parasitology*, Amsterdam and Oxford: North Holland Publishing Company.

Kennedy, C.R. (1984), The use of frequency distributions in an attempt to detect host mortality induced by infections of diplostomatid metacercariae, *Parasitology*, **89**, 209–20.

Keusch, G.T. (1985), Nutrition and immune function, in Warren, K.S. and Mahmoud, A.A.F., (Eds.), *Tropical and Geographical Medicine*, New York: McGraw-Hill Book Company.

Klein, J. (1982), *Immunology, The Science of Self–Nonself Discrimination*, New York and Chichester: John Wiley and Sons.

Lackie, A.M. (Ed.) (1986), *Immune mechanisms in invertebrate vectors. Symposia of the Zoological Society of London*, **56**, Oxford: Clarendon Press.

Lester, R.J.G. (1977), An estimate of the mortality in a population of *Perca flavescens* owing to the trematode *Diplostomum adamsi*, *Canadian Journal of Zoology*, **55**, 288–92.

Lively, C.M. (1987), Evidence from a New Zealand snail for the maintenance of sex by parasitism, *Nature*, **328**, 519–21.

Long, P.L. and Jeffers, T.K. (1986), Control of chicken cocciodsis, *Parasitology Today*, **2**, 236–40.

MacKenzie, C.D. (1986), Pathology in Tropical Medicine, *Parasitology Today*, **2**, 261–3.

Marchalonis, J.J. (1977), *Immunity in Evolution*, London: Edward Arnold.

May, R.M. and Anderson, R.M. (1978), Regulation and stability of host-parasite population interactions. II. Destabilising processes, *Journal of Animal Ecology*, **47**, 249–67.

Michel, J.F. (1976), The epidemiology and control of some nematode infections in grazing animals, *Advances in Parasitology*, **14**, 355–97.

Michel, J.F. (1985), Strategies for the use of anthelmintics in livestock and their implications for the development of drug resistance, *Parasitology*, **90**, 621–8.

Mitchell, G.F. and Handman, E. (1986), The glycoconjugate derived from a Leishmania major receptor for macrophages is a suppressogenic, disease-promoting antigen in murine cutaneous leishmaniasis, *Parasite Immunology*, **8**, 255–63.

Mitchison, N.A. and Oliveira, D.B.G. (1986), Chronic infection as a major force in the evolution of the suppressor T-cell system, *Parasitology Today*, **2**, 312–3.

Partono, F. (1985), Diagnosis and treatment of lymphatic filariasis, *Parasitology Today*, **1**, 52–7.

Phillips, S.M. and Lammie, P.J. (1986), Immunopathology of granuloma formation and fibrosis in schistosomiasis, *Parasitology Today*, **2**, 296–301.

Potts, G.R., Tapper, S.C. and Hudson, P.J. (1984), Population fluctuations in red grouse: analysis of bag records and a simulation model, *Journal of Animal Ecology*, **53**, 21–36.

Price, P.W. (1977), General concepts on the evolutionary biology of parasites, *Evolution*, **31**, 405–20.

Rau, M.E. (1983a), Establishment and maintenance of behavioural dominance in male mice infected with *Trichinella spiralis*, *Parasitology*, **86**, 319–22.

Rau, M.E. (1983b), The open-field behaviour of mice infected with *Trichinella spiralis*, *Parasitology*, **86**, 311–8.

Rau, M.E. (1984), Loss of behavioural dominance in male mice infected with *Trichinella spiralis*, *Parasitology*, **88**, 371–3.

Read, A.F. (1987), Comparative evidence supports the Hamilton and Zuk hypothesis on parasites and sexual selection, *Nature*, **328**, 68–70.

Read, A.F. (1988), Sexual selection and the role of parasites, *Trends in Ecology and Evolution*, **3**, 97–101.

Read, A.F. and Harvey, P.H. (1989), Reassessment of comparative evidence for Hamilton and Zuk theory on the evolution of secondary sexual characters, *Nature*, **339**, 618–9.

Saumier, M.D., Rau, M.E. and Bird, D.M. (1986), The effect of *Trichinella pseudospiralis* infection on the reproductive success of captive American kestrels (*Falco sparverius*), *Canadian Journal of Zoology*, **64**, 2123–5.

Smyth, J.D. (1976), *Introduction to Animal Parasitology*, London: Hodder & Stoughton, London.

Sprent, J.F.A. (1962), Parasitism, immunity and evolution, in Leeper, G.W. (Ed.), *The Evolution of Living Organisms*, pp. 149–65, Melbourne University Press.

Stanley, S. (1979), *Macroevolution: Pattern and Process*, San Francisco: Freeman.

Stephenson, L.S. (1987), *The impact of helminth infections on human nutrition*, London: Taylor & Francis.

Thomas, R.J. (1974), The role of climate in the epidemiology of nematode parasitism in ruminants, in Taylor, A.E.R. and Muller, R. (Eds.), *The Effects of Meteriological Factors upon Parasites, Symposia of the British Society for Parasitology* **12**, pp. 13–32, Oxford: Blackwell Scientific Publications.

Trent, S.C. (1963), Re-evaluation of World War II veterans with filariasis acquired in the South Pacific, *American Journal of Tropical Medicine and Hygiene*, **12**, 877–87.

Wakelin, D. (1976), Host responses, in Kennedy, C.R. (Ed.), *Ecological Aspects of Parasitology*, Amsterdam: North Holland Publishing Company.

Ward, P.I. (1988), Sexual dichromatism and parasitism in British and Irish freshwater fish, *Animal Behaviour*, **36**, 1210–5.

Whitfield, P.J. (1979), *The Biology of Parasitism: An Introduction to the Study of Associating Organisms*. London: Edward Arnold.

Wilson, G.R. and Wilson, L.P. (1978), Haematology, weight and condition of captive red grouse (*Lagopus lagopus scoticus*) infected with caecal threadworms (*Trichostrongylus tenuis*), *Research in Veterinary Science*, **2**, 331–6.

Wilson, R.S. (1971), The decline of a roach *Rutilus rutilus* (L.) population in Chew Valley lake, *Journal of Fish Biology*, **3**, 129–37.

Wolff, R.J. (1985), Mating behaviour and female choice: their relation to social structure in wild caught house mice (*Mus musculus*) housed in a semi-natural environment, *Journal of Zoology, London*, **207**, 43–51.

2. Parasite antigens

R.M. Maizels

2.1. INTRODUCTION

The presence of a parasite in a mammalian host inevitably provokes an immune response to the many hundreds or thousands of antigenic macromolecules produced by that parasite, even though relatively few of those antigens may be of critical importance to the outcome of the parasitism. Thus, although a myriad of antigenic proteins and glycoconjugates have now been identified across the whole range of parasitic organisms, only a fraction of these have been shown to be relevant in the context of host-parasite confrontation. In this chapter, the general characteristics of parasite antigens will be discussed before illustrating our current understanding by describing a small number of well-defined antigens from major parasitic pathogens.

The relevance of a parasite antigen may be considered at two levels, that of the function it performs on behalf of the parasite and that of the response it elicits in the infected host (Anders *et al.*, 1982). A central focus where these aspects meet is at the parasite surface, whether that surface takes the form of the protozoal plasma membrane, the trematode tegument or the nematode extracellular cuticle. Two considerations have reinforced this theme: that surface antigens may act as excellent targets for antibody and other neutralizing responses and that these molecules must fulfil functions (such as nutrient uptake) upon which the parasite depends. However, the realization that internal antigens may be expressed on host cell surfaces and thereby act as effective targets for T-cell responses (Townsend *et al.*, 1986) requires us also to take account of many cytoplasmic or somatic antigenic constituents in both intracellular and extracellular parasites.

A parallel interest over many years has been in the antigens secreted, released or excreted by parasites (Lightowlers and Rickard, 1988). Again, the rationale has been that these may perform some role which can usefully be blocked by the immune system, and there has certainly been a diagnostic benefit in infections where sufficient antigen is secreted for parasite products to be directly detected in the serum of a parasitized host.

Parasite and host must always be understood as an interacting pair, and it is no less important to consider the relevance of parasite antigens in terms of host-immune responsiveness and the consequences of that response. An ideal parasite antigen may be one that evokes a protective immune response, by the production of antibodies or specific T cells which can recognize one specific antigen from the parasite and thereby eliminate it. However, single protective antigens are proving to be somewhat elusive for these complex eukaryotic organisms, and increasingly combinations or 'cocktails' of antigens are envisaged as necessary for the induction of protective immunity.

In fact, many parasite products appear to have evolved in such a way that protective immune responses are unlikely to be mounted. Rather, the immune system may concern itself with 'diversionary' antigens or even decoy determinants within any one antigen, and it frequently reacts in a manner which results in the immune response itself generating damaging lesions and pathology. More detrimental still are the instances where some antigenic similarity, either coincidental or contrived, between parasite and host results in the initiation of a destructive auto-immune response against host tissue.

2.2. PROTOZOAL ANTIGENS

Trypanosome variant surface glycoproteins (VSG)

The most clearly understood parasite antigen known today is the variant surface glycoprotein (VSG) of blood-form African trypanosomes. These protozoa follow relatively uncomplicated life cycles, multiplying in the extracellular vascular fluid and undergoing differentiative changes only through the tsetse fly (*Glossina*) vector (Figure 2.1). Although the immune system is thus confronted with a single life cycle stage as a target, it singularly fails to eliminate the parasitism because of the extraordinary antigenic variation of the surface glycoprotein (Donelson and Turner, 1985).

The surface of African trypanosomes (*Trypanosoma brucei brucei, T. b. rhodesiense, T.b. gambiense, T. congolense*) is covered almost completely by a dense coat, 12–15 nm thick, composed of 7×10^6 molecules of a 59 kDa glycoprotein, the VSG. This coat is first expressed by metacyclic trypanosomes in the tsetse fly as they develop to the infective stage. The VSG is a major cell product of the parasite, accounting for 10% of the trypanosome protein synthesis, and it accordingly provokes a strong humoral antibody response. This antibody successfully kills all but a tiny proportion of the parasites, but those that survive do so because they express a different antigenic form of the surface glycoprotein. The surviving trypanosomes can therefore continue to divide at a rapid rate (approximately every six hours) so that within a few days they can again produce a full-scale infection.

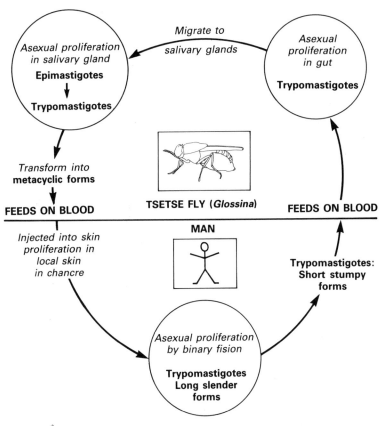

Figure 2.1. The life cycle of African trypanosomes.

The ability of trypanosomes to vary their VSG is now understood at the genetic level. The organisms have around 1000 alternative VSG genes (a major portion of the total gene complement of these parasites), all bar one of which are effectively silent in any one cell. For a gene to be utilized by the trypanosome, a copy of the genomic sequence is synthesized (an 'expression-linked copy') and inserted into a specific expression site located near the telomere of the chromosome, displacing any copy previously present at that site (Borst and Cross, 1982). This switch occurs at a relatively frequent rate of 10^{-5}–10^{-6}, ensuring that in any trypanosome infection, a few variants are always present to avoid an antibody response to the predominant type.

The molecular structure of the VSG is also known in some detail (Figure 2.2). The molecule has 540 amino acids with a general design in which the N-terminal shows little homology between variants while the C-terminal, anchored in the trypanosome membrane, contains both variable and conserved stretches (Metcalf *et al.*, 1987). All VSG variants possess at least one N-linked carbohydrate, as well as an interesting glycosylphosphatidylinositol (gpi)

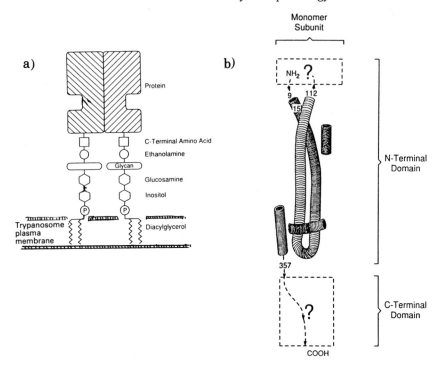

Figure 2.2. Structure of the variant surface glycoprotein (VSG) of *Trypanosoma brucei*. (A) A schematic representation of the trypanosome membrane. The shaded sections represent the protein portion of the molecule to which is attached a glycan containing a Cross Reactive Determinant, and a diacylglycerol anchor by which the whole molecule is attached to the plasma membrane. (B) Sketch of the structure of one of the subunits of the VSG (serodeme/variant MITat 1.2) based on crystallographic evidence. Only the α-helical segments, marked as cylinders, are shown for simplicity. No data are available on the three-dimensional structure at the 'top' of the molecular, nor on the C-terminal domain (Turner, 1988).

structure attached to the C-terminal amino acid (Ferguson *et al.*, 1988). None of the carbohydrate components of the VSG are exposed in the intact parasite.

The glycosylphosphatidylinositol contains a myristic acid tail imparting the VSG with a hydrophobic anchor into the cell membrane. Trypanosomes also contain a specific phospholipase which will cleave the gpi structure between its carbohydrate and fatty acid, thus rendering the VSG water soluble. Cleavage appears to occur *in vivo* and results in the C-terminal carbohydrate group becoming exposed to the immune system with consequent antibody production. This antibody will bind to all VSG variants that have been cleaved, demonstrating that the same anchor is used in all cases and defining a cross-reactive determinant (CRD); however the same antibody fails to bind intact parasites and cannot play a protective role in the infected host (Holder, 1985).

Leishmania glycoconjugates

Like African trypanosomes, protozoal parasites of the *Leishmania* genus carry a dominant surface glycoprotein, but the similarity does not extend much further (Figure 2.3). Among the fundamental differences are that *Leishmania* is an intracellular parasite and probably as a consequence T-cell mediated immune responses are far more important. The parasites undergo development in the insect (sandfly) vector to a promastigote stage and at that time a major surface protein of 62–65 000 mol. wt. appears in all species studied; it is significantly glycosylated and has hence been termed gp63. Like the VSG, gp63 is inserted into the parasite membrane via a glycophospholipid anchor, but there are fewer molecules (5×10^5) per cell, and gp63 production represents a lower although substantial (1%) proportion of total cell protein than in trypanosomes. Interestingly, the *Leishmania* surface antigen has been ascribed a functional role: it is an active protease (Etges *et al.*, 1986), and

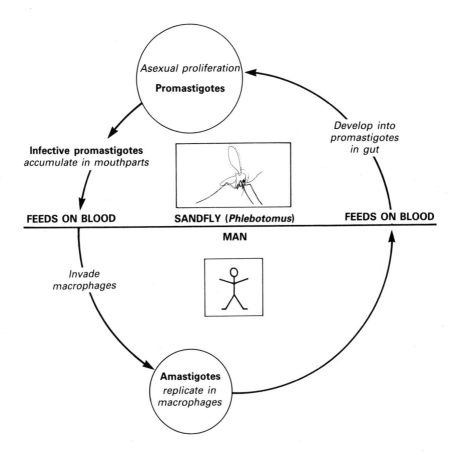

Figure 2.3. The life cycle of *Leishmania*.

this activity may to some extent enable promastigotes released from the sandfly to counteract humoral immune responses until completing entry into host macrophages for the intracellular phase of the life cycle. However, the same surface glycoprotein is also involved in binding host serum complement C3 in order to be taken up into the macrophage through its C3 receptor phagocytosis mechanism (see Chapter 7, Figure 7.5).

Leishmania parasites carry an equally prominent surface antigen, a complex phosphorylated oligosaccharide also with a lipid anchor, the LPG (lipophosphoglycan). This is a small molecule (around 9 000 mol. wt.) with a repeating phosphorylated oligosaccharide structure and an unusual heptasaccharide leading to the fatty acid terminus (Turco, 1988) which is readily secreted by parasites in culture. The LPG has a bifunctional interaction with the host immune system. If the intact LPG is purified it will elicit a protective immune response on vaccination; however, cleavage of the lipid produces a carbohydrate group with a potent suppressive effect on the same immune response (Mitchell and Handman, 1986) and capable of enhancing pathology when included in a vaccine (termed, disease-promoting antigen).

Malaria stage-specific antigens

The analysis of malaria antigens, like that of most other parasites, has had to encompass each of the life cycle stages which independently express stage-specific antigens: the invading sporozoite, the reproducing blood forms such as the merozoite, and the sexual stages which are responsible for mosquito transmission, the gametocytes. From the earliest immunological studies, it was apparent that these stages were almost completely distinct in their antigenic presentation. Thus, animals can be vaccinated effectively with an injection of radiation-attenuated sporozoite stages taken from mosquito, and this immunization will prevent any development by a subsequent virulent inoculation. However, the same animals remain fully susceptible to malaria induced by transfer of post-sporozoite blood-stage forms. Moreover, the blood-stage forms themselves display remarkable antigenic diversity both through the development cycle within the red cell, but also among genetically different isolates or strains of each species of *Plasmodium*. Consequently, the malaria parasites present an enormous spectrum of diverse and unusual antigens, and even though they have been studied in more detail than any other parasite, the importance of each major antigen in infection, immunity or pathogenesis remains controversial.

The sporozoite antigen: CSP

In many ways, the simplest stage in antigenic terms is the mosquito-derived sporozoite. The successful use of irradiated sporozoites as a vaccine correlated well with the production of antibody in vaccinees which formed precipitates around the parasite, termed the circumsporozoite precipitation reaction; the same antibody was then shown to react with an abundant surface protein (the circumsporozoite protein or CSP) of molecular weight 58–67 000 in *Plasmodium falciparum* (Nussenzweig and Nussenzweig, 1984). Much excitement centred around this molecule, which was one of the first human malaria proteins to be cloned and also showed a most unusual structure: a 4 amino acid set (Asn-Ala-Asn-Pro) was repeated up to 41 times in a block, constituting an immunodominant, repetitive epitope or repetope (Godson, 1985) (Figure 2.4). Moreover, while this repetope was species-specific, other malaria parasites all contained repetopes of totally unrelated composition; for example *P. vivax* contains a block of 9 amino acids repeated 19 times (Arnot *et al.*, 1985).

From first appearances, these repeats seemed to be responsible for stage- and species-specific protective immunity. However, cloned products or synthetic peptides corresponding to the repeats have compared poorly with living, irradiated parasites as vaccines against sporozoite infection. Consequently, the possibility is raised that these immunodominant sequences expressed in abundance on the parasite surface have evolved to perform a more diversionary function and that the protective effect of immunizing with attenuated parasites may be directed to antigens expressed once the sporozoites have entered their first host cell, the hepatocyte. This view has found recent confirmation with the demonstration that cytotoxic T cells are necessary for protective immunity to malaria (Schofield *et al.*, 1987) and the definition of T cell epitopes on the CSP and other early malaria antigens (Good *et al.*, 1988).

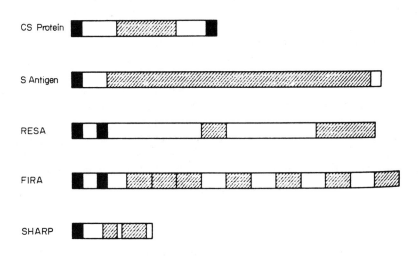

Figure 2.4. Schematic diagram of the genes for several *Plasmodium falciparum* antigens containing repetitive sequences. ▨ repetitive sequences. ■ hydrophobic sequences which function as signal peptides of transmembrane segments (Anders, 1986).

Blood-stage malaria antigens

Once sporozoites have established themselves in the liver, they multiply over the course of 4–6 weeks before a massive release of a new stage, the merozoite, which can penetrate its host's red blood cells. The molecules used by the merozoite stage to recognize, adhere to and invade the erythrocyte have therefore been closely studied, although there is no consensus about antigens critical for penetration and therefore potential targets for a protective immune response (Figure 2.5). Merozoites, taken from blood-stage parasites after schizogony of previously infected red cells, carry a high-molecular weight surface antigen in each of the species so far examined; in falciparum malaria this has been designated Pf195. Biosynthetic studies have established that this antigen is synthesized by the schizont before red cell rupture, but while on the surface of the newly-released merozoite, the molecule is physiologically processed. Its products are a 45 kDa C-terminal protein which remains anchored on the surface and a 150 kDa antigen which is liberated and further broken down; the functions of either the intact high molecular weight proteins or of their breakdown products remain unknown. However, it has been established that, like the CSP, Pf195 contains repeated blocks of amino acids and that these blocks vary between clones and isolates of parasites (Holder

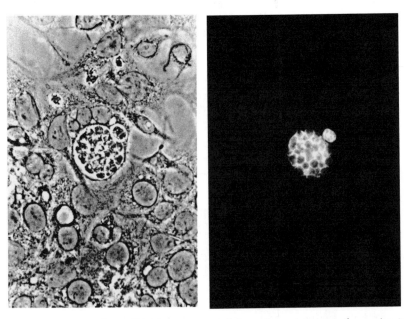

Figure 2.5. Expression of PMMSA (the precursor of the major merozoite surface antigen) on schizonts of *Plasmodium berghei*. (a) Bright field photograph of a mature *P. berghei* schizont following exoerythrocytic development in liver cells. (b) The same field seen under UV light, showing fluorescence on the schizont and on individual merozoites within the schizont. The preparation was incubated in a monoclonal antibody to PMMSA and developed to show sites of antigen expression via standard indirect immunofluorescent techniques (Suhrbier *et al.*, 1989).

et al., 1985). Thus, although purified Pf195 can exert a protective effect in experimental monkeys, it seems to be discounted as a potential vaccine because of its strain specificity.

One major malarial product which may also be involved in invasion is Pf155, or RESA, a protein first found on the surface of ring-stage (newly-invaded) erythrocytes but probably derived from the microneme, a specialized organelle within the merozoite which is involved in red cell invasion. On contact with the red cell, the merozoite appears to disgorge the contents of the micronemes and associated structures, and this antigen may then be inserted into the host cell membrane apparently by binding to the cytoskeleton. Fascinatingly, the RESA/Pf155 also contains repeating sequences, in this case two blocks of varying length in distinct domains of the protein (Cowman *et al.*, 1984).

The RESA/Pf155 molecules from different isolates of *Plasmodium falciparum* so far analysed appear to be closely conserved, and monoclonal antibody to this antigen can inhibit invasion of red cells *in vitro*. Vaccine trials have therefore been conducted to establish whether immunity against blood-stage malaria can be induced *in vivo*. A strong contrast to RESA is seen in another major product from this stage, the heat-stable, soluble, serum (S) antigens shown to be highly strain-specific in the original demonstration of antigenic diversity in malaria. These S antigens show a range of repeat structures so unrelated to one another that their molecular evolution is hard to explain. Thus, one isolate from Papua New Guinea has an S antigen with an 11 amino acid sequence repeated 100 times, while a Ghanaian isolate has a repetitive 8 -mer alternating at one position throughout the protein (Cowman *et al.*, 1985). Clearly, such repetopes must sow confusion in the immune system and may be responsible for the short-lived, strain-specific phenomena of natural immunity to malaria (Figure 2.4). The recent finding of immunological cross reactions between repetopes from CSP, RESA and S antigens has led to the suggestion that these determinants maintain high levels of low-affinity (cross-reactive) antibody in malaria infection to the exclusion of specific, high-affinity and possibly protective antibody (Anders, 1986; see Sections 4.8 and 13.3).

Sexual stage antigens involved in transmission

Only a small proportion of the blood-form parasites are capable of passing the disease on through the mosquito vector. These are a distinct cell population, the gametocytes, which are able to transform within the mosquito to motile sexual gametes which fuse within the blood meal to form successively a zygote, an ookinete and finally an oocyst in the insect gut wall. Because the gametes are extracellular forms within the mosquito blood meal, the products they express during sexual differentiation in the blood meal are unlikely to be exposed in the human host and therefore represent an additional important generation of antigenic targets. Their potential lies in the fact that in the absence of human antibody responses, there should not have been any selective pressure for antigenic variation of mosquito-stage parasites. However, if antibody is deliberately induced in a malaria carrier, the mosquito vector will ingest both gametocytes and antibody to the gametes and this antibody has been shown to effectively kill the potentially transmitting stages within the mosquito (Targett and Sinden, 1985).

The free-living gametes express several surface proteins, and antibody to two antigens of 48 and 45 kDa will, when administered *in vitro*, effectively reduce parasite survival in the insect thereby reducing or blocking transmission (Vermeulen *et al.*, 1985). However, these antigens are present in blood stream gametocytes, and antibody can be found in naturally infected humans; thus the gamete antigens may have evolved to minimize antibody-induced effects in the mosquito. In contrast, the subsequent stages arising from gamete fusion (the zygote and ookinete) share a newly-expressed surface protein of 25 kDa in *P. falciparum* and slightly less in mouse malarias. This antigen has now been sequenced (Kaslow *et al.*, 1988), and it and its murine homologues are being tested as potential transmission blocking vaccines.

2.3. HELMINTH ANTIGENS

Helminth worms are multicellular organisms which therefore bring the additional complexity of specialized tissues and organs to the host-parasite relationship. The only sense in which helminths are immunologically less complicated than the protozoa is that nearly all follow extracellular modes of life rather than sequestering within a host cell. Nevertheless, the spectrum of potential antigens from these parasites is immense, and as with malaria, research has been pursued in parallel on a wide range of antigenic products from the three helminth groups, trematodes, cestodes and nematodes. The trematode schistosomes have been the most intensely studied helminth parasites, but a range of interesting approaches have also been followed with tissue-dwelling nematodes, such as the filariae, and the highly prevalent gut nematodes, such as hookworms.

Schistosome antigens

The major focus of research in schistosome antigens has been on the products expressed by invading schistosomulae as irradiated schistosomules can stimulate a degree of protective immunity. A large array of antigens have been described (Simpson and Cioli, 1987), but among the most promising are two surface-associated products, a 38 kDa glycoprotein from schistosomules and a 26–28 kDa set of proteins shown recently to be isomers of the enzyme glutathione-S-transferase. However, it is becoming clear that either internal or secreted antigens of *Schistosoma* can play a critical part in determining whether host recognition has a protective outcome (Figure 2.6).

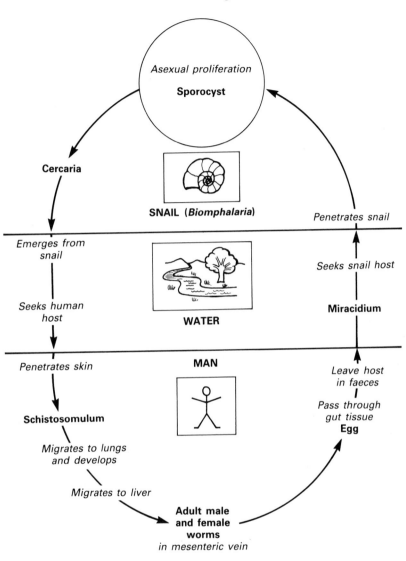

Figure 2.6. Life cycle of *Schistosoma mansoni*.

The first of these selected antigens, the 38 kDa, is a glycoprotein found on the brush-border-like tegument of the schistosomule stage, the earliest developmental form in human infection. It is clear that this antigen contains a large carbohydrate component which is immunologically most important from several standpoints. Firstly, a monoclonal antibody to a glycan epitope can protect mice from schistosomiasis, but in many natural and experimental situations this antigen elicits ineffective isotypes of antibody ('blocking antibody') rather than protective antibodies (see p. 376, Capron *et al.*, 1987).

Secondly the carbohydrate determinant is shared with a 200 kDa surface antigen and, most unexpectedly, with products from molluscs including species which may act as vectors for schistosomiasis (Dissous *et al.*, 1986). This carbohydrate group has been the subject of successful anti-idiotype mimicking in which a second antibody is raised to the binding site of the monoclonal to the carbohydrate. The second antibody selected has a similar three-dimensional profile to that of the original carbohydrate epitope and will successfully stimulate animals to produce antibodies reactive with the 38 kDa antigen (Gryzch *et al.*, 1982).

A quite separate antigen was found by correlating differences in antibody responses and resistance to infection in genetically distinct mouse strains. In particular, the 129/J mouse was found to be relatively resistant to *Schistosoma japonicum* infection and to react strongly with a 26 kDa protein (Sj 26) in parasite extracts. Once this antigen had been cloned, its amino acid sequence determined from the DNA sequence was found to be homologous to a mammalian enzyme, glutathione-S-transferase (Smith *et al.*, 1986), which acts as a general detoxification enzyme by donating electrons to potentially damaging electrophiles. Thus, an immune response to Sj 26 may impair the normal repair mechanism of the schistosome, accentuating other host defence effects and leading to parasite death (Mitchell, 1989).

Both the 38 kDa and GST are prominent surface antigens of schistosomes. However, certain internalized proteins are now known to provoke significant immune responses of a potentially protective nature. For example, paramyosin, an invertebrate muscular protein of 97 kDa, elicits a strong delayed-type hypersensitivity response in mice which helps eliminate schistosomes even though there appears to be no direct access to this antigen in the intact parasite (James *et al.*, 1985), and immunization of mice with paramyosin induces a level of protection approximately equivalent to that evoked by schistosome glutathione-S-transferase (Pearce *et al.*, 1988).

A further set of schistosome antigens of related interest are the proteolytic enzymes produced at various stages of the life cycle. For example, the free-swimming cercariae of *S. mansoni* contain preacetubular glands prepackaged with an elastase enzyme which is released on contact with human skin lipids (McKerrow *et al.*, 1985). This 30 kDa mol. wt. protease is required for penetration through the dermal layers to the underlying vasculature, and monoclonal antibodies raised against it mediate complement-dependent cytotoxicity to cercariae *in vitro* (Pino-Heiss *et al.*, 1986). Immunologically, this and other enzymes produced by later life cycle stages of *Schistosoma* are attractive antigens to study because an immune response to them may functionally block the parasite leading to immobilization of migratory stages and perhaps greater susceptibility to host-killing mechanisms (see McKerrow and Doenhoff, 1988).

Nematode antigens

There is an extensive range of nematode parasites which cause widespread disease in man and domestic animals, but in only a few species have the parasite antigens been analysed in any depth. Generally, attention has focused on secreted antigens, an important issue for all helminths (Lightowlers and Rickard, 1988), and on surface components. The surface of nematodes differs from the other helminths in being composed of a thick extracellular cuticle external to the plasma membrane of the outermost cell tissue, the hypodermis. Although the cuticle is rich in structural collagens, these generally are not exposed to the host-immune system; interest has mainly been focused on the non-collagenous components of the epicuticle, the effective surface coat of these parasites. Three groups of nematodes have been studied in comparative detail: *Trichinella*, the filarial nematodes (*Brugia* and *Onchocerca*) and the intestinal nematodes (such as *Ancylosotoma*, *Necator* and *Toxocara*).

In *Trichinella*, studies established some time ago that each stage of the parasite life cycle expressed exclusively different surface antigens as it moved from newborn larvae in the gut to the muscle (infective larvae) form and then again on ingestion by a new host to adult worm in the gut (see fig. 13.15, Philipp *et al.*, 1981). Of these stages, the muscle larvae are perhaps the most interesting, and a 48 kDa antigen appears particularly prominent in these parasites. This antigen is packed in quantity into granules within specialized cells (the stichocytes), but is also found in various processed forms on the larval epicuticle. Moreover, the 48 kDa is secreted by parasites *in vitro*, and if purified, provides an effective level of immunization against subsequent challenge (Silberstein and Despommier, 1984).

Filarial nematodes, like *Trichinella*, can survive for many years in mammalian hosts, but clearly do not show exclusively stage-specific antigens; for example, surface antigens are found which are expressed by more than one stage and cross-reactive epitopes are present on molecules of different molecular weight from different stages (Maizels *et al.*, 1983). Although these parasites are macroscopic, growing up to 50 cm length in the case of *Onchocerca*, the epicuticle seems to carry only one or a few proteins: a prominent 20 kDa surface glycoprotein from the *Onchocerca* group (Philipp *et al.*, 1984), and a glycoprotein of 29 kDa on the lymphatic filarial worms, *Brugia* and *Wuchereria* (Devaney, 1988; Maizels and Selkirk, 1988). Unlike related malaria species, which have evolved dramatically different antigenic structures, different filarial species show strong structural and immunological conservation of these major antigens (Figure 2.7). This implies both that the surface glycoproteins perform an essential role for the parasite and that the infected host is unable to make an adequate or appropriate immune response to these antigens.

One prominent antigen to which nematode-infected hosts have no problem making antibody responses is phosphorycholine (PC), a small haptenic epitope attached to polysaccharides or proteins. PC is found in a broad range of

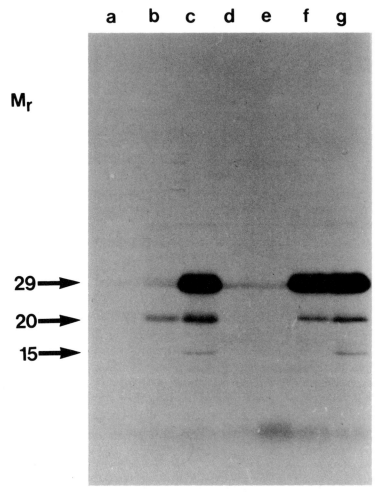

Figure 2.7. Antibodies to surface antigens cross react between filarial species. Surface proteins from *Brugia pahangi* (mol. wt. 15, 20 and 29 kDa) are precipitated by antibodies from (b) human cases of onchocerciasis; (c) *W. bancrofti*; (f) *Loa loa*; (g) *Mansonella perstans*. The non-filarial nematode, *Strongyloides stercoralis*, does not generate anti-*Brugia* reactive antibody (d, e); and no reactivity is present in normal non-endemic sera (a) (Maizels *et al.*, 1985).

nematode species (Péry *et al.*, 1974), including filariae, and the antibody provoked by it cross reacts between the stages of filarial life cycle and between the many different species which express PC (Gualzata *et al.*, 1986). This is one reason why immunological diagnosis of nematode infections, aimed at detecting antibody in human patients to extracts from parasites, has been so poorly specific, and conversely why recombinant parasite peptides, free from PC side chains and free from potentially cross-reactive oligosaccharide groups, offer excellent diagnostic reagents.

The gastrointestinal nematodes are major parasites of man and other vertebrates and can be subdivided by their mode of entry into the host. Some, like *Toxocara* and *Ascaris*, are ingested orally and enter a tissue cycle before returning to the gut. The invasive form of *Toxocara canis*, which zoonotically infects humans, actively secretes a set of glycoproteins which bear both species-specific and cross-reactive carbohydrate determinants (Maizels *et al.*, 1987). These secretions are known to have functional protease activity which, as with the schistosomula, are presumed to facilitate the migrations of developing larvae within the body.

Similar proteolytic enzymes are also major components from hookworms which gain entry to their final host by direct penetration through the skin, again presumably employing these proteases. Interestingly, such enzymes are also secreted by the adult hookworms established within the intestine; in this case they function to minimize coagulation around the adult worms feeding on the host blood supply (Hotez *et al.*, 1987). The principal protease in the dog hookworm, *Ancylostoma caninum*, is 37 kDa, while that of the human parasite, *Necator americanus*, is smaller at 33 kDa (Pritchard, 1986). These secreted enzymes may make ideal vaccine antigens, eliciting a response which impairs parasite feeding thus reducing worm burdens and preventing the blood loss which is the major pathological consequence of hookworm infection.

2.4. ANTIGENS CONSERVED BETWEEN HOST AND PARASITE

One of the general findings emerging from recent molecular studies is that conserved structural or 'housekeeping' proteins may be highly antigenic even though they retain close homology to proteins found in their mammalian hosts. Thus, structural components like myosin, actin, collagen and tubulin are common constituents of both helminth and human tissue, and tubulin is also found in intracellular structures of protozoa. Several of these shared proteins have been examined in detail.

One such shared component is Hsp-70, a stress protein termed 'heat shock protein', of 70 kDa. In schistosomes, these are abundant products with close homology to the mammalian host protein, yet the host immune system nevertheless recognizes Hsp-70 from the parasite product as foreign (see Newport *et al.*, 1988). Significantly, antibody is aimed at the C-terminal regions which have diverged in the course of evolution rather than reacting to the conserved sequences towards which the immune system maintains a state of tolerance (Hedstrom *et al.*, 1988). Closely related Hsp molecules have been found in malaria and in filarial nematodes. Despite the close homology, however, host antibody appears to discriminate between Hsp-70 products from different species because it recognizes only the species-specific

C-terminal epitopes. A very similar restriction of antigenicity appears in the recognition of schistosome myosin (Newport *et al.*, 1987).

A further example of an antigenic 'housekeeping' product, but one with an immunogenic potential, is the aldolase enzyme recently found to equate with the 41 kDa antigen of *Plasmodium falciparum*. Pf41 was originally identified as a protective antigen located in the merozoite rhoptry, a polar organelle involved in red cell invasion. Aldolase is an attractive target for vaccination as its functional constraints should counteract the tendency for antigenic variation in malaria. Indeed, so far no polymorphism has been found in pf41, which is now being tested as an experimental vaccine against malaria (Certa *et al.*, 1988). However, because of the close homology (60% by amino acid sequence) with host aldolase, it is difficult to predict whether the immune system would maintain high levels of antibody to this product without frequent reimmunization.

2.5. RECOMBINANT DNA AS A SOURCE OF PARASITE ANTIGEN

It will be clear from the above discussion of individual parasite antigens, that recombinant DNA clones have been a key to the structural and functional analysis of parasite products. These clones have been isolated from gene banks derived from parasite genomic DNA (which would include all potential parasite genes but also an unnecessarily large proportion of non-coding DNA) or from cDNA synthesized as complementary to mRNA isolated from a particular stage (and thus restricted entirely to authentic genes actively being transcribed by that stage).

Although DNA cloning is generating primary structural information at an accelerating rate, the conceptual problems remain of relating these structures to antigen function and immune recognition. Generally, the critical choice for the immune system is not whether to respond to a given antigen, but which epitopes to recognize and which arms of the immune system to select to counteract infection. Thus, within the broad biological range of parasitic infections are many examples of antigens containing non-protective 'decoy' epitopes (such as the malaria CSP), antigens to which antibody of one isotype will eliminate parasites but of another will block killing (the schistosome 38 kDa; see Chapter 13), and antigens to which either a T-cell or B-cell response is preferentially desirable. If it is appreciated that such hierarchies will be an individual characteristic of each organism, then it becomes clearer that a broad range of studies have still to be undertaken before we understand parasite antigens.

REFERENCES

Anders, R.F. (1986), Multiple cross-reactivities amongst antigens of *Plasmodium falciparum* impair the development of protective immunity against malaria, *Parasite Immunology*, **8**, 529–39.

Anders, R.F., Howard, R.J. and Mitchell, G.F. (1982), Parasite antigens and methods of analysis, in Cohen, S. and Warren, K.S. (Eds.), *Immunology of Parasitic Infections*, pp. 28–73, Oxford: Blackwell.

Arnot, D.E., Barnwell, J.W., Tam, J.P., Nussenzweig, V., Nussenzweig, R.S. and Enea, V. (1985), Circumsporozoite protein of *Plasmodium vivax*: cloning of the immunodominant epitope, *Science (Washington)*, **230**, 815–8.

Borst, P. and Cross, G.A.M. (1982), Molecular basis for trypanosome antigen variation, *Cell*, **29**, 291–303.

Capron, A., Dessaint, J.P., Capron, M., Ouma, J.H. and Butterworth, A.E. (1987), Immunity to schistosomes: progress toward vaccine, *Science (Washington)*, **238**, 1065–72.

Certa, U., Ghersa, P., Döbeli, H., Matile, H., Kocher, H.P., Shrivastava, I.K., Shaw, A.R. and Perrin, L.H. (1988), Aldolase activity of a *Plasmodium falciparum* protein with protective properties, *Science (Washington)*, **240**, 1036–8.

Cowman, A.F., Saint, R.B., Coppel, R.L., Brown, G.V., Anders, R.F. and Kemp, D.J. (1985), Conserved sequences flank variable tandem repeats in two S-antigen genes of *Plasmodium falciparum*, *Cell*, **40**, 775–83.

Cowman, A.F., Coppel, R.L., Saint, R.B., Favaloro, J., Crewther, P.E., Stahl, H.D., Bianco, A.E., Brown, G.V., Anders, R.F. and Kemp, D.J. (1984), The ring-infected erythrocyte surface antigen (RESA) polypeptide of *Plasmodium falciparum* contains two separate blocks of tandem repeats encoding antigenic epitodes that are naturally immunogenic in man, *Molecular Biology and Medicine*, **2**, 207–21.

Devaney, E. (1988), The biochemical and immunochemical characterisation of the 30 kilodalton surface antigen of *Brugia pahangi*, *Molecular and Biochemical Parasitology*, **27**, 83–92.

Dissous, C., Gryzch, J.M. and Capron, A. (1986), *Schistosoma mansoni* shares a protective oligosaccharide epitope with freshwater and marine snails, *Nature (London)*, **323**, 443–5.

Donelson, J.E. and Turner, M.J. (1985), How the trypanosome changes its coat, *Scientific American*, **252**, 32–9.

Etges, R., Bouvier, J. and Bordier, C. (1986), The major surface protein of *Leishmania* promastigotes is a protease, *Journal of Biological Chemistry*, **261**, 9078–101.

Ferguson, M.A.J., Homans, S.W., Dwek, R.A. and Rademacher, T.W. (1988), Glycosylphosphatidylinositol moiety that anchors *Trypanosoma brucei* variant surface glycoprotein to the membrane, *Science (Washington)*, **239**, 753–9.

Godson, N.G. (1985), Molecular approaches to malaria vaccines, *Scientific American*, **252**, 32–9.

Good, M.F., Berzofsky, J.A. and Miller, L.H. (1988), The T cell response to the malaria circumsporozoite protein: An immunological approach to vaccine development, *Annual Review of Immunology*, **6**, 663–88.

Gryzch, J.M., Capron, M., Lambert, P.H., Dissous, C., Torres, S. and Capron, A. (1982), An anti-idiotype vaccine against experimental schistosomiasis, *Nature (London)*, **316**, 74–6.

Gualzata, M., Weiss, N. and Heuser, C.H. (1986), *Dipetalonema viteae*: Phosphorylcholine and non-phosphorylcholine antigenic determinants in infective larvae and adult worms, *Experimental Parasitology*, **61**, 95–102.

Hedstrom, R., Culpepper, J., Schinski, V., Agabian, N. and Newport, G. (1988), Schistosome heat-shock proteins are immunologically distinct host-like antigens, *Molecular and Biochemical Parasitology*, **29**, 275–82.

Holder, A.A. (1985), Glycosylation of the variant surface antigens of *Trypanosoma brucei*, *Current Topics in Microbiology and Immunology*, **117**, 57–75.

Holder, A.A., Lockyer, M.J., Odink, K.G., Sandhu, J.S., Riveros-Moreno, V., Nicholls, S.C., Hillman, Y., Davey, L.S., Tizard, M.L.V., Schwartz, R.T. and Freeman, R.R. (1985), Primary structure of the precursor to the three major surface antigens of *Plasmodium falciparum* merozoites, *Nature (London)*, **317**, 270–3.

Hotez, P.J. Le Trang, N. and Cerami, A. (1987), Hookworm antigens: the potential for vaccination, *Parasitology Today*, **3**, 247–9.

James, S.L., Pearce, E.J. and Sher, A. (1985), Induction of protective immunity against *Schistosoma mansoni* by a non-living vaccine. I. Partial characterization of antigens recognized by antibodies from mice immunized with soluble schistozome extracts, *Journal of Immunology*, **134**, 3432–8.

Kaslow, D.C., Quakyi, I.A., Syin, C., Raum, M.G., Keister, D.B., Coligan, J.E., McCutchan, T.F. and Miller, L.H. (1988), A vaccine candidate from the sexual stage of human malaria that contains EGF-like domains, *Nature (London)*, **333**, 74–6.

Lightowlers, M.W. and Rickard, M.D. (1988), Excretory-secretory products of helminth parasites: effects on host immune responses, *Parasitology*, **96**, S123–66.

Maizels, R.M. and Selkirk, M.E. (1988), Antigens of filarial nematodes, *ISI Atlas of Science: Immunology*, **1**, 143–8.

Maizels, R.M., Partono, F., Oemijati, S., Denham, D.A. and Ogilvie, B.M. (1983), Cross-reactive surface antigens on three stages of *Brugia malayi*, *B. pahangi* and *B. timori*, *Parasitology*, **87**, 249–63.

Maizels, R.M., Sutanto, I., Gomez-Priego, A., Lillywhite, J. and Denham, D.A. (1985), Specificity of surface molecules of adult *Brugia* parasites: cross-reactivity with antibody from *Wuchereria*, *Onchocerca* and other human filarial infections, *Tropical Medicine and Parasitology*, **36**, 233–7.

Maizels, R.M., Kennedy, M.K., Meghji, M., Robertson, B.D. and Smith, H.V. (1987), Shared carbohydrate epitopes on distinct surface and secreted antigens of the parasitic nematode *Toxocara canis*, *Journal of Immunology*, **139**, 207–14.

McKerrow, J.H. and Doenhoff, M.J. (1988), Schistosome proteases, *Parasitology Today*, **4**, 334–40.

McKerrow, J.H., Jones, P., Sage, H. and Pino-Heiss, S. (1985), Proteinases from invasive larvae of the trematode parasite *Schistosoma mansoni* degrade connective-tissue and basement-membrane macromolecules, *Biochemical Journal*, **231**, 47–51.

Metcalf, P., Blum, M., Freymann, D., Turner, M. and Wiley, D.C. (1987), Two variant glycoproteins of *Trypanosoma brucei* of different sequence classes have similar 6Å resolution X-ray structures, *Nature*, **325**, 84–6.

Mitchell, G.F. (1989), Glutathione S-transferases — potential components of anti-schistosome vaccines, *Parasitology Today*, **5**, 34–7.

Mitchell, G.F. and Handman, E. (1986), The glycoconjugate derived from a *Leishmania major* receptor for macrophages is a suppressogenic, disease-promoting antigen in murine cutaneous leishmaniasis, *Parasite Immunology*, **8**, 255–63.

Newport, G., Culpepper, J. and Agabian, N. (1988), Parasite heatshock proteins, *Parasitology Today*, **4**, 306–12.

Newport, G.R., Harrison, R.A., McKerrow, J., Tarr, P., Kallestad, J. and Agabian, N. (1987), Molecular cloning of *Schistosoma mansoni* myosin, *Molecular and Biochemical Parasitology*, **26**, 29–38.

Nussenzweig, R.S. and Nussenzweig, V. (1984), Development of sporozoite vaccines, *Philosophical Transactions of the Royal Society Series B*, **307**, 117–28.

Pearce, E.J., James, S.L., Hieny, S., Lanar, D.E. and Sher, A. (1988), Induction of protective immunity against *Schistosoma mansoni* by vaccination with schistosome paramyosin (Sm97), a nonsurface parasite antigen, *Proceedings of the National Academy of Sciences (USA)*, **85**, 5678–82.

Péry, P., Petit, A., Poulain, J. and Luffau, G. (1974), Phosphorylcholine-bearing components in homogenates of nematodes, *European Journal of Immunology*, **4**, 637–9.

Philipp, M., Taylor, P.M., Parkhouse, R.M.E. and Ogilvie, B.M. (1981), Immune response to stage-specific surface-antigens of the parasite nematode *Trichinella spiralis*, *Journal of Experimental Medicine*, **154**, 210–5.

Philipp, M., Gómez-Priego, A., Parkhouse, R.M.E., Davies, M.W., Clark, N.W.T., Ogilvie, B.M. and Beltrán-Hernández, F. (1984), Identification of an antigen of *Onchocerca volvulus* of possible diagnostic use, *Parasitology*, **89**, 295–309.

Pino-Heiss, S., Pettit, M., Beckstead, J.H. and McKerrow, J.H. (1986), Preparation of mouse monoclonal antibodies and evidence for a host immune response to the preacetabular gland proteinase of *Schistosoma mansoni* cercariae, *American Journal of Tropical Medicine and Hygiene*, **35**, 536–43.

Pritchard, D.I. (1986), Antigens of gastrointestinal nematodes, *Transactions of the Royal Society for Tropical Medicine and Hygiene*, **80**, 728–34.

Schofield, L., Villaquiran, J., Ferreira, A., Schellekens, H., Nussenzweig, R. and Nussenzweig, V. (1987), γ Interferon, CD8+ T cells and antibodies required for immunity to malaria sporozoites, *Nature (London)*, **330**, 664–6.

Silberstein, D.S. and Despommier, D.D. (1984), Antigens from *Trichinella spiralis* that induce a protective response in the mouse, *Journal of Immunology*, **132**, 898–904.

Simpson, A.J.G. and Cioli, D. (1987), Progress towards a defined vaccine for schistosomiasis, *Parasitology Today*, **3**, 26–8.

Smith, D.B., Davern, K.M., Board, P.G., Tiu, W.U., Garcia, E.G. and Mitchell, G.F. (1986), M$_r$ 26,000 antigen of *Schistosoma japonicum* recognized by resistant WEHI 129/J mice is a parasite glutathione-S-transferase, *Proceedings of the National Academy of Sciences (USA)*, **83**, 8703–7.

Suhrbier, A., Holder, A.A., Wiser, M.F., Nicholas, J. and Sinden, R.E. (1989), Expression of the precursor of the major merozoite surface antigens during the liver stage of malaria, *American Journal of Tropical Medicine and Hygiene*, **40**, 19–23.

Targett, G.A.T. and Sinden, R.E. (1985), Transmission blocking vaccines, *Parasitology Today*, **1**, 155–8.

Townsend, A.R.M., Bastin, J., Gould, K. and Brownlee, G.G. (1986), Cytotoxic T lymphocytes recognize influenza haemagglutinin that lacks a signal sequence, *Nature (London)*, **324**, 575–7.

Turco, S.T. (1988), Proposal for a function of the *Leishmania donovani* lipophosphoglycan, *Biochemical Society Transactions*, **16**, 259–61.

Turner, M.J. (1988), Trypanosome Variant Surface Glycoprotein, in Englund, P.T. and Sher, A. (Eds.), *The Biology of Parasitism*, pp. 349–69, New York: Alan R. Liss, Inc.

Vermeulen, A.N., Ponnudurai, T., Beckers, P.J.A., Verhave, J.-P., Smits, M.A. and Meuwissen, J.H.E.T. (1985), Sequential expression of antigens in sexual stages of *Plasmodium falciparum* accessible to transmission-blocking antibodies in the mosquito, *Journal of Experimental Medicine*, **162**, 1460–76.

3. Differentiation of bone marrow cells into effector cells

D.M. Haig and H.R.P. Miller

3.1. INTRODUCTION

Histological analysis of inflammatory reactions in the vicinity of parasites has, over the last 100 years, revealed some fairly characteristic lesions which, with the development of *in vitro* techniques for analysis of haemopoiesis and for the study of cytokines, can begin to be understood at the molecular level. For example, it has long been known that infection with most helminth parasites is associated with eosinophilia and with a local accumulation of eosinophils within the vicinity of the helminth. Similarly, nematodiasis is frequently associated with the accumulation of mast cells and basophils in the gastrointestinal mucosa (Miller, 1984). Typically, schistosome egg granulomas are populated by eosinophils, macrophages and giant cells, the precursors of which originate from haemopoietic tissue (adult bone marrow and spleen). Inflammation can vary at different stages of infection; for example, invasion of the intestine by the protozoan *Eimeria* sp. is initially associated with neutrophil accumulation and, subsequently, with infiltration of mononuclear cells, eosinophils and mast cells (Rose and Hesketh, 1982). In effect, different species of parasites are associated with variant local inflammatory responses as assessed by histological and immunocytochemical techniques.

Because all of the recruited cells, including lymphoid elements, in an inflammatory focus are originally derived from bone marrow, the function of this organ is vital to the development of parasite-induced inflammation. Experiments in laboratory animals over the last decade have clearly established that bone marrow is itself a vital element of the protective response against parasites and that individual populations of bone marrow-derived cells, such as macrophages, can play a crucial role in the protective response against many parasites. Whilst it is generally recognized that inflammatory cells derived from bone marrow are non-specific effectors, they are specifically targeted against the parasite by elements of the immune response, most notably through the production of inflammatory cytokines by T cells and, also, by the arming of inflammatory cells with antigen-specific and often cytophilic immunoglobulins.

3.2. BONE MARROW AS THE SOURCE OF INFLAMMATORY CELLS

Eight major families of haemopoietic cells have been identified: neutrophil, eosinophil and basophil granulocytes, monocytes, mast cells, megakaryocytes, erythroid cells and lymphocytes. The first seven families have no antigen specificity but, as will be discussed below, many bone marrow-derived cells can be directed to sites of inflammation by the specific interaction of parasite antigens with B cell immunoglobulins or with T cells which then produce a variety of lymphokines. However, a crucial element in the build up of this inflammatory response is the capacity of the stem cells in bone marrow to self replicate and to generate differentiated progenitor cells.

Haemopoietic cells in adult bone marrow or spleen can be divided into three major groups (Lord, 1983; Metcalf, 1984; Figure 3.1): (i) Pluripotential stem cells (PSC), with the capacity to generate all the blood cell lineages, are capable of self renewal and are detected by their ability to protect lethally irradiated animals following adoptive transfer (Till and McCulloch, 1961). A small number of these cells (0.05–0.2% of normal mouse bone marrow cells) are maintained throughout life by dividing to produce one daughter cell of the parental phenotype and the other capable of differentiating further. Multipotential stem cells are the immediate progeny of PSC and can generate several hundred progenitor cells within 7–10 days. (ii) Progenitor cells comprize approximately 1.5% of normal mouse bone marrow cells and are generally committed to one or at most two cell lineages while not yet morphologically identifiable. One progenitor cell can, over 7–14 days, generate up to 10 000 mature progeny. Multipotential stem cells and progenitor cells can be detected *in vitro* by their ability to form colonies of cells in semi-solid agar or methycellulose (Bradley and Metcalf, 1966). (iii) Differentiated cells are generated from progenitor cells and are morphologically recognizable as the immediate precursors of mature myeloid and erythroid cells. Under normal conditions of haemopoiesis progenitor cells cannot revert to stem cells and, similarly, differentiated cells cannot revert to the progenitor compartment.

There are two forms of haemopoiesis: (i) Steady-state haemopoiesis is the maintenance of controlled numbers of all the blood cell types; the turnover of stem/progenitor cells is stimulated by contact with accessory bone marrow/ haemopoietic stromal cells (Dexter *et al.*, 1977; Lord, 1983). Stromal cells

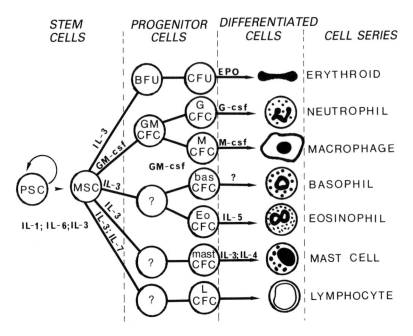

Figure 3.1. Haemopoietic cell compartments
Abbreviations: PSC = pluripotential stem cell; MSC = multipotential stem cell; BFU = erythroid burst forming unit; GM-CFC = granulocyte-macrophage colony forming cell; G-CFC = granulocyte (neutrophil)-CFC; M-CFC = monocyte-CFC; bas = basophil; Eo = eosinophil; L = lymphocyte; EPO = erythropoietin. For cytokine abbreviations, see Table 1.

may secrete several substances to maintain haemopoiesis including the cytokines IL-6, GM-CSF, IL-7, G-CSF, M-CSF (see Table 3.1). During steady state haemopoiesis small numbers of progenitor cells are distributed to various tissues around the body. (ii) Induced haemopoiesis occurs after certain traumas [e.g. blood loss (shock) and irradiation] or can be mediated by the immune response. Under these conditions haemopoiesis is drastically accelerated. Immune response-induced haemopoiesis is effected primarily by activated T cells producing IL-6, IL-3 and GM-CSF on contact with antigen (Miyajima *et al.*, 1988; see below). The term induced haemopoiesis will be used hereafter to refer to immune response induced haemopoiesis.

Neutrophil, eosinophil and basophil lineages develop within the bone marrow although, during infection, foci of haemopoiesis may also occur in other tissues. The bone marrow acts as a reservoir of granulocytic cells (particularly neutrophils). By contrast, monocytes are seeded into lymphoid and other tissues where they differentiate into mature macrophages often displaying morphological and functional heterogeneity depending on tissue localization. Similarly, mast cell progenitors are seeded into the tissues and develop their characteristic morphology *in situ*. Like macrophages, mast cells are heterogeneous, with two major subpopulations in mucosal and connective tissues. What is not yet clear is whether there are committed progenitors for each subset of macrophage and mast cell or whether the local tissue environment drives the development of single, committed progenitor cell lines into several functionally distinct subsets of mature cells.

3.3. CONTROL OF BONE MARROW DIFFERENTIATION

Cytokines drive bone marrow cell differentiation

Cytokines are secreted regulatory molecules which include the interleukins, interferons, and colony-stimulating factors. Interleukins (IL) are now defined as secretory glycoproteins of leucocytes that are involved in inflammatory responses (see *Immunology Today*, 1986; Paul, 1988). Interferons (IFN) are a family of proteins defined by their ability to induce an anti-viral state in target cells (Stewart, 1980), and colony-stimulating factors (CSF) are glycoproteins which stimulate the survival, proliferation and differentiation of haemopoietic cells (Metcalf, 1984). Recently a large number of human and mouse cytokine genes have been cloned and recombinant products studied. Those reactive on bone marrow cells are listed in Table 3.1. The following additional and often complicating observations have been made.

Spectrum of activity

Individual cytokines can often stimulate a wide spectrum of activities *in vitro* by interacting with a variety of target cells. Examples are IL-1 and IL-6. The IL-1 gene products are amongst the most pleiotropic molecules known, targeting a range of cells including T cells, B cells, macrophages, fibroblasts, certain brain and liver cells (Dinarello, 1984) as well as stimulating haemopoietic stem cells (Mochizuki *et al.*, 1987; Leary *et al.*, 1988).

Cytokine-cytokine stimulatory interactions

Certain cytokines can stimulate the secretion of other cytokines from target cells. For example IL-1 stimulates production of IL-6 by fibroblasts and IL-6 has activities that are similar to those of IL-1 so it is often difficult to determine which is the functional cytokine under these circumstances (Wong and Clark, 1988).

Synergistic interactions between cytokines

Some cytokines which, on their own, have little or even no detectable activity will synergize with other functionally active cytokines to augment the overall response. Murine IL-4, for example, will augment IL-3-driven mast cell

Table 3.1 Recombinant human and mouse cytokines active on bone marrow cells.

Name (Alternatives)	Source	Principal activity on bone marrow cells	Other functions
1. Interleukin-1 (IL-1; Haemopoietin-1; Lymphocyte activating factor)	Monocytes/macrophages; Fibroblasts	Increases stem cell responses by synergizing with GM-CSF and M-CSF No activity alone.	Many. Fever induction; B-cell stimulation; T-cell activation
2. Interleukin-6 (IL-6; IFN-B2; B-cell stimulating factor-2; Hybridoma growth factor)	T cells; Monocytes/macrophages; Fibroblasts; Mast cells*	Induces IL-3 receptor Expression on stem cells Synergizes with IL-3 to expand stem blast cell numbers.	Anti-viral activity; stimulation of B-cell differentiation and IgG production. T-cell activation factor.
3. Interleukin-3 (IL-3; Multi-CSF; Mast cell growth factor; Burst promoting activity)	Helper T-cells; Mast cells*	Multipotential stem cell growth/differentiation factor. Growth factor for mast cells and early eosinophils, monocytes and megakaryocytes.	Induces 20-α-OH steroid dehydrogenase and THY-1 (in mice)
4. Granulocyte-macrophage colony stimulating factor (GM-CSF)	Helper T-cells Monocytes/macrophages; Fibroblasts; Mast cells*	Growth/differentiation factor for stem/progenitor cells, especially neutrophil and monocyte lineages	Activates neutrophil and macrophage phagocytic and cytotoxic functions

5. Granulocyte-colony stimulating factor (G-CSF)	Monocytes/macrophages; Fibroblasts; Endothelial cells	Neutrophil growth factor.	Activates neutrophil function
6. Monocyte-colony stimulating factor (M-CSF)	Monocytes/macrophages; Fibroblasts; Embryonic Tissue	Monocyte/macrophage growth factor	Activates macrophage function
7. Interleukin-4 (IL-4); B-cell stimulating factor-1; T-cell growth factor-II	Helper T-cells; Mast cells*	Mast cell growth factor Synergizes with IL-3 (in the mouse)	Stimulates IgG_1 and IgE production. Augments MHC class II antigen expression on B cells. T-cell growth factor.
8. Interleukin-5 (IL-5); B-cell growth factor II; T-cell replacing factor	Helper T-cells; Mast cells*	Eosinophil differentiation factor	Stimulates IgA production, B-cell proliferation and eosinophil function
9. Interferon-γ (IFN-γ; macrophage activating factor)	Helper T-cells	Antagonistic effect on IL-3/GM-CSF induced stem cell proliferation	Stimulates IgG_{2a} production; anti-viral activity; induces/augments MHC Class II molecule on a variety of cells.
10. Erythropoietin (EPO)	Kidney epithelial cells	Stimulates full development of red blood cells	
11. Interleukin-7 (IL-7)	Bone marrow stromal cells	Supports pre-B lymphocyte growth *in vitro*	

* Plant *et al.*, 1989; Wodnar-Filipowicz *et al.*, 1989.

proliferation (Smith and Rennick, 1986), and IL-1 will synergize with GM-CSF and M-CSF to augment haemopoietic stem/progenitor cell proliferation and differentiation (Cosman, 1988). Figure 3.1 shows the activities of the various cytokines on the different haemopoietic cell compartments. This is highly schematic and should be studied with the above points in mind.

Cytokines act in a hierarchical way to control haemopoiesis

When IL-3 occupies its receptor on stem/progenitor cells, GM-CSF, M-CSF and G-CSF are inhibited from binding. This is not due to competition for receptor but is a consequence of receptor down modulation induced by occupancy of IL-3 receptors (Walker *et al.*, 1985). GM-CSF down modulates G- and M-CSF receptors, whereas G-CSF only down modulates M-CSF receptors when present at high concentration. Thus a hierarchical down modulation (and hence activation) of CSF receptors can be demonstrated. This parallels the known cell lineage stimulating potential of the four growth factors. For the maximum possible expansion of, for example, neutrophils, haemopoietic cells must first be stimulated by IL-1/IL-6 followed by IL-3 followed by GM-CSF and finally G-CSF. The continuous presence of G-CSF and GM-CSF during the initial stages of stem cell division encourages synergistic interactions between the cytokines, further expanding granulopoiesis. Similarly a maximal eosinophil response would require IL-3/GM-CSF and IL-5, and a maximal mast cell response (in the mouse) IL-3 and IL-4. As helper T cells are a major source of IL-3, GM-CSF, IL-4 and IL-5 (Miyajima *et al.*, 1988; Table 3.1 and see below), the above are examples of inductive haemopoiesis. In contrast, steady-state haemopoiesis is less well understood and probably depends on stochastic mechanisms of stem cell cycling and the availability of cytokines from bone marrow stromal cells, e.g. IL-7 (for B-cell precursor development, Hunt *et al.*, 1987; Goodwin *et al.*, 1989), GM-CSF and possibly IL-6.

Differential production of cytokines by helper T-cell subsets

Two helper T-cell subsets (Th$_1$ and Th$_2$) have been identified in the mouse on the basis of the cytokines that they are capable of secreting on contact with specific antigen or mitogen (Mosmann and Coffman, 1987; Bottomly, 1988; Figure 3.2). Note that both types secrete the haemopoietically active molecules IL-3 and GM-CSF thereby underlining the importance of a progenitor pool expansion in the generation of inflammatory responses. Th$_1$ cells mediate DTH reactions and, through the activities of IFN-γ on B cells, select for the IgG$_2$ antibody responses typically seen in the development of immunity to certain bacteria and intracellular parasites. On the other hand Th$_2$ cells produce IL-4 which stimulates mast-cell development and acts on B cells to select IgE and IgG$_1$ responses and IL-5 which stimulates eosinophil differentiation and selects for IgA antibody responses typically seen in

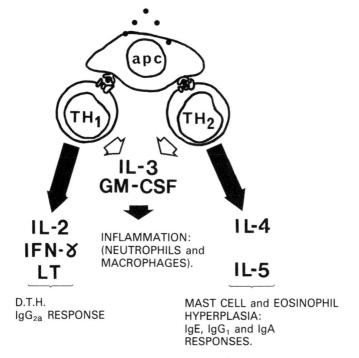

Figure 3.2. Helper T-cell subsets in the mouse.
apc = antigen presenting cell; D.T.H. = Delayed type hypersensitivity reaction; LT = Lymphotoxin see Table 1 for cytokine abbreviation code. Both TH$_1$ and TH$_2$ subsets secrete cytokines in response to activation by antigen presented by an apc.

helminth parasite infections. In a study on cutaneous Leishmaniasis in mice, it has been shown that Th$_1$ cells raised against one antigen of *Leishmania major* could confer protection against infection, whereas Th$_2$ cells recognizing a different antigen from the pathogen exacerbated an infection (Scott *et al.*, 1988). Comparable subdivision of T-helper cells may not occur in man because individual T-cell clones can produce IL-2, IFN-γ and IL-4.

Control of cytokine production

Cytokines are generally biologically active in the 10^{-11} to 10^{-13} molar range. Molecules like IL-3 could have devastating effects if allowed to act either systemically or over a long period of time. Consequently many regulatory mechanisms exist to control cytokine production.

Local secretion

Cytokines are produced and act locally at the site of stimulation. Most have a short half life in serum where they are inhibited by factors which discourage systemic activity.

Transient mRNA production

Cytokine messenger RNA is only transiently produced and at very low concentrations, often not exceeding 0.1% of the total message in the cell. Nucleotide sequences rich in AU in the 3' region of the mRNA ensure its rapid degradation (Miyajima *et al.*, 1988).

Coordinated control of gene activity

Certain cytokine genes are coordinately controlled. For example the genes for IL-2 and IFN-γ in the mouse are adjacent on chromosome 11 and share a highly conserved regulatory region 5' of the coding regions (Fujita *et al.*, 1986). IFN-γ has potent anti-proliferative effects and will suppress over-exuberant responses induced by IL-2. Similarly IL-3 and IL-4 lie on adjacent genes on mouse chromosome 11, and IL-4 may regulate IL-3-induced responses as a consequence of the down modulation of receptors discussed above. IFN-γ has potent antagonistic effects on responses induced by IL-3 and IL-4 and will prevent stem/progenitor cell replication during inductive haemopoiesis (Raefsky *et al.*, 1985).

The complex interactions between the cytokines described above are almost certainly over-simplified since it has been known for some years that 'regulator cascades' in which one cytokine induces the production or release of another can occur in cells as disparate as bone marrow stromal cells, Kupffer cells in the liver, and T cells. Clearly, therefore, the production of cytokines by individual cells within organs as complex as gut, brain, or reproductive tract, will need to be measured before the effect of 'tissue environment' on constitutive and induced haematopoiesis can be fully determined.

3.4. DIFFERENTIATED CELLS OF BONE MARROW ORIGIN

Monocytes/macrophages

Macrophages which are derived from circulating monocytes are widely distributed in the body and are part of a family of mononuclear phagocytes which include Kupffer cells in the liver, dendritic cells, Langerhans cells and veiled cells in peripheral lymph. They contain a variety of lysosomal hydrolytic enzymes (e.g. glycosidases) and proteinases and secrete non-lysosomal enzymes including plasminogen activator, collagenases and elastase (Jessup *et al.*, 1985). Also, when stimulated, they produce sulfidopeptide leukotrienes, prostaglandins, oxygen metabolites and platelet activating factor acether (PAF). Therefore, when activated, they are potent inflammatory cells.

The principal cytokines which activate macrophages once they have left bone marrow are IFN-γ and GM-CSF. For example, *Trypanosoma cruzi*, a flagellated protozoan parasite which normally replicates in many different nucleated host cells including phagocytes, is rapidly killed by macrophages activated by IFN-γ (Murray *et al.*, 1983). Similarly, GM-CSF induces macrophages to produce greater levels of secreted and membrane-bound IL-1 and to express immunoglobulin receptors (FcR) facilitating the phagocytosis of opsonized pathogens (Morrissey *et al.*, 1988).

Macrophages are probably also involved in the elimination of tissue helminths since *in vitro* experiments have established that antibody-dependent cellular cytotoxicity against helminth targets is mediated by IgG and IgE and that macrophages, like other bone marrow-derived inflammatory cells carry $Fc_\gamma R$ and $Fc_\epsilon R$ as well as receptors for complement (Dessaint *et al.*, 1979). Complexes of IgE or IgG on the surface of the parasite, when bound to the appropriate FcR on the surface of the macrophage, trigger the release of lysosomal enzymes and oxygen metabolites as well as the generation of membrane-derived sulfidopeptide leukotrienes, prostaglandins, and PAF (Rouzer *et al.*, 1982). Many of these products are either directly toxic to the parasite or are in some way detrimental to their survival.

Macrophage cytotoxicity is greatly enhanced against helminths in the presence of parasite-specific IgE (Capron *et al.*, 1986). Unlike mast cells which bind IgE through the classical high-affinity (Ka $10^9 M^{-1}$) receptor $Fc_\epsilon RI$, macrophages carry a lower affinity (Ka $10^7 M^{-1}$) $Fc_\epsilon RII$ (Capron *et al.*, 1986). The affinity of the $Fc_\epsilon RII$ increases ten-fold in the presence of IgE aggregates or complexes and, in parasitic infection associated with raised IgE titres, a high proportion of inflammatory cells selectively bind IgE. Once bound, either as dimer or as a multimer to the $Fc_\epsilon RII$, IgE triggers the release of mediators from the macrophages (Capron *et al.*, 1983).

One additional recent observation on the effect of interleukins on macrophage function *in vitro* has established that stimulation with IL-4 causes fusion of macrophage membranes and the formation of giant cells (McInnes and Rennick, 1988). Parasitic granulomas, particularly around schistosome eggs, are rich in mononuclear phagocytes and T cells, and a feature of this, and many other similar lesions, is the formation of giant cells. Whether such giant cells are the consequence of localized production of IL-4 by T cells has yet to be determined.

Overall, macrophages respond to and produce a variety of cytokines. Importantly, several of the cytokines increase the capacity of the cell to produce inflammatory mediators and/or kill pathogens and parasites. The T-cell derived cytokines and cytophilic immunoglobulins are crucial factors in arming macrophages against parasites.

Mast cells

Mast cell precursors are generated in bone marrow and seed out into the tissues, probably as progenitor cells, where they differentiate and develop granules and the functional properties of mast cells (Jarrett and Haig, 1984). They are an apparently less heterogeneous population than the mononuclear

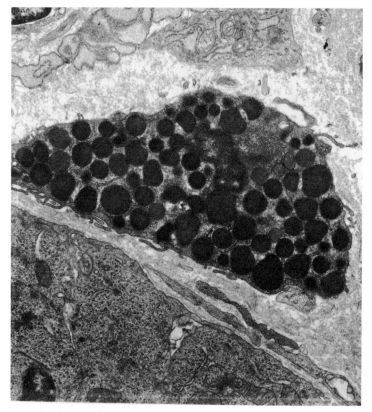

Figure 3.3. Ultrastructure of a rat mucosal mast cell *in situ* in the jejunum.
Note large electron dense cytoplasmic granules which store the mast cell mediators.
Mag: ×11,250

phagocytes, comprizing two major subsets which reside in the mucosae and connective tissues, respectively. The mucosal mast cell differs from connective tissue mast cells in its granule content of highly basic proteinases, acidic proteoglycans and histamine (Enerback, 1987; Figure 3.3). Mucosal mast cells in the rat and mouse contain highly soluble chymotrypsin-like enzymes and, in man, they lack an insoluble chymase which is found in skin mast cells. Both cell types store large amounts of inflammatory mediators which can be released instantaneously.

The differentiation of mast cell progenitors is highly T-cell dependent in the rat and mouse and, in these species as well as man there is evidence to suggest that mucosal mast cells remain T-cell dependent even after they have reached maturity (Miller, 1987). The extensive mast cell proliferation in response to helminth infection (see Chapter 10), occurring particularly in the parasitized intestine is also highly T-cell dependent since it can be adoptively transferred with T cells, does not occur in the athymic or immunocompromized

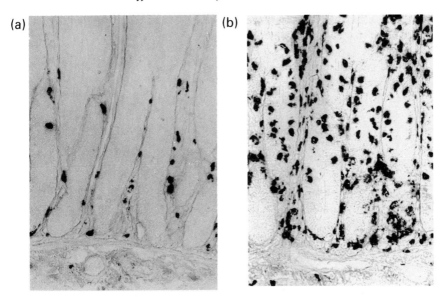

Figure 3.4a. Localization of mast cells in uninfected rat jejunum.
Mag: ×430

Figure 3.4b. Localization of mast cells in rat jejunum 11 days after *N. brasiliensis* infection.
Mag: × 930

host and can be suppressed by drugs which suppress T-cell function (Figure 3.4a, b; Miller, 1987).

Mast-cell recruitment and differentiation in rodents is IL-3 dependent, but it is not yet certain whether IL-3 is responsible for mast cell growth in man because it has proved very difficult to grow human mast cells *in vitro*. In the mouse, IL-4 is implicated in the further differentiation and maturation of cultured mast cells (Miyajima *et al.*, 1988). Since IL-4 also induces synthesis of IgG_1 and IgE by B cells (Miyajima *et al.*, 1988), it is likely that this one lymphokine is of general importance in the control of murine immediate hypersensitivity reactions although it is not yet clear whether IL-4 has a comparable role in man and other species.

The role of lymphokines in the maintenance and recruitment of connective tissue mast cells has yet to be defined, but these cells are very long-lived, unlike the mucosal subset which survive in the rat for only 30–40 days (Enerback, 1981). Both mast cell subsets probably act as sentinel cells and, like macrophages, are in place in skin or enteric mucosa awaiting the arrival of the pathogen or parasite. In a host which has yet to experience the pathogen, the mast cells may be directly activated by secreted parasite products or indirectly through cross-reacting antigens against which the host has already mounted an antibody response. In the sensitized host, the presence of IgE on the mast cell surface or of other anaphylactic immunoglobulins (IgG_4 in man, IgG_{2a} in rat, IgG_1 in guinea pig) in the vicinity of mast cells causes a much more efficient and instantaneous release of the large store of mediators such that, if parasite antigens are injected intravenously into a sensitized animal, it will suffer severe and often fatal anaphylactic shock.

Mast cell surface membranes carry the $Fc_\epsilon RI$ which has a very high affinity for IgE (Metzger, 1983). Therefore when injected intradermally into a non-sensitized host, IgE remains bound to the surfaces of skin mast cells for a number of days in contrast to the anaphylactic IgG immunoglobulins which remain for only a few hours (Ovary, 1981). Mast cells also have receptors for the C3b complement fragment. Once activated via surface receptors, the mast cell releases the contents of the granule to the exterior. Histamine diffuses rapidly away from the granule and causes local increases in vascular permeability; so too does 5-hydroxytryptamine in those species in which mast cells store this monoamine. However, the insoluble chymase and heparin proteoglycan of the connective tissue mast cell probably do not diffuse any distance and exert a local inflammatory effect in the immediate vicinity of the cells (Schwartz and Austen, 1984). By contrast the soluble chymases of mucosal mast cells diffuse readily like histamine and are detected in blood and lymph within minutes of challenge (Miller *et al.*, 1983).

Importantly, mast cells are potent sources of the membrane-derived lipid mediators leukotriene (LT) C_4, D_4 and E_4, platelet activating factor acether, and prostaglandin (PG)D_2. Again, depending on the species, the mucosal mast cells produce LTC_4 and PGD_2 whereas those in the connective tissues generate PGD_2 alone (Brodie *et al.*, 1988).

The precise role mast cells play in parasitic infection has not been fully defined but it is now clear that the recruitment of large numbers of mucosal mast cells to parasitized enteric tissue is associated with activation and release of granule and membrane-derived mediators (Miller, 1984). These may exert a direct toxic or disorienting effect on the parasite and probably cause increased vascular permeability, shedding of epithelial cells, and secretion of mucus, all of which may be detrimental to the survival of enteric parasites (Miller, 1987). Finally, mast cells are thought to release chemotactic peptides which recruit both eosinophils and neutrophils to the site of mast cell activation (Schwartz and Austen, 1984). *In vitro* studies suggest that the mast cell-derived eosinophil chemotactic factor of anaphylaxis (ECFA) increase the ability of eosinophils to kill parasites (Kay, 1988).

Eosinophils

An increase in the number of eosinophils in tissue lesions or in blood is a common feature of parasitic helminthiasis (Miller, 1984; see Chapter 10). However, unlike mast cells and macrophages, the eosinophil differentiates to a morphologically recognizable form within the bone marrow or within foci of haemopoiesis associated with parasitic infection before it is recruited to the tissues as a differentiated cell (Figure 3.5). Eosinophils, like mast cells, contain numerous granules in which the preformed mediators, major basic

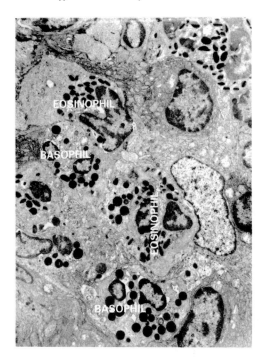

Figure 3.5. Section through rat intestine showing eosinophils and basophils DII following *N. brasiliensis* infection.
Mag: × 3,000

protein, eosinophil cationic protein, eosinophil peroxidase and eosinophil-derived neurotoxin are located. Again, like mast cells and macrophages, stimulated eosinophils generate lipid membrane-derived mediators, LTC_4 and PAF (Kay, 1988).

Whilst the CSFs induce formation of colonies *in vitro* of macrophages and granulocytes, including eosinophils, from bone marrow stem cells, IL-5 in its own right is a potent stimulator of eosinophil growth and differentiation (Silberstein and David, 1987; see Table 3.1). Where IL-5 is produced *in vivo* is not known, nor is it clear whether IL-5 produced at sites distant from bone marrow is active on cells within the marrow or only in the immediate vicinity of the T cell itself.

The mechanisms which promote mediator release by eosinophils have been partially characterized and involve the triggering of membrane components such as the FcγR and complement (CR1 and 3) receptors (Kay, 1985). An extensive study of the role of IgE in the activation of eosinophils has shown that eosinophils have the low-affinity $Fc_εRII$ (Capron *et al.*, 1981). This receptor is more abundant on eosinophils from patients with eosinophilia than on eosinophils from normal patients. Similarly, eosinophils which, on separation through density gradients, are hypodense and thought to be activated express more Fc$_\epsilon$RII than normodense eosinophils (Wardlaw and Kay, 1987). The hypodense eosinophils are abundant in tissues and blood of parasitized hosts, and comparative studies *in vitro* show that these cells kill parasites more effectively than normodense eosinophils. Although Fc$_\epsilon$R and CR1 and 3 are involved in cell activation, expression of these receptors is not increased on hypodense eosinophils and it would seem that interaction with IgE through Fc$_\epsilon$RII is the most crucial event in stimulating mediator release (Wardlaw and Kay, 1987).

The granule constituents which are released following activation comprise (i) major basic protein from the crystalline granule core which is highly toxic to parasites. The release of major basic protein is observed when eosinophils are in contact with IgG-coated schistosomulae. Similarly, this highly toxic protein kills the larvae of *Trichinella spiralis* as well as the protozoan parasite *Trypanosoma cruzi*. (ii) Eosinophil cationic protein from the granule matrix which is again toxic to parasites and, amongst its other properties, causes secretion of histamine by mast cells. (iii) Eosinophil granule peroxidase which is also a matrix protein and in the presence of hydrogen peroxide and halide is microbicidal, and also induces histamine release from mast cells (Wardlaw and Kay, 1987).

In addition to preformed granule mediators, the eosinophils generate large quantities of leukotriene C$_4$ and platelet activating factor (Wardlaw and Kay, 1987). It is not certain whether these mediators affect parasite survival, but they are potent stimulators of smooth muscle, causing, for example, bronchoconstriction, and they also mediate the release of mucus. Intestinal and respiratory tract mucus secretion is enhanced in the presence of both enteric and lung worms, and studies have shown that leukotriene C$_4$ is generated in the intestines of rats infected with *T. spiralis* (Moqbel *et al.*, 1987). It is not known whether leukotriene C$_4$ is generated from mast cells, eosinophils or other bone marrow-derived inflammatory cells (see Chapter 10).

Because eosinophils are recruited to lesions as mature cells, it has been possible to characterize potential chemoattractants by *in vitro* studies of peripheral blood eosinophils. A wide variety of chemotactic molecules have been identified and include ECF-A from mast cells and other peptides formed by the complement system (Wardlaw and Kay, 1987). Histamine, under certain circumstances is chemotactic, but the most potent chemoattractants are the lipid, membrane-derived mediators, leukotriene B$_4$ and platelet activating factor. Many of these chemoattractants are produced by other bone marrow-derived cells after activation.

Overall, therefore, the eosinophil is predominantly a mature effector cell which can be recruited very rapidly from the bone marrow or from other inflammatory foci of haemopoiesis and which has potential to damage or kill tissue-dwelling parasites.

Basophils

Like eosinophils, basophils mature within the bone marrow and are recruited to the inflammatory foci as mature, rapidly available effector cells. They have, however, a number of properties which are similar to those of the mast cell, especially in the number, size and content of their granules. Histamine, proteoglycans and proteinases are present in the granules, so too are major basic protein and lypophospholipase, both of which are also present in eosinophils (Ackerman *et al.*, 1982). Unlike mast cells, basophils are polymorphonuclear leukocytes, have a very short life span of 8–12 days and normally comprize 0.5–1% of the circulating leukocytes and 0.3% of nucleated cells in the bone marrow.

Factors which exclusively support basophil growth have yet to be determined although a proportion of foetal human cord blood leukocytes, grown long term in the presence of IL-3, develop the characteristics of basophils (Ogawa *et al.*, 1983). Again, the cytokines GM-CSF and IL-3 support the growth of colonies in which basophils are detected in low numbers. However, there is ample evidence from adoptive transfer studies in laboratory animals to suggest that basophil recruitment is highly T-cell dependent, although the lymphokines involved have not been characterized (Askenase, 1980).

The histamine content of human basophils is low, and they produce negligible levels of prostaglandin D_2 when compared with pulmonary mast cells. Basophils generate leukotriene C_4 and D_4 at relatively low concentrations and their granules contain serine esterases which have yet to be characterized (Schulman *et al.*, 1983).

Basophils are recruited in large numbers to cutaneous lesions following intradermal sensitization and challenge with soluble antigen (cutaneous basophil hypersensitivity reactions; Askenase, 1980). Similar lesions are seen following viral infection of skin, contact sensitivity reactions, and tumour or allograft rejection. Importantly, massive infiltration of skin lesions with basophils occurs in guinea pig dermis during infection with arthropod parasites (ticks) where the host has been sensitized by previous exposure (Askenase, 1980; see Chapter 8). Experiments in which T cells have been adoptively transferred have shown that such cells are crucial to the development of the basophil reaction, and the selective elimination of basophils from the host by injecting anti-basophil antisera suggests that basophils play an important role in the protective response against ticks in the guinea pig model. Similarly, there is circumstantial evidence to indicate that basophils have some role in the immune elimination of intestinal helminths (Askenase, 1980). The principal immunoglobulin isotypes responsible for activation of basophils in the guinea pig are IgE and IgG_1, and it is possible, using trace amounts of purified IgG_1 to passively transfer cutaneous basophil hypersensitivity reactions to naive guinea pigs (Galli *et al.*, 1984). In man, the basophil membrane bears the high affinity $Fc_\epsilon RI$, like the mast cell, and activation of the basophil is mediated through crosslinking of surface receptors (Galli *et al.*, 1984).

In summary, the basophil matures in bone marrow and is recruited to inflammatory foci as an adult, competent cell. If activated through membrane receptors it will, like the mast cell, release histamine and other granule mediators. These may have a direct anti-parasitic effect as shown against ticks or they may promote local inflammatory changes which are detrimental to parasite survival.

Neutrophils

As members of the granulocyte series, neutrophils are non-dividing granulated cells which mature in the bone marrow and have a short life span in the circulation. Primary (azurophilic) granules within the neutrophil cytoplasm contain myeloperoxidase, neutral proteases, acid hydrolases, cationic proteins and lysozyme. The secondary (specific) granules contain lactoferrin, collagenase, lysozyme, enzymes which cleave C5, and a protein which binds vitamin B_{12} (Kay, 1988).

Differentiation of neutrophils occurs in the bone marrow but the cells are recruited to sites of inflammation by a variety of chemotactic substances. These include the well-known bacterial products but also the lipid mediators leukotriene B_4 and PAF and mast cell-derived chemotactic substances. The chemotactic factors activate the neutrophil membrane causing mobilization of receptors with increased expression of receptors for complement and IgG (Kay, 1988).

Activated neutrophils are cytotoxic for complement-coated helminths (Moqbel *et al.*, 1983) and, importantly, neutrophils comprize the earliest recruited cell population in IgE-mediated immediate hypersensitivity responses. The neutrophil response tends to be transient and is superseded by immigration of other inflammatory cell populations.

In summary, the neutrophil tends to be a rapid response cell, recruited within minutes or a few hours of IgE-mediated allergic reactions. Their function in parasitic infection has yet to be fully understood.

3.5.   CONCLUSIONS

Despite the well-known association between parasitism and the recruitment of bone marrow-derived effector cells, it is only in the last two to three years that the molecular mechanisms involved have begun to be understood. Pluripotential stem cells which undergo continuous self-renewal within the bone marrow also produce daughter multipotential stem cells which, in turn, produce progenitor cells committed to one or at most two cell lineages. Under normal conditions these events occur within adult bone marrow and spleen and are driven in a hierarchical fashion by cytokines produced locally by stromal and other cells in a process known as constitutive haemopoiesis.

Infection of the host with bacteria or parasites and consequent generation of antigen-specific T cells is associated with induced (inflammatory) haemopoiesis. In addition to increased cell production in bone marrow, progenitor cells seeded out from bone marrow differentiate into mature inflammatory cells as a consequence of cytokine production by activated T cells. Rapid production and recruitment to the site of infection of inflammatory cells, such as macrophages, mast cells, eosinophils and basophils, are, therefore, highly T-cell dependent events. Similarly, neutrophil production is influenced by T cells, although recruitment of neutrophils is primarily via chemotactic peptides.

Once recruited, bone-marrow derived inflammatory cells are thought to serve anti-parasitic functions either through direct actions of secreted preformed or generated mediators (from mast cells, eosinophils, basophils and sometimes neutrophils) on the parasite, or through enhanced phagocytosis and killing of unicellular parasites after activation of macrophages, neutrophils or eosinophils. Some of the cytokines responsible for recruitment also participate in activation of inflammatory cells with increased receptor expression, increased synthesis of mediators and increased phagocytic or killing properties.

In effect the bone marrow-derived inflammatory cell is substantially influenced by the local production of T cell-derived cytokines, and the T cell, responding to parasitic antigens, is central to the inflammatory responses that have been so long identified with parasite infection.

REFERENCES

Ackerman, S.J., Weil, G.J. and Gleich, G.J. (1982), Formation of Charcot-Leyden crystals by human basophils, *Journal of Experimental Medicine*, **155**, 1597.

Askenase, P.W. (1980), Immunopathology of parasitic diseases: involvement of basophils and mast cells, *Springer Seminars in Immunopathology*, **2**, 417.

Bottomly, K. (1988), A functional dichotomy in CD4$^+$ T lymphocytes, *Immunology Today*, **9**, 268–74.

Bradley, T.R. and Metcalf, D. (1966), The growth of mouse bone marrow cells *in vitro*, *Australian Journal of Experimental Biology and Medical Science*, **44**, 287–300.

Broide, D.H., Metcalf, D.D. and Wasserman, S.I. (1988), Functional and biochemical characterization of rat bone marrow-derived mast cells, *Journal of Immunology*, **141**, 4298.

Capron, M., Capron, A., Dessaint, J.P., Torpier, G., Johansson, S.G.O. and Prin, L. (1981), Fc receptors for IgE on human and rat eosinophils, *Journal of Immunology*, **126**, 2087.

Capron, M., Jouault, T., Prin, L., Joseph, M., Ameisen, J.C., Butterworth, A.E., Papin, J.P., Kusnierz, J.P., and Capron, A. (1986), Functional study of a monoclonal antibody to IgE Fc receptor of eosinophils, platelets and macrophages (Fc$_\epsilon$R$_2$), *Journal of Experimental Medicine*, **164**, 72–89.

Capron, A., Dessaint, J.P., Capron, M., Joseph, M., Ameisen, J.C. and Tonnel, A.B. (1988), From parasites to allergy: The second receptor for IgE (Fc$_\epsilon$R2), *Immunology Today*, **7**, 15–8.

Cosman, D. (1988), Colony stimulating factors *in vivo* and *in vitro*, *Immunology Today*, **9**, 97–8.

Dessaint, J.P., Torpier, G., Capron, M., Bazin, H. and Capron, A. (1979), Cytophilic binding of IgE to the macrophage 1. Binding characteristics of IgE on the surface of macrophages in the rat, *Cellular Immunology*, **46**, 12–23.

Dexter, T.M., Allen, T.D. and Lajtha, L.G. (1977), Conditions controlling the proliferation of haemopoietic stem cells *in vitro*, *Journal of Cell Physiology*, **91**, 335–42.

Dinarello, C.A. (1984), Interleukin-1, *Reviews of Infectious Diseases*, **6**, 51–95.

Enerback, L. (1981), The gut mucosal mast cell, *Monographs in Allergy*, **17**, 222.

Enerback, L. (1987), Mucosal mast cells in the rat and man, *International Archives of Allergy and Applied Immunology*, **82**, 249.

Fujita, T., Shibuya, H., Ohashi, T., Yamanishi, K. and Taniguchi, T. (1986), Regulation of Interleukin-2 gene: functional DNA sequences in the 5′ flanking region for the gene expression in activated T lymphocytes, *Cell*, **46**, 401–7.

Galli, S., Dvorak, A.M. and Dvorak, H.F. (1984), Basophils and mast cells: morphologic insights into their biology, secretory patterns and function, *Progress in Allergy*, **34**, 1.

Goodwin, R.G., Lupton, S., Schmierer, A., Hjerrild, K.J., Jerzy, R., Clevenger, W., Gillis, S., Cosman, D. and Namen, A.E. (1989), Human interleukin-7; molecular cloning and growth factor activity on human and murine B-lineage cells, *Proceedings of the National Academy of Sciences (USA)*, **86**, 302–7.

Hunt, P., Robertson, D., Weiss, D., Rennick, D., Lee, F. and Witte, O.N. (1987), A single bone marrow-derived stromal cell type supports the *in vitro* growth of early lymphoid and myeloid cells, *Cell*, **48**, 997–1007.

Immunology Today (1986), **Vol. 7, No. 11 (Nov.)**, 321–2.

Jarrett, E.E.E. and Haig, D.M. (1984), Mucosal mast cells *in vivo* and *in vitro*, *Immunology Today*, **5**, 115–9.

Jessup, W., Leoni, P. and Dean, R.T. (1985), The macrophage in inflammation, in Venge, P. and Lindborn, A. (Eds.), Inflammation: Basic mechanisms, tissue injuring principle and clinical models, pp. 161–86, Stockholm: Almquist and Wiksell.

Kay, A.B. (1985), Eosinophils as effector cells in immunity and hypersensitivity disorders. *Clinical and Experimental Immunology*, **38**, 294.

Kay, A.B. (1987), Allergy and Inflammaton, London: Academic Press.

Kay, A.B. (1988), Mechanisms in allergic and chronic asthma which involve eosinophils, neutrophils, lymphocytes and other inflammatory cells, *Bailliere's Clinical Immunology and Allergy*, **2**, 1.

Leary, A.G., Ikebuchi, K., Hirai, Y., Wong, G.G., Yang, Y-C., Clark, S.C. and Ogawa, M. (1988), Synergism between interleukin-6 and interleukin-3 in supporting proliferation of human haematopoietic stem cells: Comparison with interleukin-1, *Blood*, **71**, 1759–63.

Lord, B.I. (1983), in Potten, C.S. (Ed.), Stem Cells: Their identification and characteristics, pp. 118–43, Edinburgh: Churchill-Livingstone.

McInnes, A. and Rennick, D.M. (1988), Interleukin-4 induces cultured monocytes/macrophages to form giant multinucleated cells, *Journal of Experimental Medicine*, **167**, 598–612.

McLaren, D.J., McKean, J.R., Olsson, I., Venge, P. and Kay, A.B. (1981), Morphological studies on the killing of schistosomula of *Schistosoma mansoni* by human eosinophil and neutrophil cationic proteins *in vitro*, *Parasite Immunology*, **3**, 359–73.

Metcalf, D. (1984), The hemopoietic colony stimulating factors, Amsterdam: Elsevier.

Metzger, H. (1983), The receptor on mast cells and related cells with high affinity for IgE, *Contemporary Topics in Molecular Immunology*, **9**, 115.

Miller, H.R.P. (1984), The protective mucosal response against gastrointestinal nematodes in ruminants and laboratory animals, *Veterinary Immunology and Immunopathology*, **6**, 167.

Miller, H.R.P. (1987), Immunopathology of nematode infestation and expulsion, in Marsh, M.N. (Ed.), Immunopathology of the small intestine, pp. 177.

Miller, H.R.P., Woodbury, R.G., Huntley, J.F. and Newlands, G.F.J. (1983), Systemic release of mucosal mast cell protease in primed rats challenged with *Nippostrongylus brasiliensis*, *Immunology*, **49**, 471.

Miyajima, A., Miyatake, S., Schreurs, J., De Vries, J., Arai, N., Yokota, T. and Arai, K. (1988), Co-ordinate regulation of immune and inflammatory responses by T cell-derived lymphokines, *FASEB Journal*, **2**, 2462–73.

Mochizuki, D.Y., Eisenman, J.R., Conlon, P.J., Larsen, A.D. and Tushinski, R.J. (1987), Interleukin-1 regulates haemopoietic activity, a role previously ascribed to hemopoietin-1. *Proceedings of the National Academy of Sciences (USA)*, **84**, 5267–71.

Moqbel, R., Sass-Kuhn, S.P., Goetzl, E.J. and Kay, A.B. (1983), Enhancement of neutrophil- and eosinophil-mediated complement-dependent killing of Schistosmula of *Schistosoma mansoni in vitro* by leukotriene B4. *Clinical and Experimental Immunology*, **52**, 519.

Moqbel, R., Miller, H.R.P., Wakelin, D., MacDonald, A.J. and Kay, A.B. (1987), Leukotrienes and intestinal worms, in Kay, A.B. (Ed.), Allergy and Inflammation, p. 367, London: Academic Press.

Morrissey, P.J., Bressler, L., Charrier, K. and Alpert, A. (1988), Response of resident murine peritoneal macrophages to *in vivo* administration of granulocyte-macrophage colony-stimulating factor. *Journal of Immunology*, **140**, 1910.

Mosmann, T. and Coffman, R. (1987), Two types of mouse helper T-cell clone: implications for immune regulation. *Immunology Today*, **8**, 223–7.

Murray, H.W., Rubin, B.Y. and Rotherwel, C.D. (1983), Killing of intracellular *Leishmania donovani* by lymphokine-stimulated human mononuclear phagocytes. Evidence that interferon is the activating lymphokine, *Journal of Clinical Investigation*, **72**, 1506.

O'Garra, A., Umland, S., De France, T. and Christiansen, J. (1988), B cell factors are pleiotropic, *Immunology Today*, **9**, 45–58.

Ogawa, M., Nakahata, T., Leary, A.G., Sterk, A.R., Ishizaka, K. and Ishizaka, T. (1983), Suspension culture of human mast cells/basophils from umbilical cord mononuclear cells. *Proceedings of the National Academy of Sciences (USA)*, **80**, 4494.

Ovary, Z. (1981), IgE production and suppression in mice, *International Archives of Allergy and Applied Immunology*, **66**, supplement 1, 8–18.

Paul, W.E. (1988), Lymphokine nomenclature, *Immunology Today*, **9**, 366–7.

Plant, M., Pierce, J.H., Watson, C.J., Hanley-Hyde, J., Nardan, R.P. and Paul, W.E. (1989), Mast cell lines produce lymphokines in response to cross linkage of $Fc_\epsilon R1$ or to calcium ionophores, *Nature*, **339**, 64–7.

Raefsky, E.L., Platanias, L.C., Zoumbas, N.C. and Young, N.C. (1985), Studies of interferon as a regulator of haemopoietic cell proliferation, *Journal of Immunology*, **135**, 2507–11.

Rose, M.E. and Hesketh, P. (1982), Coccidiosis: T-lymphocyte-dependent effects of infection with *Eimeria nieschulzi* in rats. *Veterinary Immunology and Immunopathology*, **3**, 499–508.

Rouzer, C.A., Scott, W.A., Hamil, A.L., Liu, F.T., Katz, D.H. and Cohn, Z.A. (1982), Secretion of leukotriene C_4 and other arachidonic acid metabolites by macrophages challenged with immunoglobulin E immune complexes, *Journal of Experimental Medicine*, **156**, 1077–86.

Schulman, E.S., MacGlashan, D.W. Jr., Schleimer, R.P., Peters, S.P., Kagey-Sobotka,

A., Newball, H.H. and Lichtenstein, L.N. (1983), Purified human basophils and mast cells: current concepts of mediator release, *European Journal of Respiratory Diseases*, **64**, supplement 128, 53.

Schwartz, L.B. and Austen, K.F. (1984), Structure and function of the chemical mediators of mast cells, *Progress in Allergy*, **34**, 271.

Scott, P., Natovitz, P., Coffman, R.L., Pearce, E. and Sher, A. (1988), Immunoregulation of cutaneous Leishmaniasis. T cell lines that transfer protective immunity or exacerbation belong to different T helper subsets and respond to distinct parasite antigens. *Journal of Experimental Medicine*, **168**, 1675–84.

Silberstein, D.S. and David, J.R. (1987), The regulation of human eosinophil function by cytokines, *Immunology Today*, **8**, 380.

Smith, C.A. and Rennick, D.M. (1986), Characterisation of a murine lymphokine distinct from IL-2 and IL-3 possessing a TCGF activity and a MCGF activity that synergises with IL-3, *Proceedings of the National Academy of Sciences (USA)*, **83**, 1857–61.

Stewart, W.E. (1980), Interferon nomenclature, *Nature*, **286**, 110.

Till, J.E. and McCulloch, E.A. (1961), A direct measurement of the radiation sensitivity of normal mouse bone marrow cells, *Radiation Research*, **14**, 215–22.

Walker, F., Nicola, N.A., Metcalf, D. and Burgess, A.W. (1985), Hierarchical down-modulation of haemopoietic growth factor receptors, *Cell*, **43**, 269–76.

Wardlaw, A.J. and Kay, A.B. (1987), The role of the eosinophil in the pathogenesis of asthma, *Allergy*, **42**, 321.

Wodnar-Filipowicz, A., Heusser, C.H. and Moroni, C. (1989), Production of the haemopoietic growth factors GM-CSF and interleukin-3 by mast cells in response to IgE receptor mediated activation, *Nature*, **339**, 150–2.

Wong, G.G. and Clark, S.C. (1988), Multiple actions of interleukin-6 within a cytokine network, *Immunology Today*, **9**, 137–9.

4. Antigen uptake, processing and presentation

R.K. Grencis

4.1. INTRODUCTION

Parasites gain entry to the body along a variety of routes, through the skin, across mucosal surfaces, or by injection directly into the blood stream, and during their life cycles often utilize a combination of all three. A consequence of this is that parasites present not only a complex and considerable antigenic load to the host but also confront it at a number of distinct sites around the body. It follows, therefore, that the way in which the body handles antigen will depend upon the molecular nature of the antigen, the site of antigen exposure, the cell types which encounter the antigen and the immunoregulatory mechanisms operating in the host at that particular time.

Much of what is known about antigen handling mechanisms during immune responses has come from *in vitro* studies using cells from animals and man. Moreover, the majority of investigations have utilized cell populations from donors undergoing an immune response against defined injected proteins or bacterial or viral infections.

Based upon observations and experimental data from such studies, this chapter will attempt to present a concise overview of current theories concerning the mechanisms involved in the handling of foreign molecules by the body and relate them to antigen handling during infection by protozoan and metazoan parasites. Investigations into antigen presentation during parasitic infection will not only be crucial for effective vaccine design but may also provide insights into the variation in disease susceptibility observed between individuals infected with parasites under natural conditions.

4.2. ANTIGEN HANDLING – SETTING THE SCENE

During the 1950s early theories concerning antigen handling by cells involved in immunity (most notably potential antibody-producing cells) were centred around the idea of antigen entering a cell and acting as a template for the subsequent production of complementary antibody. This hypothesis was modified in the 1960s following experiments which demonstrated that antibody-producing cells contained little if any antigen, whereas antigen was found in abundance in other cell types, namely macrophages. These observations stimulated research into the involvement of the macrophage in antigen handling and gave rise to the view that macrophage-associated or macrophage-treated antigen was much more effective in the generation of antibody-producing cells than native antigen (Unanue, 1981).

Following the categorization of lymphocytes into the T- and B-cell compartments, the next major step towards understanding the mechanisms involved in antigen handling was the demonstration of the importance of self-molecules in the presentation of antigen to T lymphocytes (see Zinkernagel and Doherty, 1979). These observations had a profound effect upon the way in which investigators thought about antigen-T-cell interactions. This was particularly important as acquired immunity is essentially dependent upon the effective presentation of antigen to T lymphocytes. In contrast, therefore, to B cells which are capable of recognizing and binding free antigen through their surface immunoglobulin molecules, T cells possess a specific receptor which only recognizes foreign antigen in conjunction with self-molecules expressed on the surface of other cells. The self-molecules are those encoded by genes of the major histocompatibility complex (MHC), i.e. Class I molecules and most notably the Class II or Ia molecules. The cells of the body which express such molecules, and therefore present antigen to T cells, have been termed 'antigen presenting cells' (APCs). Over the last 15 years extensive *in vitro* and *in vivo* research (see Moller, 1987) has shown that cells which can present antigen to T cells encompass a heterogeneous group which are seeded throughout the body and exhibit a variety of morphological types (Table 4.1). However, there is one important feature which is common to all these cell types; in order to present antigen effectively, the cells must constitutively express or be induced to express MHC Class II molecules on their surface.

4.3. ANTIGEN PROCESSING – AN OVERVIEW

Early studies demonstrated that T cells were unable to discriminate between native and denatured protein antigens, and more recent studies have largely confirmed this view. The conversion from a native to a non-native form has been termed 'antigen processing'. Antigen processing is carried out by the antigen-presenting (MHC Class II expressing) cell.

Table 4.1. Cell types which have been characterized as antigen-presenting cells.

Macrophages (peritoneal, splenic thymic, alveolar, bone marrow derived, liver)	T cells
	Gut epithelial cells
Dendritic cells	Astrocytes
Langerhans cells	Fibroblasts
B lymphocytes	Endothelial cells

An insight into the nature of antigen processing was first reported by Ziegler and Unanue (1982) in the early 1980s. They were studying *in vitro* the antigen-presenting properties of macrophages to specific T cells using the bacterial antigens of *Listeria monocytogenes*. Their experiments examined the effects of fixation of the antigen-presenting macrophages prior to or after addition of the bacterial antigen, and the effects of various lysomotropic agents or metabolic inhibitors upon antigen presentation to *Listeria*-specific T cells taken from immunized animals. Their experiments allowed two main conclusions to be drawn: (1) antigens require an obligatory period of time for interaction with the APC before effective T-cell stimulation can occur; and (2) antigens may require an acidic intracellular micro-environment for subsequent effective antigen presentation and proteolytic mechanisms may be involved.

Such studies were extended by those of Shimonkovitz *et al.* (1983) using the protein antigen ovalbumin. They observed that proteolytic digests of ovalbumin (i.e. peptide fragments after treatment of ovalbumin with trypsin) but not the native molecule were effectively presented by lightly fixed APC to T cells taken from animals immunized with ovalbumin. Effective presentation only occurred in an MHC-restricted manner. These studies established that a processed form of the antigen could interact with the MHC molecules upon the surface of an APC and stimulate T cells to divide and when taken together with observations by Ziegler and Unanue (1982), Unanue (1981) and others, suggested that intracellular processing of antigen would require internalization of antigen molecules by the APC.

Most eukaryotic cells are constantly sampling their environment by endocytic activity and internalization of the cell membrane. This may be via pinocytic activity in which fluid and small particles are internalized. Phagocytosis tends to be restricted to specialized cell types in the body, such as monocytes and macrophages. In both kinds of endocytosis most efficient internalization follows interaction of antigenic molecules with membrane receptors. This interaction may be via specific receptors for antigen, or more likely by receptors normally involved in membrane-receptor-mediated endocytosis which fortuitously binds particular regions of antigen molecules. Such internalization mechanisms help to explain how antigen gains access to intracellular processing compartments (Figure 4.1). Studies using the inhibitor monensin – a cationic ionophore which prevents recycling of cell surface receptors through endosomes (via clathrin-coated pits) and then through the

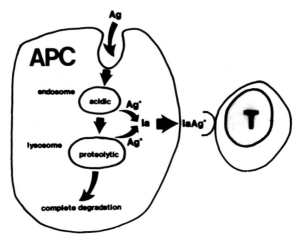

Figure 4.1. Basic mechanism of intracellular antigen processing. Upon endocytosis by an antigen-presenting cell (APC), the antigen (Ag) may be altered by the acidic conditions within the endosome or by proteolytic activity within the lysosome. This 'processed' antigen (Ag*) then will or will not possess an affinity for MHC Class II molecules (Ia). Those antigens which have a strong affinity will interact and will be able to participate in the presentation process to T cells at the membrane surface. Presumably, those antigens which have a low affinity for MHC molecules will be completely degraded by the cell, e.g. within the lysosome.

golgi apparatus – strongly suggest that APCs require an intracellular pathway for processing antigen (see Berzofsky *et al.*, 1987).

Although this concept may hold true for strongly phagocytic cell types such as macrophages, it is clear from a number of studies that other potent antigen-presenting cells (e.g. dendritic cells) exhibit poor phagocytic activity. Whilst a limited degree of processing may occur in these cells in intracellular sites, an alternative site for processing may occur within or close to the cell membrane. In this context protease activity has been demonstrated in the cell membranes of both the APC and the T cell (Figure 4.2; Kramer and Simon, 1987). Such proteases could play an important role in antigen-processing mechanisms at the cell surface. However, other experiments controlling for surface protease activity suggest that certain antigens do not require processing at all. This may indeed be the case, if aspects of the molecular configuration of an antigen predispose it to direct interaction with MHC Class II molecules. In addition, small changes may occur in the antigen upon association with membrane constituents prior to presentation to T cells, which may be sufficient to render an unprocessed 'native' antigen presentable.

Examination of a variety of protein molecules has enabled Allen (1987) to propose three basic types of antigen classified according to their processing requirements (Table 4.2). Certain antigens seem to require little in the way of processing; some require unfolding of the tertiary structure (denaturation such as alkylation of disulfide bonds) and some require proteolytic cleavage,

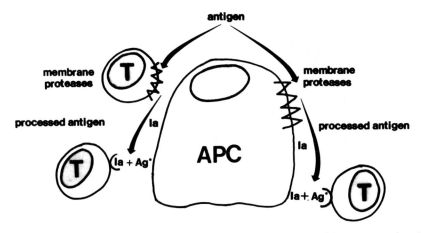

Figure 4.2. Antigen processing by surface proteases. Antigen is processed by proteases found in close association with the membranes of either the antigen-presenting cell or T cell. Once processed the antigen (Ag*) may associate with surface Class II molecules on the APC and be presented to T cells.

Table 4.2. Classification of antigens according to the metabolic processing required to enable recognition by T cells (taken from Allen, 1987, with permission).

	Processing requirements	Examples
Type I	None	Fibrinogen, listerial proteins
Type II	Unfolding	Lysozyme, myoglobin ribonuclease
Type III	Cleavage	Lysozyme, myglobin, ovalbumin, insulin

before they are recognized by T cells. Recent work has begun to examine the endopeptidases involved in cleaving antigenic proteins during processing. Using a series of proteinase-specific inhibitors to interfere with the presentation of myoglobin to T-cell clones, Takahashi and colleagues (see Berzofsky *et al.*, 1988) have demonstrated that thiol proteinases, such as capthepsin B and L, were necessary and sufficient to process myoglobin enabling presentation of three distinct antigenic determinants (epitopes) to T cells. Capthepsin B is present in endosomes of APCs and preferentially cleaves after pairs of basic amino acids, e.g. Lys-Lys or Arg-Arg. Increasing information about the specificity of such proteinases will help in prediction of potential T-cell epitopes, although it remains to be determined whether other enzymes, such as the serine proteinases, play important roles in processing other protein antigens.

A further degree of complexity arises from studies of antigen presentation in viral infections. Differences in presentation of viral-derived antigens to T cells can arise from the differences in site of viral antigen production and subsequent processing (Mills, 1986). In common with mechanisms described earlier, viral antigens acquired exogenously may be processed with endocytic or membrane mechanisms. However, in addition, the intracellular habitat of the virus provides a new site of endogenously produced viral antigens. Antigen-presenting studies involving influenza haemagglutinin produced exogenously and endogenously have provided important information concerning the intracellular pathways used in processing antigens and the interaction of the antigens with the two classes of MHC molecules (see Germain, 1986).

Endogenously produced antigens may proceed along different processing pathways (possibly missing interaction with lysosomes) and as a result may show differential affinity for certain MHC molecules (Class I rather than Class II molecules). Indeed, there may be differences between MHC Class I and Class II molecules in terms of the mechanisms used to transport them to the cell surface which may impose constraints upon the 'type' of antigen with which they interact. These observations may be of particular importance to studies of processing and presentation of antigens from intracellular-dwelling parasitic protozoa.

4.4.   ANTIGEN-MHC INTERACTION

Following investigations of antigen (insulin) presentation by macrophages to insulin-specific T cells, Rosenthal (1978) coined the term 'determinant selection' to describe the observation that the MHC could determine which part of a protein antigen was preferentially recognized by a T cell. This view can now be extended to encompass the idea that whatever the configuration of antigen, in order to become immunogenic it will need to form an association with an MHC molecule and make contact with the T-cell receptor. With regard to the APC, the first critical component is for the antigen to make an association with an MHC molecule. Depending on the antigen this may occur at the cell surface or occur within the APC prior to expression on the membrane. In this latter case, the antigen may combine with newly synthesized MHC molecules or with MHC molecules expressed on the originally internalized membrane (Figure 4.3).

Our understanding of the antigen binding interaction was furthered by the demonstration of a direct association between an antigenic peptide and purified MHC molecules by Babbitt *et al.* (1985) employing the technique of equilibrium-binding analysis. Equal concentrations of a fluorescently labelled immunogenic peptide (from hen egg lysozyme) were placed on either side of a semi-permeable membrane. To one side of the membrane a solution was added which contained purified MHC Class II molecules (small antigen molecules may pass through the membrane unlike the larger MHC molecules). After a period of equilibrium, it was possible to measure an increased concentration of labelled antigen on the side of the membrane containing

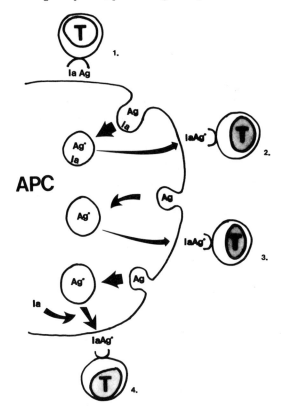

Figure 4.3. Interaction between MHC Class II molecules and antigen. 1. Antigen (Ag) in native form has sufficient affinity to interact with Ia molecules expressed at the cell surface leading to effective presentation to T cells. 2. During endocytosis membrane-bound Ia is internalized with antigen. Following processing the antigen (Ag*) is re-expressed at the membrane surface in association with the original membrane-bound Ia molecule. 3. Antigen is endocytosed, processed and re-expressed at the surface of the cell, where it now associates with the Ia molecule also expressed upon the membrane. 4. Antigen is endocytosed, processed and prior to arrival at the membrane surface associates with newly synthesized Ia molecules within the cell. The Ia molecule and antigen complex is now expressed at the surface and presented to T cells. APC = antigen-presenting cell.

MHC molecules. Moreover, the binding was demonstrated to be saturable, homogeneous and dependent upon a particular form (polymorphism) of the MHC molecule.

Bearing in mind the vast array of antigens that the immune system confronts and therefore, by definition, may bind to MHC molecules, it is reasonable to suggest that the MHC molecule must have many separate binding sites. However, there are only a few polymorphic MHC molecules expressed in each individual. Moreover, observations from competition studies using different antigenic peptide fragments which bind the same MHC molecule suggested that only one antigen binding site was present. The answer to this conflicting state of affairs has come from the recent description of the three-

Figure 4.4. Schematic diagram of the top surface of an MHC molecule (Class I) binding site (presumably the view of the site that would come into contact with the T-cell receptor). The site consists of a groove-shaped structure bounded by α-helices at the sides and β-pleated sheets at the base (based on Bjorkman *et al.*, 1987).

dimensional structure of the antigen binding site of the MHC Class I molecule by Bjorkman *et al.* (1987) and by modelling of the Class II molecule-binding site by Brown *et al.* (1988). The described structure exhibits a single potential binding site which is located on the top surface of the molecule facing away from the membrane. The site is envisaged as a groove bounded by α-helices at the side and β-strands at the base (Figure 4.4). The groove is approximately 2.5 nm long, 1.0 nm wide and 1.1 nm deep. Such a size would enable binding of peptides of approximately 8–20 amino acids in length depending on their configuration. Of course it is possible that larger molecules could be recognized with the non-active sites protruding out of the groove. If one site binds many antigens, then some structural homology might be expected between antigens. This has been observed with a number of peptides, although the site is large enough for a peptide to bind over different overlapping positions. The above description is for Class I MHC molecules; however, because of the similarities in recognition of antigens by Class I- and Class II-restricted T cells (Townsend *et al.*, 1986), many of the binding site features described for Class I molecules also apply to Class II molecules (see Brown *et al.*, 1988).

The molecular construction and hence shape and charge of a protein antigen will determine its affinity for a certain MHC molecule. Once bound to an MHC molecule, it is thought that such a complex is presented to the T-cell receptor after expression on the surface of the APC. The site on the antigen molecule which makes contact with the T-cell receptor is termed the epitope; the site which makes contact with the MHC molecule is termed the agretope

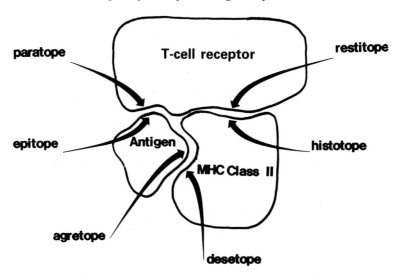

Figure 4.5. Terminology for interacting sites between antigen, the T-cell receptor and Ia molecules (after Schwartz, 1986).

(Figure 4.5). It is now clear from a number of studies that, for protein antigens at least, the agretope and epitope are composed of a small number of amino acid residues interspersed in the primary sequence of the peptide. These residues have been identified for an immunogenic peptide of hen egg lysozyme (Allen *et al.*, 1987). Using modelling analysis it has been proposed that this particular antigenic peptide would assume an α-helical conformation, with the agretope on one face of the helix and the epitope on the other face. There is evidence that a number of well-described, T-cell stimulating peptides do conform to the α-helix configuration. Moreover, many of the helices are amphipathic, that is contain both hydrophilic and hydrophobic regions. Such amphipathic molecules could orientate themselves easily in the lipid bilayer of the APC. This would place the molecule in an ideal position to associate with MHC molecules (Allen *et al.*, 1987).

Using predictive algorithms it has been possible to successfully identify areas of molecules which would be particularly good candidates for T-cell stimulating sites. Again important characteristics seem to be regions of hydrophobicity adjacent or near to alternate areas of hydrophilic residues (see Berzofsky *et al.*, 1987). Peptides of approximately 10–20 amino acids are good T-cell stimulating antigens, and this size fits in well with current theories concerning the size of MHC molecule antigen binding sites (see above). Thus, even though native protein antigens may not contain helical regions in known 'immunodominant' sites, once freed from the tertiary structure by antigen processing mechanisms, the small peptide fragments may then be able to fold into an amphipathic structure.

A comparative analysis of the primary structure of a number of known T-cell epitopes has led to the discovery of common amino acid sequence patterns. A large percentage contain a linear pattern composed of a charged residue or glycine followed by two hydrophobic residues, although the minimum size necessary for full stimulation is 8–12 amino acids. Further analyses of residues around this central pattern have revealed subpatterns of residues which are specific for different allelic forms of MHC molecules. Moreover, when the primary structure is allowed to assume a helical conformation, the residues comprizing the MHC allelic-specific subpattern become juxtaposed (Rothbard and Taylor, 1988). It must be remembered, however, that although amphipathic α-helices may make particularly good T-cell antigens, not all good T-cell antigens will need to be of this particular configuration, and a number of investigations have suggested that other types of molecule, such as β-pleated strands, may be involved.

4.5. THE T-CELL ANTIGEN RECEPTOR

Investigations into defining the T-cell antigen receptor were based on the hypothesis that T cells express structures on their surface that are involved in specific recognition of antigen. Early studies were based upon the production of monoclonal antibodies specific for molecules on the surface of T-cell hybridomas, T-cell lymphomas and isolated T-cell clones. Important observations were that the antibodies were 'clone' specific, and immunoprecipitation studies of a number of different clones revealed the precipitation of similar structures. This was a disulfide bonded heterodimer of approximately 80–90 kDa. (It is interesting to note the comparison with the major disulfide bonded heterodimer on the surface of B cells – the antigen-specific receptor, i.e. the immunoglobulin molecule.) The T-cell antigen receptor is composed of two glycoprotein chains and occurs as two distinct types. The most common is composed of one α and one β chain (αβ receptor), the less common composed of one γ and one δ chain (γδ receptor). Each has constant and variable regions, in a manner similar to that found in immunoglobulin molecules (Figure 4.6). Indeed, a number of structural features that provide the framework of the antibody-combining sites are also present in the T-cell antigen receptor. This can be inferred from studies of the sequence variability profiles of the genes coding for the variable regions of the alpha and beta chains of the T-cell receptor. A comparison with the profiles of the genes coding for the variable regions of the immunoglobulin molecule shows many similarities and suggests that the T-cell antigen receptor binding site is essentially the same as that of the immunoglobulin molecule (see Marrack and Kappler, 1987).

However, whereas the immunoglobulin molecule is responsible for recognizing antigen alone, the T-cell receptor is responsible for both MHC and antigen recognition. Moreover, another fundamental difference between T- and B-cell recognition of antigen is the involvement of other cell surface

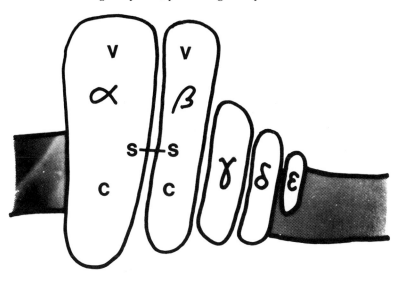

Figure 4.6. Diagrammatic representation of the T-cell antigen receptor within the cell membrane, showing α and β chains (containing variable (v) and constant regions (c) and disulphide bond). Also shown are the accessory chains γ, δ and ε (CD3 complex).

molecules in T-cell activation which are non-antigen specific in nature. Among these are included the CD3 complex (comprizing γ, δ and ε polypeptides) involved in the transduction of signal from the APC to the T cell *via* enzyme systems, notably protein kinase C. Another glycoprotein complex, termed LFA-1, is thought to be involved in cell-cell adhesion. A 50 kDa glycoprotein molecule, termed CD2, found on T cells is thought to interact with a 50–70 kDa glycoprotein (LFA-3) which is expressed upon almost all haemopoietic cells (including, therefore, APC). In addition to providing a stabilizing influence upon cell to cell contact, these structures are thought to convey activation signals to both T cell and APC leading possibly to lymphokine and cytokine secretion. The CD4 and CD8 membrane glycoproteins expressed on T-cell subsets are thought to be involved in interactions with MHC molecules expressed upon APCs (see Figure 4.7). All these molecules serve to focus and stabilize the APC-T-cell interaction at the membrane surface (Unanue, 1984).

4.6. SECONDARY SIGNALS

In addition to the interactions at the membrane surface between cells as mentioned above, other secondary membrane and intracellular functions are induced by the T cell-APC interaction. Two processes have been identified and studied in depth. The first mechanism concerns the regulation of MHC Class II (Ia) molecules and the second concerns the regulation and production of the cytokine interleukin 1 (IL-1).

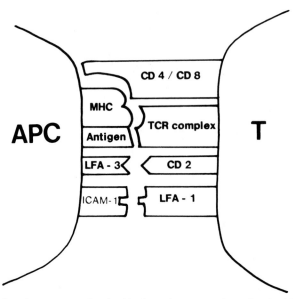

Figure 4.7. Cell surface structures involved in the antigen-presenting cell (APC/T cell interactions; see text for details).

Considerable attention has been devoted to the control of expression of Class II molecules upon macrophages (see Oppenheim *et al.*, 1986). Class II molecule expression on macrophages is not constitutive as on other APC such as dendritic cells. In general a basal level of Class II expression in macrophages is maintained, partly by self-regulation involving the production of prostaglandins which inhibit Class II expression. Under certain conditions, such as those accompanying infection of the host, macrophages change the levels of Class II expressed. The only well-established factor known to induce Class II expression is -interferon (IFN-γ).

This lymphokine is produced primarily by T cells during the antigen presenting process. Once bound to the APC, IFN-γ induces new expression of messenger RNA for Class II molecules, and shortly afterwards, increased levels of Class II are observed on the membrane surface (Figure 4.8). Indeed, exposure of many different cell types to IFN-γ induces increased levels of Class II expression on their surfaces. Bearing in mind the importance of MHC molecules in restricting antigen presentation, it is evident how T cells can regulate the antigen-presenting capabilities of APC.

The APC (particularly macrophages) can also influence the T cell via the production of the cytokine IL-1 (Oppenheim *et al.*, 1986). Two basic forms of IL-1 have been described, one secreted and one found at the membrane surface. It is thought that the membrane-associated type is important in antigen-presenting mechanisms. Direct cell to cell contact has been found to induce the membrane-associated form of IL-1 possibly via signals through

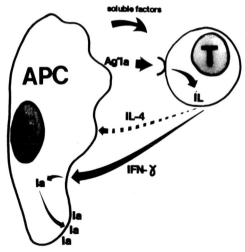

Figure 4.8. Induction of MHC Class II molecules (Ia) by T-Cell lymphokines. Following antigen (Ag*) presentation and T-cell activation, lymphokines (IFN-γ and possibly IL-4) are secreted. These induce an increase in Ia production and expression in the antigen-presenting cell (APC).

the MHC/antigen/T-cell receptor complex. Furthermore T cells are thought to release an IL-1 inducing protein upon interaction with macrophages although the identity of this molecule is as yet unknown. It is still unclear whether or not IL-1 is an absolute requirement for the activation of T cells although it is known that T cells bear high-affinity receptors for IL-1 and that IL-1, in conjunction with T-cell mitogens, can enhance the transcription of a number of lymphokine genes. Alternatively, it has been suggested that IL-1 may exert its effect upon T cells by an indirect stimulation of dendritic cells. For example, it is recognized that IL-1 enhances the clustering of dendritic cells with T cells in a unique manner not involving antigen or MHC molecules.

4.7. APC CHARACTERISTICS AND CELLULAR COOPERATION

As mentioned earlier, many different cell types are capable of functioning as antigen-presenting cells. It has been suggested that any cell type which expresses Class II molecules or can be induced to express them, can, under the right conditions, present antigen or fragments of antigen to T cells. Indeed a number of detailed studies examining the role of Class II molecules in T-cell recognition of antigen have been able to demonstrate effective T-cell stimulation by artificially constructed liposomes containing just MHC molecules and antigen. However, several specific cell types are thought to dominate the antigen-presenting role *in vivo*. These are cells of the macrophage/

monocyte line, dendritic cells, B lymphocytes and possibly gut epithelial cells. Not all the mechanisms of antigen processing and presentation discussed in the previous section will apply to all, but the following brief discussion of the characteristics of each of these major APC types will serve to highlight important differences between APC and to introduce current ideas concerning cell to cell interactions in the immune response, particularly in relation to the mechanisms of T-B co-operation.

Macrophages

Macrophages and monocytes of the mononuclear phagocyte system are the major components of the reticuloendothelial system and as such are sequestered at various sites around the body. These bone marrow-derived cells participate in a vast array of immunological and inflammatory responses and have been the subject of intense investigation (see reviews Adams and Hamilton, 1984; Unanue and Allen, 1987; see Chapter 3). As may be expected from a highly active cell type expressing multiple functions, populations of macrophages exhibit a marked heterogeneity in terms of surface markers, biochemistry, secretion and morphology. One common feature of macrophages is the capacity of many of their various properties to be up or down regulated particularly during the process of activation. Another is their strong phagotocytic activity mediated through a variety of cell surface receptors, and a third is their ability to express Class II molecules upon their surface membrane.

The regulation of Class II molecules upon the macrophage surface has been extensively studied particularly in connection with their antigen-presenting capacity (see above). The resting macrophage expresses low levels of Class II molecules which may be regulated by locally produced prostaglandins. Under certain conditions, the Class II expression may be significantly increased often accompanied by an increase in the capacity of the cell to present antigen. The major factor responsible for the increase in Ia expression is action of the lymphokine IFN-γ released by T cells (or in some cases natural killer cells). Macrophages are also potent producers of IL-1 and have been observed in both the secreted and membrane-associated forms. Indeed it has been suggested that the expression of membrane-associated IL-1 is a key feature of the antigen-presenting capacity of the macrophage and is intimately involved in this process during direct cell to cell contact. However, the subtle mechanisms involved in macrophage/lymphocyte and macrophage/APC interactions remain to be fully described.

Dendritic cells

The term 'dendritic cell' encompasses a variety of cell types described in the literature and includes interdigitating cells, veiled cells, Langerhans cells, follicular dendritic cells and interstitial cells. First described by Steinman and

Cohn (1973), lymphoid dendritic cells show several common features. They are irregularly shaped and as their name suggests, produce an array of cell processes including dendrites, pseudopods and veils. They are found at many anatomical sites of the body including the skin (Langerhans cells), connective tissue (interstitial cells), lymph (veiled cells), Peyers patches, spleen and lymph nodes (interdigitating cells, follicular dendritic cells). Most are derived from precursors found in the bone marrow (with the exception of follicular dendritic cells which are present in the B-cell areas of organized lymphoid tissue and are involved in the capture of immune complexes).

Dendritic cells are poorly endocytic, and this is reflected in the paucity of endosomes and lysosomes in their cytoplasm. Many of the surface markers found on other APC types including Fc receptors, macrophage markers, immunoglobulin, NK markers and T- and B-cell markers are lacking. However, they do express complement receptor markers, the leucocyte common antigen and MHC Class I molecules. Furthermore, they constitutively express MHC Class II molecules.

Another notable difference between dendritic cells and macrophages is that dendritic cells do not appear to produce IL-1 (i.e. they do not express mRNA for IL-1 production). Notwithstanding these differences, dendritic cells are potent antigen-presenting cells, on a cell for cell basis much more potent than Class II-expressing macrophages. A unique feature of dendritic cells which may be crucial to their effect is the ability to physically associate or cluster with T cells in an antigen- and MHC-independent manner (Austyn, 1987). This clustering is thought to be essential for T-cell activation. It is therefore becoming increasingly apparent that dendritic cells play a very important role in the generation of primary immune responses. Moreover, in the light of recent evidence it may be reasonable to suggest that other APC types, such as B cells and macrophages, are mostly involved in presenting antigen to primed T cells. It is also becoming clear that there is a complex interplay between APC types, e.g. macrophages may influence the T-cell clustering ability of dendritic cells by the release of IL-1.

B lymphocytes

Although it has been known for many years that to make antibody against the majority of antigens a B cell requires T cell help, the exact mechanisms which are involved in this interaction are only recently becoming clear. Following on from the antigen-bridging concepts of Mitchison (1971) and subsequently Katz *et al.* (1973), a major step forward came from the observation that APC, such as macrophages, take up antigen and partially degrade it before presentation to T cells. By analogy it was hypothesized that a B cell could only be specifically helped to produce antibody by a T cell if antigen was processed and presented to that T cell in a similar manner. Furthermore, it was proposed that the immunoglobulin expressed on the B-cell surface could serve as a specific receptor for native antigen prior to processing.

Such a hypothesis has largely been borne out by a number of experiments utilizing B-cell hybridomas and MHC-restricted, antigen-specific T-cell clones (Howard, 1985). It was shown that B cells could bind antigen molecules specifically via surface immunoglobulin and that such B cells could efficiently present antigen to T cells in an MHC-restricted manner. Also B cells were shown to endocytose antigen in a non-specific manner and re-present effectively to T cells. The next step came from experiments in which non-transformed B cells from immunized animals were shown to present antigen to T cells at very low antigen concentrations. This suggested that antigen capture by Ig led to very effective T-cell stimulation (see Howard, 1985). However, the demonstration that Ig-captured antigen is processed before presentation remained unproven until recently. A series of elegant experiments by Lanzavecchia (1986) utilizing human T- and B-cell clones specific for the same antigen from the same donor individual helped resolve this problem. By separating in time the binding of antigen to Ig and the presentation of antigen to T cells, it was shown that following a brief pulse of antigen, B cells could effectively present antigen even in the presence of a blocking antibody specific for the surface-antigen-specific Ig. Incubation of the antibody with the cells during the antigen pulse period effectively prevented antigen presentation. Moreover, using metabolic inhibitors, it was shown that internalization of antigen, followed by a processing step, was required (Figure 4.9).

More recently, the direct interaction of helper T cells with B cells has been visualized within the microtubule organizing centre with the T cell becoming orientated to face the area of cell-cell contact (Kupfer *et al.*, 1986). This step is accompanied by the co-ordinated re-orientation of the Golgi complex of the cell, the latter being most likely involved in the transport of lymphokines and growth factors. It has also been shown that the T-cell derived factor, interleukin-4, is capable of inducing Class II expression on resting B cells, whereas T-cell derived IFN-γ can modulate this expression (Sanders *et al.*, 1987). Moreover, IL-4 is thought to enhance the ability of antigen-specific B cells to form conjugates with T cells. Thus, the release of lymphokines by T cells can subtly regulate the APC capacity of B cells.

Under the influence of these and other T-cell derived factors, the B cell is also able to proceed along the pathway of differentiation, maturing into an antibody-producing plasma cell. The specificity of the antibody will be identical to that of the surface Ig of the B cell used to capture the native antigen. All these observations lead to the conclusion that the Class II, Ig-bearing B cell can play an important role in antigen presentation. Indeed recent evidence from experiments using B-cell depleted mice have suggested that the B cell may play a major role in antigen presentation and clonal expansion of activated T cells within lymph nodes (Ron and Sprent, 1987).

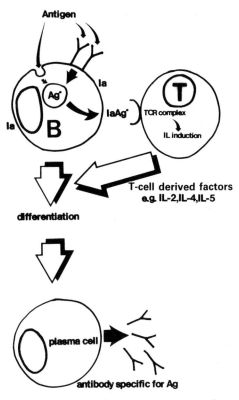

Figure 4.9. Antigen presentation by B cells. Following internalization of antigen (Ag) specifically recognized by surface immunoglobulin (and to a lesser extent non-specifically by pinocytosis) antigen is processed (Ag*) and re-expressed at the membrane surface in association with Ia molecules. Upon recognition by the T-cell receptor complex (TCR), the T cell is stimulated to produce interleukins (IL) which act upon the activated B cell and may cause it to differentiate into a plasma cell. This antibody-producing cell secretes immunoglobulin specific for the antigenic epitope originally recognized by the antigen-presenting B cell.

Gut epithelial cells

Entry of antigen through mucosal surfaces and its subsequent processing and presentation may occur through a number of routes (Nicklin, 1987). Antigen may gain access to the lymphoid compartment by diffusing through epithelial discontinuities, such as those found at the villus tip, or via transportation through epithelial cells or microfold (M) cells. M cells are modified epithelial cells found in areas of organized lymphoid tissue in the small intestine called Peyers patches. Antigen is taken up from the gut lumen by phagocytosis in the M cell. Observations suggest that the resultant phagosomes containing antigen do not fuse with lysosomes (M cells contain few) but pass through the cytoplasm and fuse with the lateral cell membrane to discharge their contents into the lateral space. Because M cells are closely associated with lymphoid cells and APC such as dendritic cells and macrophages, antigen may be processed and presented by these latter type cells at this site. Thus the M cell does not seem to play an antigen-presenting role but more of an antigen acquisition and transport role.

Another prominent cell type within the mucosa which may have an APC role is the gut epithelial cell. Several studies have demonstrated the presence of Class II molecules on the surface of gut epithelial cells of many host species. It has been shown that Class II positive epithelial cells can process and present antigen to antigen-specific T cells. Moreover these APCs seem to selectively stimulate CD8-positive T cells, i.e. T cells which characteristically exhibit suppressor or cytotoxic functions (Mayer and Shlien, 1987). This may be a factor contributing to the observed 'suppressed' nature of the gut immune system with regard to many orally ingested antigens (e.g. food commensals, etc.).

4.8. ANTIGEN HANDLING DURING PARASITIC INFECTION – THE STORY SO FAR

Although the complexity and variety of parasite antigens will undoubtedly present a number of unique features to the host immune system, it is reasonable to use the mechanisms described above as working hypotheses for the handling of parasite antigens (see review by Kaye, 1987). Indeed, there is a wealth of data concerning serological and cellular responses to protozoan and metazoan parasite infections which indicate that a multitude of parasite molecules are recognized as foreign, processed and then presented to the immune system. The immune system can respond with proliferation of T cells, production of antibodies and their interaction with other inflammatory and immune cell types. However, the very nature of the parasitic association shows that the immune responses generated against parasite molecules are often ineffective in eliminating the infection. The mechanisms underlying immune evasion by parasites are areas of intense research. In this context, one focus of attention will be an analysis of antigen presentation (see Chapter 13).

An impressive beginning has been made in the definition of parasite antigens in terms of molecular weights, their biochemical properties and purification, serological and cellular responses against them, and their capacity to stimulate protective or suppressive responses (see Chapter 2). Nevertheless, for the vast majority of species, it will be some time before the approaches used in the analysis of APC mechanisms for defined peptides (as described earlier) will be feasible for parasite antigens. However, with the increased effort now being directed towards the cloning and expression of genes coding for parasite antigens, such appropriate studies will become a reality. A beginning has already been made, highlighting the fact that the approaches used for 'artificial' antigens can be successfully applied to the analysis of immune mechanisms in parasitic infections.

The site of antigen presentation

The variation observed in the immune responses generated by parasite infection are a reflection of the anatomical location of the parasite within the host, the type of APC locally present and the nature of the antigens (secreted *versus* membrane bound). For example, during the skin-penetrating phase of infection by parasites, such as schistosomes or hookworms, APC, such as Langerhans cells, may be important in the presentation of parasite antigens secreted during passage through the skin; for intracellular pathogens, such as malarial parasites, antigens found within red cell membranes may be important in presentation by splenic macrophages; for intestinal dwelling tapeworms or nematodes, Class II bearing epithelial cells or macrophages and dendritic cells of the Peyers patches may be important in presenting both secreted and parasite-bound antigens. Furthermore, as the parasite migrates around the body, e.g. following infection by a skin-penetrating helminth, a whole array of potential APC may be involved (Figure 4.10).

The variation in the subsequent immune response generated by a different APC at different sites is evidenced by a number of vaccination studies. In the murine model of *Schistosoma mansoni* infection, animals can be protected with a schistosomula antigen given intradermally with an adjuvant but not

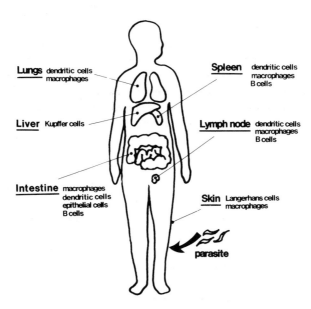

Figure 4.10. Diagram to show the variety of cell types which may be involved in presenting parasite antigens following infection by a skin-penetrating helminth.

intravenously (see Section 8.4). The difference in protective immunity is reflected in the types of response generated. In the former (intradermal) the components of T-cell mediated immunity are promoted and a marked cutaneous delayed hypersensitivity is evident in the absence of an antibody response. In the latter (intravenous), a strong antibody response is generated in the relative absence of a cell-mediated immune response (James, 1986). However, as might be expected, this particular situation does not apply to all parasite infections. Indeed, Liew (1986) has shown in the *Leishmania major* system that immunization of mice with attenuated (irradiated) promastigotes given intravenously leads to the generation of a population of protective T cells which can help cure the disease, whereas the same antigen given subcutaneously leads to the generation of T cells which lead to exacerbation of the disease.

It is clear from both of these examples that the characteristics of APC of different types and sites can determine the type of immune response generated. One way in which parasites may do this has recently been proposed by Anders (1986) and is based upon the presentation of particular parasite antigens by B cells. Critical to his hypothesis is the demonstration of a number of immunodominant parasite protein antigens which exhibit repetitive sequences of amino acids, a characteristic of *Plasmodia*, *Trypanosoma* and *Leishmania* antigens (see Section 2.2). Anders has proposed that in the malaria sporozoite, the repetitive protein molecule presents the host with a complex network of cross-reacting epitopes. During natural infection these epitopes would generate B cells with a great variety of combining-site affinities through their surface Ig molecules. Because of the continued antigenic supply during natural infection, the B cells specific for a particular epitope and low affinity will continue to interact with antigen, process it and present it to T cells which will, in turn, provide help for production of low-affinity antibody. This will continue to occur even with antigens for which the affinity is high due to their cross-reactive nature with low-affinity epitopes. Thus, there will be an excessive increase of specific low-affinity antibody produced during infection resulting in a low-key overall response and weak immunity (see Sections 2.2 and 13.3).

Differences in responses may also result from the capacity of certain parasite molecules to modulate a normally protective anti-parasite immune response. Some parasite infections have been shown to induce changes in the proportion of Class II molecules expressed upon certain cell types although this effect is usually as a result of T-cell produced IFN-γ. *Leishmania donovani* infected macrophages exhibit low levels of Class II molecules on their surface, and the levels cannot be increased after stimulation with IFN-γ. Although the parasite molecules involved in this modulation are unknown, it has been shown that the effect is mediated by host-produced inflammatory mediators such as prostaglandins (Reiner *et al.*, 1987).

Another mechanism whereby parasites may exert a modulatory effect at the level of the APC is at interleukin-1 production. As mentioned earlier in this chapter, IL-1, particularly its membrane-associated form, has been implicated as an important component in T-cell activation. Interference with this step may be one way the parasite can reduce effective antigen presentation and hence the subsequent generation of host-protective immune responses. Indeed, *L. donovani* has been shown to infect macrophages without inducing the production of IL-1 (Reiner, 1987).

The presentation of defined parasite antigens

Detailed studies concerning the mechanisms of antigen presentation during parasitic infection have suffered from lack of sufficient quantities of highly purified antigens. However, the genes coding for a number of parasite antigens have already been cloned (see Section 2.5), and this has enabled analysis of host-parasite interactions at the molecular level.

In the case of malaria, the methodologies developed using artificial antigens have been successfully applied to the study of the processing of parasite antigens. The obvious aim of this type of research is the production of a vaccine. One target in the malaria life cycle which has been extensively studied is the stage that infects the liver cells – the sporozoite. A detailed molecular analysis of the antigens of the sporozoite revealed an immunodominant region (with regard to recognition by antibodies) on the immunogenic circumsporozoite protein (CS protein). Antibodies to this epitope (composed of a tandemly repeated sequence of amino acids) have been shown to prevent infection of hepatocytes by sporozoites *in vitro* and protect mice *in vivo* from a challenge infection of murine malaria. However, the repeating epitope was not widely recognized by T cells, at least not in murine models.

The whole amino acid sequence of the CS protein has been analyzed by Good *et al.* (1987, 1988) using an algorithm for amphipathicity to predict areas of the molecule which should be efficient at stimulating T cells (see Section 4.4). This algorithm predicts regions of protein in which the amino acid sequence folds into an α-helix. A segment from residues 323 to 349 was chosen as the most likely candidate for good T-cell stimulation. The amphipathicity of this segment was apparent when the helix was viewed as a spiral (Figure 4.11) with the hydrophobic regions separated from the hydrophilic regions. This region was thus identified as a major peptide located at the carboxy terminal side of the repetitive region. Interestingly, the repetitive region would not be predicted to form an amphipathic helix using this algorithm. A synthetic peptide corresponding to the predicted T-cell stimulating region was shown to prime T helper cells which could generate a secondary antibody response specific for the repeat region, and *in vitro* stimulate lymph node cells from an immunized mouse to divide. Moreover, when this peptide was conjugated to the repetitive B-cell site and injected into mice, it was found to be highly immunogenic when compared to immunization with the B-cell epitope alone.

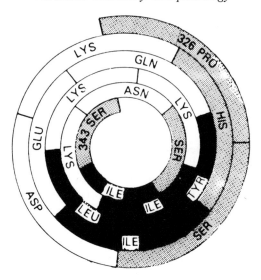

Figure 4.11. Diagrammatic representation of the T-cell epitopes of the circumsporozoite protein from residues 326 proline to 343 serine. The view is looking down the α-helix. The amphipathic nature of this structure can be seen in that the hydrophobic areas (dark shading) are separated from the hydrophilic areas (open) by non-polar residues (stippled areas) (taken from Good *et al.*, 1987; copyright 1987 by the AAAS).

Of course, the protective immune response to parasite infections, including malaria, are known to be highly complex and in many situations are largely undefined. It has been demonstrated that functional host immunity to parasites involves both antibody and antibody independent-mechanisms, and the discovery of further T-cell epitopes important in generating cell-mediated immune effector mechanisms in malaria and other species is eagerly awaited. Nevertheless, regardless of the suitability of the malaria T- and B-cell epitopes so far described in terms of vaccination, this example serves to illustrate the immense potential and value of such approaches when applied to complex parasite antigens.

Whilst the above discussion has concentrated upon protein antigens, it is becoming increasingly clear that a number of important parasite antigens are carbohydrate in nature. This has been shown to be true for both protozoan (*Leishmania*) and metazoan (*Schistosoma*) parasites. Vaccination of mice with a glycolipid from *L. major* was found to protect against the development of cutaneous lesions. The presence of the lipid moiety was essential for protection; vaccination with the lipid-free carbohydrate increased the severity of lesion development (Handman *et al.*, 1987). The nature of antigen handling of these molecules remains to be described but must be taken into account when investigating parasite-antigen-induced responses. A variety of polysaccharides and polyanionic molecules are known to be retained by macrophages and interfere with antigen presentation. Certainly with regard to intracellular bacteria, the intact organism or dominant carbohydrate-containing moieties have been shown to strongly inhibit presentation. Inhibition is thought to result from interference in the intracellular processing chemistry or traffic of endosomes within the APC (Harding *et al.*, 1988).

4.9. SUMMARY

As is evident from the above discussion, antigen handling in terms of processing and presentation is a highly complex phenomenon. The type of immune response subsequently generated will be dependent upon many factors, most notably the molecular form of the antigen and the nature of the APC. Other factors briefly mentioned here, such as genetic disposition and immunocompetence of the host, are likely to assume a central importance when antigen presentation is considered in the context of infection by a parasite. As a consequence, the detailed analysis of antigen handling during parasite infection is a formidable programme of investigation bearing in mind the complex molecular nature of protozoan and metazoan organisms and their intricate parasite life cycles. However, based upon observations made from presentation of 'artificial' antigens, such analysis has begun and a rational approach can now be applied to studies of parasite antigens. The advances in molecular immunology and molecular parasitology will pave the way for definitive experimentation in this area. Such studies will undoubtedly provide a fascinating insight into the crucial aspects of the host-parasite relationship. In nature, this relationship is strongly in favour of the parasite. A detailed knowledge of the handling of parasite molecules by the host may enable manipulation of the relationship so that the parasite is not so favourably positioned.

ACKNOWLEDGEMENTS

I would like to thank Professor I.V. Hutchinson for reviewing the manuscript, Mrs J. Crosby for photography and Mrs J. Jolley for excellent secretarial assistance.

REFERENCES

Adams, D.O. and Hamilton, T.A. (1984), The cell biology of macrophage activation, *Annual Review of Immunology*, **2**, 283–318.

Allen, P.M. (1987), Antigen processing at the molecular level, *Immunology Today*, **8**, 270–3.

Allen, P.M., Matsueda, G.R., Evans, R.J., Dunbar, J.B. Jr., Marshall, G.R. and Unanue, E.R. (1987), Identification of the T cell and Ia contact residues of a T cell antigenic epitope, *Nature (London)*, **327**, 713–5.

Anders, R.F. (1986), Multiple cross-reactivities amongst antigens of *Plasmodium falciparum* impair the development of protective immunity against malaria, *Parasite Immunology*, **8**, 529–39.

Austyn, J.M. (1987), Lymphoid dendritic cells, *Immunology*, **62**, 161–70.

Babbitt, B.P., Allen, P.M., Matsueda, G., Haber, E. and Unanue, E.R. (1985), Binding of immunogenic peptides to Ia histocompatibility molecules, *Nature (London)*, **317**, 359–61.

Berzofsky, J.A., Brett, S.J., Streicher, H.Z. and Takahashi, H. (1988), Antigen processing for presentation to T lymphocytes: Function, Mechanisms and Implications for the T-cell repertoire, *Immunological Reviews*, **106**, 5–31.

Berzofsky, J.A., Cease, K.B., Cornette, J.L., Spouge, J.L., Margalit, H., Berkower, I.J., Good, M.F., Miller, L.H. and de Lisi, C. (1987), Protein antigenic structures recognised by T cells. Potential applications to vaccine design, *Immunological Reviews*, **98**, 9–52.

Bjorkman, P.J., Saper, M.A., Samraoui, B., Bennett, W.S., Strominger, J.L. and Wiley, D.C. (1987), The foreign antigen binding site and T cell recognition regions of class I histocompatibility antigens, *Nature (London)*, **329**, 512–8.

Brown, J.H., Jardetzky, T., Soper, M.A., Samraouri, B., Bjorkman, P.J. and Wiley, D.C. (1988), A hypothetical model of the foreign antigen binding site of Class II histocompatibility molecules, *Nature*, **322**, 845–50.

Germain, R.N. (1986), The ins and outs of antigen processing and presentation, *Nature (London)*, **322**, 687–9.

Good, M.F., Berzofsky, J.A. and Miller, L.H. (1988), The T cell response to the malaria circumsporozoite protein: an immunological approach to vaccine development, *Annual Review of Immunology*, **6**, 663–88.

Good, M.F., Maloy, W.L., Lunde, M.N., Margalit, H., Cornette, J.L., Smith, G.L., Moss, B., Miller, L.H. and Berzofsky, J.A. (1987), Construction of synthetic immunogen: Use of new T-helper epitope on malaria circumsporozoite protein, *Science*, **235**, 1059–62.

Handman, E., McConville, M.J. and Goding, J.W. (1987), Carbohydrate antigens as possible parasite vaccines. A case for the *Leishmania* glycolipid, *Immunology Today*, **8**, 181–5.

Harding, C.V., Leyva-Cobian, F. and Unanue, E.R. (1988), Mechanisms of antigen processing, *Immunological Reviews*, **106**, 77–92.

Howard, J.C. (1985), Immunological help at last, *Nature (London)*, **314**, 494–5.

James, S.L. (1986), Induction of protective immunity against *Schistosoma mansoni* by a non-living vaccine. III. Correlation of resistance with induction of activated larvacidal macrophages, *Journal of Immunology*, **136**, 3872–7.

Katz, D.H., Hamaoka, M.E., Dorf, M.E. and Benacerraf, B. (1973), Cell interactions between histoincompatible T & B lymphocytes. The H-2 gene complex determined successful physiologic lymphocyte interactions, *Proceedings of the National Academy of Sciences (USA)*, 2624–8.

Kaye, P.M. (1987), Antigen presentation and the response to parasitic infection, *Parasitology Today*, **3**, 292–9.

Kramer, M.D. and Simon, M.M. (1987), Are proteinases functional molecules of T lymphocytes? *Immunology Today*, **8**, 140–2.

Kupfer, A., Swain, S.L., Janeway Jr, C.A. and Singer, S.J. (1986), The specific direct interaction of helper T cells and antigen presenting B cells. *Proceedings of the National Academy of Sciences (USA)*, **83**, 6080–3.

Lanzavecchia, A. (1986), Antigen presentation by B lymphocytes. A critical step in T-B collaboration, *Current Topics in Microbiology and Immunology*, **130**, 65–78.

Liew, F.Y. (1986), Cell mediated immunity in experimental cutaneous Leishmaniasis, *Parasitology Today*, **2**, 264–70.

Marrack, P. and Kappler, J. (1987), The T cell receptor, *Science*, **238**, 1073–9.

Mayer, L. and Shlien, R. (1987), Evidence for function of Ia molecules on gut epithelial cells in man, *Journal of Experimental Medicine*, **166**, 1471–83.

Mills, K.H.G. (1986), Processing of viral antigens and presentation to class II-restricted T cells, *Immunology Today*, **9**, 260–3.

Mitchison, N.A. (1971), The carrier effect in the secondary response to hapten-protein conjugates II cell co-operation, *European Journal of Immunology*, **1**, 18–27.

Moller, G. (Ed.) (1987), Antigenic requirements for activation of MHC-restricted responses, *Immunological Reviews*, **98**.

Nicklin, S. (1987), Intestinal Uptake of Antigen: Immunological Consequences, in Miller, K. and Nicklin, S. (Eds.), Immunology of the Gastrointestinal Tract, pp. 87–110, Boca Raton: CRC Press.

Oppenheim, J.J., Kovacs, E.J., Matsushima, K. and Durum, S.K. (1986), There is more than one interleukin-1, *Immunology Today*, **7**, 45–56.

Reiner, N.E. (1987), Parasite accessory cell interactions in murine leishmaniasis. I. Evasion and stimulus-dependent suppression of the macrophage interleukin-1 response by *Leishmania donovani*, *Journal of Immunology*, **138**, 1919–25.

Reiner, N.E., Ng, W. and McMaster, W.R. (1987), Parasite-accessory cell interactions in murine Leishmaniasis. I. *Leishmania donovani* suppresses macrophage expression of class I and class II major histocompatibility complex gene products, *Journal of Immunology*, **138**, 1926–32.

Ron, Y. and Sprent, J. (1987), T cell priming *in vivo*: a major role for B cells in presenting antigen to T cells in lymph nodes. *Journal of Immunology*, **138**, 2848–56.

Rosenthal, A.S. (1978), Determinant selection and macrophage function in genetic control of the immune response. *Immunological Reviews*, **40**, 136–52.

Rothbard, J.B. and Taylor, W.R. (1988), A sequence pattern common to T cell epitopes, *The EMBO Journal*, **7**, 93–100.

Sanders, V.M., Fernandez-Botran, R., Uhs, J.W. and Vitetta, E.S. (1987), Interleukin 4 enhances the ability of antigen-specific B cells to form conjugates with T cells, *Journal of Immunology*, **139**, 2349–54.

Schwartz, R.H. (1986), Immune response (Ir) genes of the murine major histocompatibility complex, *Advances in Immunology*, **38**, 31–202.

Shimonkovitz, R., Kappler, J., Marrack, P. and Grey, H.M. (1983), Antigen recognition by H-2 restricted T cells. I. Cell free antigen processing, *Journal of Experimental Medicine*, **158**, 303–16.

Steinman, R.M. and Cohn, Z.A. (1973), Identification of a novel cell type in peripheral lymphoid organs of mice. I. Morphology, quantitation, tissue distribution, *Journal of Experimental Medicine*, **137**, 1142–62.

Townsend, A.R.M., Rothbard, J., Gotch, F.M., Bahadur, G., Wraith, D. and McMichael, A.J. (1986), The epitopes of influenza nucleoprotein recognised by cytotoxic T lymphocytes can be defined with short synthetic peptides, *Cell*, **44**, 959–68.

Unanue, E.R. (1981), The regulatory role of macrophages in antigenic stimulation. Part II. Symbiotic relationship between lymphocytes and macrophages, *Advances in Immunology*, **31**, 1–136.

Unanue, E.R. (1984), Antigen-presenting function of the macrophage, *Annual Review of Immunology*, **2**, 395–428.

Unanue, E.R. and Allen, P.M. (1987), The basis for the immuno-regulatory role of macrophages and other accessory cells, *Science*, **236**, 551–7.

Ziegler, K. and Unanue, E.R. (1982), Decrease in macrophage antigen catabolism by ammonia and chloroquine is associated with inhibition of antigen presentation to T cells, *Proceedings of the National Academy of Sciences (USA)*, **79**, 175–8.

Zinkernagel, R.M. and Doherty, P.C. (1979), MHC-restricted cytotoxic T cells: Studies on the biological role of polymorphic major transplantation antigens determining T cells restriction specificity, function and responsiveness, *Advances in Immunology*, **27**, 51–177.

5. Humoral and cellular effector immune responses against parasites

Kenneth Crook

5.1. INTRODUCTION

The most striking feature of the immune response to infections with parasites is the wide diversity of ways in which the response manifests itself. The involvement, or not, of antibody and the different types of cells, from T lymphocytes down to platelets, suggest a highly differentiated network of cells. From this network the various elements involved depend on the type of parasite infecting the host and the particular problems which this presents. For example, protozoan malaria parasites which spend most of their time in mammalian hosts living within circulating erythrocytes are clearly going to be seen in a different manner by the immune system to the metazoan schistosome trematodes which infect the mammalian host through the skin and migrate through the tissues to the liver via the lungs.

The purpose of this chapter is not to give a comprehensive review of immunity to parasites but is instead intended to provide an up-to-date overview of work on some of the important mechanisms involved in the defence against infection, as well as drawing particular attention to those areas which are currently in the limelight of immunological interest. A few examples of original data taken from papers quoted in the text will also be given. The choice of these has been relatively arbitrary, the purpose again not being to cover all possible aspects of research but merely to help the reader gain some feel for the hard data from which the workers' conclusions are drawn.

First of all, it is important to have some understanding of how the key cell types in any immune response (the T and B lymphocytes) develop and how they are organized within the lymphoid tissue to maximize encounter with antigen (see Chapter 4) and initiate the cascade of effector mechanisms against the invader. A more comprehensive review of this aspect, however, may be found in Male *et al.* (1987).

5.2. LYMPHOPOIESIS

Foetal development

The earliest appearance of lymphocyte-like cells in mice is that of antibody-forming cells at day nine of gestation in the yolk sac (Tyan, 1968). From about day 12 of gestation, haemopoiesis (and consequently lymphopoiesis, since all circulating blood cells are derived from the same stem cell pool) moves to the foetal liver. Foetal liver cells have been shown to be able to transfer the capability for antibody-formation (B cells; Doria *et al.*, 1962) and delayed type hypersensitivity (T cells; Tyan, 1968) to irradiated recipients.

Towards the end of gestation, around day 19, the number of stem cells in the liver drops rapidly and the site of haemopoiesis moves again, this time to the bone marrow. By this stage, however, cells have begun to seed the foetal thymus and differentiate into mature T cells. B-cell lymphopoiesis remains within the liver, and moves to the bone marrow after birth.

B-cell development

B-cell development like T-cell development, can essentially be divided into two main steps. The first is antigen independent and involves the cells becoming committed to the B- or T-lymphocyte lineage, while the second requires antigen for the cells to move on to terminal differentiation into mature effector cells.

The major distinguishing feature of B lymphocytes is their ability to express on their surface and then to secrete antibody molecules made up of immunoglobulin (Ig) heavy (H) and light (L) polypeptide chains (Figure 5.1). Each of these chains is made up of domains of about 110 amino acids in length. These domains are termed either variable (V) or constant (C) depending on the makeup of their amino acid sequences, with L chains having one V and one C domain and H chains having one V and either three or four C domains. The V regions of the H and L chains are situated at the N-terminus of the molecule and interact to form the specific antigen-combining site. At the genetic level, these chains each result from a combination of separate gene segments (Figure 5.2) which rearrange during B-cell differentiation to make a single messenger RNA molecule. This rearrangement of a few groups of gene clusters allows the generation of an antibody repertoire with a diversity of as much as 10^8 or 10^9 binding sites from a relatively small number of genes.

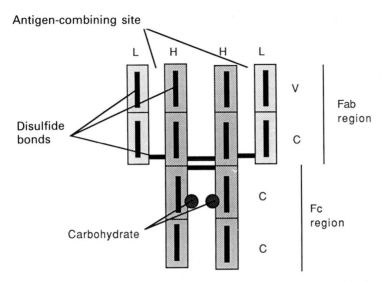

Figure 5.1. Immunoglobulin molecule. A four-chain molecule made up of two identical heavy (H) chains and two identical light (L) chains held together by disulfide bridges, represented by the dark bands. Each chain consists of domains whose tertiary structure is dictated by intradomain disulfide bridges. These domains are either variable (V) or constant (C) depending on their polypeptide sequence. The Fab region comprizes the antigen-binding sites, formed by the interaction between light and heavy chain V domains, while the Fc region is involved in interactions with accessory cells during an immune response.

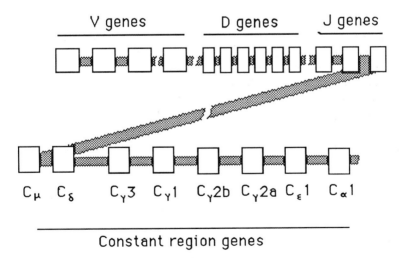

Figure 5.2. Arrangement of mouse immunoglobulin heavy chain genes in the germline DNA. Each heavy chain results from the recombination of one V gene with one D and one J gene, followed by the joining of this VDJ message to a message from one C region, depending on the class of immunoglobulin to be produced. There is a very large number of V genes as well as twelve D genes and three J genes to choose from. The breaks in the line representing the non-coding DNA indicate large lengths of intervening sequence between coding sequences (adapted from Male *et al.*, 1987).

One of the earliest indications that a cell has become committed to the B-cell lineage is the rearrangement of these gene segments. When they have been transcribed and translated, it is possible to observe L and H polypeptide chains in the cell cytoplasm. The first of the Ig C domains to be transcribed is that of the IgM isotype, and a cell with cytoplasmic L and H-mu chains is termed a pre-B cell. IgM will then begin to be expressed on the surface of such cells which are now known as B cells. This is not a mature phenotype as B cells still require stimulation by antigen to drive them on towards terminal differentiation into plasma cells.

Once a cell has expressed surface IgM, further gene rearrangements in the region of the C gene segments can lead to this being replaced by another Ig isotype. This process is known as class switching and it can be stimulated by a variety of factors such as activation by antigen or the action of lymphokines derived from T cells or accessory cells. Having expressed surface Ig, B cells move out into the blood and begin circulating around the blood and lymphatic systems ready for encounter with foreign antigen.

The final stage of B-cell development, which occurs in the periphery, follows activation by antigen which binds to the B cell via the membrane-Ig molecule and, except in the case of the so-called thymus-independent antigens, the additional action of T-cell-derived lymphokines. The cell then proceeds through several more cell cycles until the plasma cell stage is reached; this no longer expresses surface Ig but does secrete large numbers of these molecules into the surrounding tissues.

T-cell development

Commitment of pluripotent stem cells to become T cells occurs in the foetal liver before birth and afterwards in the bone marrow. These poorly characterized pre-T cells then migrate through the circulation and seed the thymus where the processes of T-cell differentiation and antigen receptor selection take place. The T-cell antigen receptor is made up of two polypeptide chains (alpha and beta) of similar size, each of which comprizes two domains very like the V and C domains of the Ig molecule (Figure 5.3). A second class of T-cell receptor has been discovered more recently, comprizing gamma and delta chains, which are only present on a very small subpopulation of peripheral T cells (around 3–5%) although they do appear to be present in much higher numbers in the skin and gut epithelium. The role of these cells is not clear (see Chapter 10), but they appear to have only a very limited specificity as defined by their usage of V-region genes and thus may represent some sort of natural killer (NK) cell population.

At the genetic level, too (Figure 5.4), there are similarities between the T-cell receptor and immunoglobulin, diversity resulting from the rearrangement of small gene segments to create a single transcript. As receptor genes are rearranged, the early T cells express two surface antigens of note—CD4 and CD8. Gradually, as cells interact with antigen-presenting accessory cells in

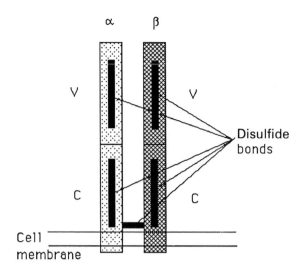

Figure 5.3. T-cell receptor. This is cell-bound heterodimer made up of two polypeptide chains, alpha and beta, held together near the cell surface by a disulfide bond. Each chain is in turn made up of one variable (V) and one constant (c) domain, and as with the Ig molecule, these domains have a tertiary structure in part dictated by an intradomain disulfide bond. There is a significant degree of sequence homology between Ig and TCR V domains and between c domains.

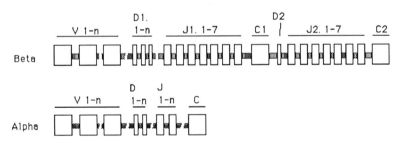

Figure 5.4. Arrangement of the mouse TCR alpha and beta genes in the germline DNA. As with the Ig genes, recombination occurs between individual V, D, and J genes. This VDJ sequence then joins with a C region to form a single transcript. With the beta genes, the V gene may recombine with D, J, and then C from one of two subsets of these genes (adapted from Male *et al.*, 1987).

the thymus, one or other of these surface markers is lost, and the resulting mature T cells become restricted to recognize foreign antigen in association with either Class I MHC molecules, in the case of CD8, or Class II MHC molecules, in the case of CD4 (see Chapter 4 for a more complete discussion of antigen recognition by T cells).

On expression of an appropriate antigen receptor, the mature T cells migrate out of the thymus and into the blood and lymphatic circulation awaiting encounter with the combination of foreign antigen plus MHC for which they are specific.

5.3. THE ORGANIZATION OF THE LYMPHOID TISSUE

The functions of the primary lymphoid organs (thymus and bone marrow) have already been briefly discussed above. This section will describe how the secondary lymphoid tissues function to maximize and optimize the interaction between antigen and T and B lymphocytes. Again, however, for a more detailed description of the organization of the lymphoid system and lymphocyte trafficking, the reader is referred to Male *et al.* (1987).

Lymph nodes constitute the central element in the peripheral immune system, acting as the sites where T and B cells are first presented with antigen and are able to interact with each other if necessary. The basic anatomy of a lymph node is shown in Figure 5.5. Antigen enters via the afferent lymphatics in one of two ways. It may be in the form of an immune complex (i.e. antigen complexed with specific antibody and complement components) in which case it will enter the follicles in the B-cell area in the cortex where it will bind to the surface of follicular dendritic cells (FDC) and be presented to and stimulate B cells. Alternatively, antigen may enter the lymph node on

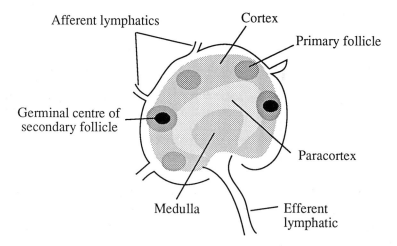

Figure 5.5. Structure of a lymph node. This is divided into distinct T- and B-cell areas (paracortex and cortex, respectively). Antigen enters by the afferent lymphatics, while lymphocytes enter via the high endothelial-walled venules in the paracortex. Once stimulated, the lymphocytes leave the node by the efferent lymphatic.

antigen-presenting cells and will migrate through the cortex to the T-cell area in the paracortex. These antigen-presenting cells may then differentiate into a cell known as an interdigitating cell (IDC) which has very extensive cellular processes, presumably to maximize presentation of antigen to the surrounding T cells.

The T cells thus activated may then associate with B cells at the cortex/paracortex border and provide them with the necessary signals to complete the activation process already begun when the B cells bound antigen. The effector T and B cells subsequently migrate out of the lymph node by way of the efferent lymphatics and towards the site of infection.

The other major peripheral lymphoid organ is the spleen where lymphocytes recirculate through specific T- and B-cell areas in a slightly different way but with a similar end result although the types of antigen passing through the spleen may be distinct from those seen in the lymph nodes.

Unstimulated and long-lived memory T and B cells continuously recirculate around the blood and lymphatics ready to encounter antigen. They may also pass through the lymph nodes and do so by crossing the high endothelial-walled venules from the blood into the paracortex. The T cells remain there while the B cells migrate into the cortex. In this way, with a pool of recirculating lymphocytes bearing specific receptors for antigen and with sites where these cells can be stimulated and interact with each other, the immune system is ready to detect and respond to stimulation by any foreign material. This stimulation is at the base of an extensive network of defensive mechanisms involving many types of cells as well as soluble molecular factors.

5.4.   B LYMPHOCYTES

There are three ways in which B cells could function as effector cells in the immune response against parasites: by means of antibody which could function directly against the parasite to assist in damaging it; antibody as a regulatory molecule; or by secreting other factors and lymphokines which may stimulate other effector cells into action.

Antibodies as effectors

The role of antibodies in the immune response is classically defined as being against unicellular organisms, especially prokaryotes, opsonizing them which allows the foreign cell to be phagocytosed by cells, such as macrophages, or activating the complement cascade which will bring about lysis of the cell by the action of the terminal complement components.

In infections by eukaryotic parasites, antibodies will largely only be active on their own against protozoa. The best example of this is in infections with malaria. The infective form, the sporozoite, spends a very short time in the blood before invading the liver and entering the hepatocytes where the next stage in the life cycle develops. It has been clearly shown that antibodies directed against the sporozoite stage can act in a blocking manner to prevent the infection of liver cells (Potocnjak *et al.*, 1980; Yoshida *et al.*, 1980) and thus prevent further development and replication of the parasite (even Fab fragments of Ig molecules can block invasion confirming that this is a physical effect rather than a cytotoxic one). This phenomenon has been exploited in the development of a vaccine against human malaria (Herrington *et al.*, 1987) although the value of such anti-sporozoite antibodies in controlling infection with malaria in the field has been questioned by Hoffman *et al.* (1987) who found no correlation between levels of anti-sporozoite antibodies and onset of disease. Clinical trials of such vaccines have led to the realization that T-cell responses must also be initiated, and it is towards this goal that malaria vaccine research has more recently been directed (see Chapter 14).

IgA is the major immunoglobulin isotype secreted at mucosal surfaces (e.g. the gut, lungs, genito-urinary tract) and in at least one case has been shown to protect against infection by a parasite. Lloyd and Soulsby (1978) transferred IgA from mice immune to the tapeworm *Taenia taeniaeformis* to naive mice and found the recipients to be immune to challenge with the same parasite. However, it is not clear how IgA is active in this situation—it may be acting in a blocking manner, as in the malaria experiments described above or it may be binding to an accessory cell to act via an antibody-dependent, cell-mediated cytotoxicity (ADCC) mechanism as described below. Immunity at mucosal surfaces is described in more detail in Chapter 10.

Less-established roles for antibody molecules include possible anti-enzyme and anti-sensory properties. Infections with the hookworm *Ancylostoma caninum* are associated with high levels of proteolytic enzymes presumably released by the worm to allow its passage through the tissues. Transfer of immune serum to dogs with *A. caninum* infections, however, led to a marked reduction in proteolytic activity of parasite extracts (Thorson, 1956). It has also been suggested, although never been proved, that immunoglobulins may interfere in some way with the surface of the parasite to prevent it either responding to its environment correctly (perhaps by blocking an important receptor) or halting its motility through tissues in the case of invasive parasites such as schistosomes or hookworms.

The main mechanism through which antibodies are effective against parasites is by antibody-dependent, cell-mediated cytotoxicity (ADCC). In this, parasite-specific antibodies bind to effector cells, such as eosinophils or macrophages, via surface receptors specific for the Fc portion of Ig molecules. These cells, which are by themselves unable to recognize antigen, are now rendered specific by the surface-bound antibody and can bind specifically to the parasite surface and exert their cytotoxic effect against the target. Examples of this will be given later in those sections pertaining to the effector cells participating in ADCC.

Antibodies as immune regulators

The ability of Ig molecules to interact with effector cells and play a role in regulating the immune response has been suggested on many occasions, but conclusive proof has usually not been forthcoming. The only possible involvement of antibodies as immunoregulatory molecules for which there has been some evidence is as part of an idiotypic/anti-idiotypic (Id/anti-Id) network. Idiotypes are antigenic determinants on Ig molecules themselves (Kunkel *et al.*, 1963; Oudin and Michel, 1963), particularly around the hypervariable regions at the antigen binding end of the molecule. These idiotypes are recognized by anti-idiotypic antibodies from the same individual, and it was suggested that this interaction might play an important role in immunoregulation (Jerne, 1974)—the Network Theory. Although these Id/anti-Id interactions may exist in many parasite systems (e.g. Gorczynski, 1987; Cesbron *et al.*, 1988), they are not well understood and the full significance, and indeed relevance, of an extensive idiotypic network in immune responses in general is not clear.

B-cell-derived lymphokines

The major lymphokine produced by B cells is interleukin-1 (IL-1) which has a stimulatory effect on a variety of other cells including T cells, NK cells, and neutrophils. In this way B cells can potentiate an immune response without actually producing antibody.

5.5. T LYMPHOCYTES

These cells are classically thought of as the controlling cell behind an immune response with the ability to stimulate and organize other effector cell types against the foreign organism. As described above T cells can be broadly divided into two subsets, functionally as well as antigenically: the T-helper (Th) cells and T-cytotoxic (Tc) cells. Th cells bear the CD4 differentiation antigen while Tc cells bear CD8 (see Chapter 4).

T-helper cells

Th cells recognize antigen in association with Class II molecules usually present on a specialized presenting cell, such as a dendritic cell or a B lymphocyte, and this acts as the initial activation step. Interleukin-2 produced either by the same T cell or by another activated T cell in the vicinity amplifies this activation giving rise to the fully functional Th cell. Such cells are then capable of producing a wide array of lymphokines including IL-2, IL-3, IL-4, IL-5 and interferon-gamma (IFN-gamma), which are important in recruiting other cells into the site of inflammation, and once there, in activating them to make their contribution in the response against the parasite (see below and Chapter 3).

The importance of T cells in determining the outcome of parasitic infections was first suggested by experiments in athymic mice (either after surgical removal or by using congenitally athymic nude mice). These mice often showed a delayed and weak, or even non-existent, immune response to the infection (e.g. Manson-Smith *et al.*, 1979). Transfer of lymphocytes from immune animals was able to restore protective immunity. Further experiments involving the depletion and restoration of cells have elucidated the identity of the critical cell type even further and shown that a T-cell-mediated response is important in the defence against many different kinds of parasite. Phillips *et al.* (1987b) have depleted T-cell subsets *in vivo* by giving rats intravenous injections of monoclonal antibodies specific for subset markers and subsequently infecting the animals with *Schistosoma mansoni*. Depletion using either a monoclonal antibody recognizing RT7.1 (a pan-T-cell marker) or W3/25 (recognizing CD4) led to an increase in initial worm development and a decrease in both the T-cell response, as measured by a delayed type hypersensitivity reaction, and in antibody levels. Depletion using OX8 (recognizing CD8) led to the opposite effects.

Similarly, Schofield *et al.* (1987b) found that although depletion of T-cell subsets had little effect on the initial development of *S. mansoni*, animals receiving a dose of a monoclonal antibody recognizing CD4 (i.e. Th cells) were less resistant to a secondary infection. Depletion of CD8 positive cells led to an increase in resistance and lower morbidity. As well as reinforcing the hypothesis that CD4 Th cells are vital in the anti-schistosome response, these results also suggest that the elements important in effecting a primary response may be different from those in a secondary response. This difference between the response to primary and challenge infections may be because certain effector mechanisms are not able to reach a stage where they are able to influence the parasite in a primary response. On secondary stimulation, however, the more rapid initiation of immune mechanisms by memory cells may allow other, possibly more effective, systems to come into play earlier. This is an important general point about all immune responses and one which reinforces the heterogeneity of the mechanisms involved.

Kelly and Colley (1988), also making use of *in vivo* T-cell subset depletions with monoclonal antibodies, showed that not only was the depleted cell subset important but the timing of the depletion could also play a role in the outcome of infection with *S. mansoni*. Depletion of CD4 positive cells before a challenge infection gave the expected decrease in resistance. However, if the administration of antibody was delayed a few days until the worms had moved out of the skin the depletion made no difference to the response against the worm, indicating that, by this stage, the Th cells had already made their contribution in inducing the activation of other effector cells.

Many workers have recently generated parasite-specific T-cell lines and clones by repeated *in vitro* antigenic stimulation of lymphoblasts taken from infected animals. Transfer of these cells to animals has been shown in many parasite systems to be as effective in conferring protection on the recipient as an unfractionated population of lymph node cells: Canessa *et al.* (1988) found that *Toxoplasma gondii*-specific, CD4-positive T-cell lines could help *in vitro* B-cell proliferation and differentiation and that supernatants from these T-cell cultures stimulated *T. gondii*-infected macrophages to kill their intracellular parasites; Lammas *et al.* (1987) showed that transfer of enhanced eosinophilia and resistance to infection with *Mesocestoides corti* could be obtained by injection of the mice with a specific T-cell line; similarly, Riedlinger *et al.* (1986) induced an accelerated expulsion of *Trichinella spiralis* from infected mice if they were first given an injection of a Th cell line raised against *T. spiralis* larval antigen (Table 5.1); Auriault *et al.* (1987) obtained protection to *S. mansoni* infection in rats with a T-cell line raised against a 28 kDa antigen; Brake *et al.* (1986, 1988) have protected mice from infection with *Plasmodium chabaudi adami* by transfer of a *P.c. adami*-specific, CD4-positive T-cell line (1986) and subsequently with a T-cell clone derived from this line (1988).

Table 5.1. Transfer of specific T-cell lines protects against *T. spiralis* infection (data adapted from Riedlinger *et al.*, 1986).

Group	Days MLNC[a] in culture	Worm nos. on day 7 Post-infection (mean)
Control	—	185
Test[b]	28	60
Control	—	202
Test	77	60
Control	—	146
Test	98	75

[a] MLNC—Mesenteric Lymph Node Cells
[b] Test mice received a dose of 2×10^6 cultured T cells.

These results with T-cell clones are a particular cause for excitement as they will eventually allow a complete elucidation of the fine T-cell specificity required at the initiation of the response as well as the precise mechanisms each individual T cell employs in its contribution to the overall host response. Examples of this have already begun to be described after the observation that murine T-helper cell clones can apparently be divided into two groups according to the lymphokines that they are able to release (see Mosmann and Coffman, 1987, for a review). These were named Th1 and Th2, and in addition to both being able to secrete certain lymphokines, Th1 can uniquely produce IL-2 and IFN-gamma, while only Th2 can secrete IL-4 and IL-5 (see Chapter 3).

This classification seemed to provide a solution to results from *Leishmania major* infections in mice which otherwise appeared to present a confusing picture of the precise role of T cells in the defense against this parasite. Liew *et al.* (1987) showed that T cells taken from mice immunized intravenously with irradiated *L. major* promastigotes could transfer protection to naive recipients which might be the predicted result. On the other hand, if the T cells were taken from mice immunized subcutaneously, on transfer to naive animals the effects of the disease were exacerbated. Thus it appeared that, in this model, T cells could act in both a protective and disease-promoting manner. In addition, Sadick *et al.* (1987) presented results which seemed to suggest that if *L. major*-susceptible strains of mice were treated with a monoclonal antibody recognizing CD4-positive cells, they behaved more like *L. major*-resistant strains of mice.

More recent experiments have exploited the Th1/Th2 subdivision to clarify the reasons for the above, apparently unexpected, results. Scott *et al.* (1988) were able to protect BALB/c mice (normally very susceptible to infection) against *L. major* infection with one T-cell line; however, this approach caused

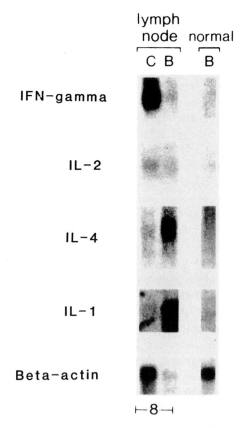

Figure 5.6. Northern blot analysis of messenger RNA obtained from lymph nodes from mice infected with *Leishmania major*. Assays were carried out eight weeks after infection in susceptible BALB/c (B) and resistant C57BL/6 (C) strains of mice. Nitrocellulose-bound mRNA was incubated with radioactively labelled nucleic acid probes specific for a variety of lymphokines (interferon-gamma, interleukin-2, interleukin-4, interleukin-1, and finally beta-actin as a positive control). Production of IL-2, IL-1, and beta-actin mRNA was visible in both B and C strains, while IFN-gamma was uniquely expressed by resistant C strain cells, and IL-4 was only produced by susceptible B strain cells (taken from Heinzel *et al.*, 1989; used with permission of the authors and Rockefeller University Press).

a more serious manifestation of the disease by transferring two other T-cell lines. They showed that the disease-exacerbating T-cell line could secrete IL-4 and IL-5 (properties of Th2 cells), but that the protective cell lines could secrete IL-2 and IFN-gamma (properties of Th1 cells). In agreement with this, Heinzel *et al.* (1989) measured the lymphokine mRNA produced in the lymphoid organs of mice resistant and susceptible to *L. major* infection (C57BL/6 and BALB/c, respectively). High levels of IL-4 message and low levels of IFN-gamma message were detected in BALB/c mice, while the inverse was found in C57BL/6 mice, as can be seen from the Northern blot presented in Figure 5.6. This seemed to confirm that Th1 cells were protective in *L. major* infection while Th2 cells aggravated the disease, and the ability of mice to resist *L. major* infection would depend on their ability to preferentially induce Th1 rather than Th2 cells. Heinzel *et al.* (1989) also carried out *in vivo* depletions of CD4-positive T cells with monoclonal antibodies which indicated that Th1 cells could recover more rapidly than Th2, providing protection for BALB/c mice and helping to explain the results of Sadick *et al.* (1987) described above.

Mosmann *et al.* (1988) have also used the Th1/Th2 subdivision to explain the high IgE production (stimulated by IL-4 and suppressed by IFN-gamma) and eosinophilia (induced by IL-5) normally found in infections with *Nippostrongylus brasiliensis*. They tested lymphokine secretion by lymph node and spleen cells and found high levels of IL-4 and IL-5 but low levels of IFN-gamma and IL-2. This suggests that preferential induction of Th2 cells is important in controlling infection.

However, it is perhaps worth pointing out that recent studies by Kelso and Gough (1988) argue that the range of lymphokines secreted by T-cell clones is a completely random affair and that no neat Th1/Th2 classification exists. Even if this is the case, though, the central message from the above experiments is not substantially altered, and we can still say that the efficacy of a T-cell response is dependent on the ability of those T cells to begin producing the appropriate combination of lymphokines on receipt of the appropriate signal.

T-cytotoxic/suppressor cells

These cells bear the CD8 surface marker and recognize foreign antigen in association with Class I MHC molecules. This is certainly true for Tc cells although it is not clear how the putative T-suppressor subpopulation recognizes antigen since it does not appear to express a receptor like other T cells. In any case, this chapter will not deal with suppression of immune responses, as this subject is addressed elsewhere in this book (see Chapter 13).

The role of Tc cells in immunity to parasites is not well established, and there are only a few examples documenting their action. Butterworth *et al.* (1979) demonstrated that Tc cells specific for alloantigens could adhere to schistosomula (remembering that schistosomes pick up host antigens during their passage through the tissues) but they do not appear to be able to damage the worms and the adherence is probably an *in vitro* artefact with no relevance to protective immunity. Tc cells taken from mice infected with *Plasmodium falciparum* are able to kill L cells (a fibroblastic cell line) transfected with the gene encoding the circumsporozoite protein (Kumar *et al.*, 1988) although the *in vivo* importance of this has not been demonstrated. *In vivo* depletion experiments with monoclonal antibodies in mice infected with malaria have given results opposite to those expected from the CD4-depletion experiments described above. If mice immune to malaria are injected with monoclonal antibodies against either CD4 or CD8 and then reinfected, only the mice depleted of their Th cells remain immune (Schofield *et al.*, 1987b; Weiss *et al.*, 1988). The mice which had lost their Tc cells are no longer able to resist a secondary infection. These results would suggest a role for Tc cells in at least secondary immunity to malaria.

Tc cells have also been found to exert a cytotoxic effect on lymphocytes infected with the protozoan parasite *Theileria parva* (Goddeeris *et al.*, 1986).

5.6.   NATURAL KILLER CELLS

These are cells of a somewhat poorly defined nature, and they may in fact represent a mixture of different cell types grouped together by virtue of their granular appearance and ability to lyse certain target cells but not others although they lack the precise specificity of Tc cells. It may be that they are an evolutionary forerunner of the Tc cell. The gamma/delta receptor-bearing T cells found in such high numbers in the skin and intestinal epithelium might be expected to fall somewhere between Tc cells and NK cells in evolutionary terms. The data supporting this is purely circumstantial however—gamma chains are the first to rearrange in ontogeny and they have a very restricted usage of V gene segments implying a limited specificity.

A role for NK cells in immunity against parasites has, as for Tc cells, received only limited attention, and the work that has been done would suggest that they are not very important. Hughes *et al.* (1987) showed that NK cells could not bind *Toxoplasma gondii* parasites and that there was no evidence for *in vitro* killing by these cells—in fact, survival of the parasites appeared to be enhanced if NK cells were present! NK activity was found in pulmonary cell populations from *Trichinella pseudospiralis*-infected mice (Niederkorn *et al.*, 1988) although again there was no functional significance ascribed to this. Spontaneous killing of *Trypanosoma musculi* and *T. lewisi* by NK-like cells was, however, demonstrated by Albright *et al.* (1988). In addition, killing was increased if parasite-specific antibody was added, suggesting that these cells could also function in an ADCC system. But on the whole it would appear that NK cells do not have a vital part in the immune response against parasites.

5.7. LYMPHOKINES

It is only really in the last ten years that people have begun to look in detail at the role of these molecules, not only as important mediators in the defense against parasites but also as a contributory element in the pathology often associated with disease. This has come about with the purification and cloning of many of the molecules allowing a standardization of the nomenclature and providing a pure source of factors for use in *in vitro* and *in vivo* studies. Up until then, different laboratories had been giving the same factor different names, leading to a rather confused picture of the situation.

Interferon-gamma has long been known to be able to induce macrophages to differentiate to a more activated phenotype, with increased MHC expression, increased membrane ruffling and enhanced phagocytosis and cytotoxicity, as well as several other characteristics associated with activation. As macrophages are an important cell type in the defense against many parasites, it seemed at least possible that IFN-gamma might increase their effect. Reed (1988) injected mice with recombinant IFN-gamma and incubated macrophages taken from these animals with *Trypanosoma cruzi*. Cells were still infected at the normal rate but killing of the intracellular parasites was greatly increased. In a human study, Salata *et al.* (1987) took lymphocytes from patients infected with *Entamoeba histolytica* and incubated them with parasite antigen. The lymphocytes produced IFN-gamma which could induce macrophages to kill *E. histolytica* trophozoites in culture. A monoclonal antibody specific for IFN-gamma removed this activity from the lymphocyte supernatants, confirming that IFN-gamma is the active factor. Schofield *et al.* (1987a) could inhibit the infection of hepatocytes by *Plasmodium berghei* sporozoites by treating the hepatocytes with recombinant IFN-gamma. This inhibition was maximal if treatment was given six hours before infection, but even six hours after infection inhibition of sporozoite entry was at a level of about 75%. Hepatocytes are able to express up to about 44 000 surface receptors each for IFN-gamma, so this may be an important mechanism of host protection.

If BALB/c mice, to be infected with the protozoan parasite *Eimeria vermiformis*, are first given a dose of a monoclonal antibody specific for IFN-gamma, the infection is much more severe than in controls (Rose *et al.*, 1989). The intracellular location of this parasite within enterocytes might suggest that IFN-gamma is able to act directly on this cell type.

Tumour necrosis factor (TNF—of which there are two forms known as alpha and beta), or cachectin, has also been found to enhance the cytotoxicity of accessory cells for parasites. Wirth and Kierszenbaum (1988) showed that macrophages had an enhanced ability to destroy *Trypanosoma cruzi* parasites, if TNF (in addition to bacterial endotoxin) was present. An extract from the U937 cell line containing a mixture of factors was shown to enhance the toxicity of human eosinophils for *Schistosoma mansoni* larvae, due in part (although not exclusively) to TNF. Recombinant TNF (alpha and beta) could also increase platelet-mediated ADCC of *Schistosoma mansoni* (Damonneville *et al.*, 1988). TNF-beta is produced by CD4-positive lymphocytes, which can also secrete IFN-gamma, and this could also render the platelets more effective. But when the two factors were mixed together, the effect was additive, suggesting that TNF and IFN-gamma can act synergistically. In a similar piece of work, Esparza *et al.* (1987) suggested that TNF and IFN-gamma could, when administered together, cause greater macrophage killing of *S. mansoni* schistosomula than either factor alone. IFN-gamma can also synergize with granulocyte/monocyte colony-stimulating factor (GM-CSF) in killing of *Leishmania donovani* by human macrophages (Weiser *et al.*, 1987). With *Trypanosoma cruzi*, GM-CSF can activate macrophages to inhibit replication of the intracellular parasite although it is not as efficient as either IFN-gamma or TNF-alpha (Reed *et al.*, 1987). Human hepatocytes in culture could inhibit the intracellular schizogony of *Plasmodium falciparum* and thus prevent blood-stage infection if the cultures were first treated with either IFN-gamma or IL-1 (Mellouk *et al.*, 1987).

In vivo studies have helped to support the *in vitro* observations, at least in the case of malaria. The *in vivo* administration of TNF into mice led to a decrease of parasitaemia in mice infected with *Plasmodium yoelii* and survival of mice given a lethal dose of the parasite (Taverne *et al.*, 1987). Similarly intraperitoneal injections of IFN-gamma or release of TNF-alpha from an osmotic pump protected mice against infection with *P. chabaudi adami* (Clark *et al.*, 1987).

As mentioned in the introduction to this section, cytokines may also act against the host by exacerbating the pathology of the disease. Clark *et al.* (1981) have argued that, among other factors, TNF may be important in causing the depression of normal haemopoiesis in bone marrow, fever, hypergammaglobulinaemia, splenomegaly and other symptoms which often accompany malarial infections. More direct proof for this comes from the work of Grau *et al.* (1987) who showed that the rapidly fatal neurological complications associated with *Plasmodium berghei* infection of CBA mice (and a model for human cerebral malaria) were almost certainly due to the action of TNF on macrophages (see Chapter 7).

5.8. ANTIBODY-DEPENDENT, CELL-MEDIATED CYTOTOXICITY

As described above this process involves specific antibody binding to phagocytic cells (including eosinophils, neutrophils, and macrophages) via their surface receptors for the Fc region of the Ig molecule. These cells (which have been drawn into the site of inflammation and activated following the chemotactic and stimulatory action by a variety of factors such as complement components, prostaglandins, histamine, interleukins, as well as other mediators released by platelets, mast cells, and basophils) can now bind specifically to the surface of the parasite and either ingest it, in the case of a protozoan, or release cytotoxic mediators on to the parasite if it is larger.

ADCC against schistosomes has been reported by several workers—Kassis *et al.* (1979), Capron *et al.* (1980), and Verwaerde *et al.* (1987) have all shown that eosinophils and macrophages are capable of significant anti-schistosomula cytotoxicity in both mice and rats. In *S. mansoni* infections in the rat, the antibody isotype most effective varied with the accessory cell, macrophages preferentially interacting with IgE and eosinophils with IgG2a (Capron *et al.*, 1980). This presumably reflects the distribution of isotype-specific Fc receptors on the surfaces of these cell types. Verwaerde *et al.* (1987) found that an important antibody in ADCC involving platelets, eosinophils and macrophages appeared to be directed against a molecule of molecular weight around 26 kDa. Passive transfer of antibody specific for this antigen could confer 40–60% protection on the recipient rats. Table 5.2 illustrates some data from this experiment.

Table 5.2. Transfer of specific IgE protects against *S. mansoni* infection (data adapted from Verwaerde *et al.*, 1987).

Experiment no.	No. of worms in control mice[a]	No. of worms in test mice[b]	% Protection
1	106	38	64
2	140	72	49
3	57	30	47

[a] Control mice received 0.5 ml saline.
[b] Test mice received 0.5 ml serum containing a monoclonal IgE specific for *S. mansoni*.

Mehta *et al.* (1980) and Kazura and Aikawa (1980) have shown that ADCC can be active against *Litomosoides carinii* microfilariae and *Trichinella spiralis* nematodes, respectively. The protozoan *Trypanosoma dionisii* (Thorne *et al.*, 1979) and *Trypanosoma cruzi* (Olabuenaga *et al.*, 1979) are also susceptible to killing by granulocytes and monocytes in combination with parasite-specific antibody, confirming the generality in the efficacy of this mechanism, at least *in vitro*. Figure 5.7 is a scanning electron micrograph of an immature schistosome under attack from a variety of accessory cells bound to its surface.

Eosinophils

These cells are similar to neutrophils in overall morphology although they seem to be especially important in the immune response against many parasites. The reasons for this are not clear but it may be that the different nature of the granules which they possess may make them more suitable in this role, while neutrophils are perhaps more effective against prokaryotes.

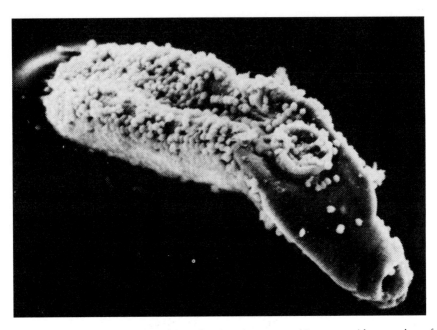

Figure 5.7. A scanning electron micrograph of an immature schistosome with a number of leucocytes bound to its surface (taken from McLaren and Terry, 1982; used with permission of the authors and Blackwell Publishers Ltd).

Alternatively it is possible that neutrophils are specialized for phagocytosing and destroying small particles and micro-organisms, while eosinophils are geared towards secreting their cytotoxic granules on to the surface of a much larger organism (such as a nematode larva) which cannot be phagocytosed. In any case, they are often very effective, and in some comparative studies (Butterworth *et al.*, 1977, with schistosomes; Kazura and Aikawa, 1980, with *Trichinella*), eosinophils have been observed to be much more cytotoxic than either neutrophils or macrophages (which were sometimes completely ineffective).

   Some workers have examined the steps leading up to cytotoxicity in infection with *S. mansoni* in humans (David *et al.*, 1980), *S. mansoni*, *T. spiralis*, and *Nippostrongylus brasiliensis* in rats (McLaren *et al.*, 1977), and *Fasciola hepatica* in rats (Davies and Goose, 1981), and all seem to come to some kind of consensus view on how this operates—eosinophils adhere to the surface of the parasite and then degranulate, releasing peroxidase either on to the parasite and causing damage or into cytoplasmic vacuoles which then discharge on to the parasite. Such a scheme is shown in Figures 5.8 (McLaren *et al.*, 1978). In these two pictures, eosinophils are seen binding to, penetrating, and finally stripping off the cuticle of *S. mansoni* schistosomulae.

(a)

(b)

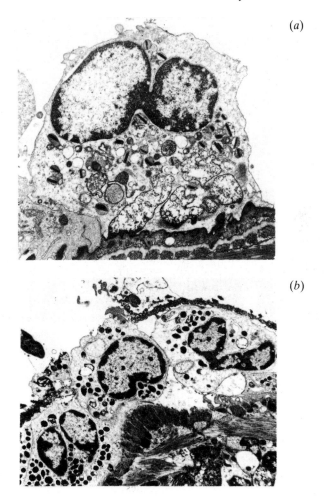

Figure 5.8. a) Eosinophil attached to the tegument of a schistosomulum. An eosinophil granule is visible opening on to the surface of the worm. b) Some eosinophils have broken through the tegument and are attached to the naked body of the worm (both figures taken from McLaren *et al.*, 1978; used with permission of the authors and Cambridge University Press).

However, it should be added that in certain systems eosinophils have been found to be unable to damage parasites; McLaren and Terry (1982) demonstrated that only older *S. mansoni* were susceptible to eosinophil-mediated cytotoxicity. Cook *et al.* (1988) found eosinophils to be completely incapable of damaging, or even adhering to, *Mesocestoides corti* parasites. These last two results merely serve to emphasize the point that the immune response against parasites is highly heterogeneous, and it is not possible to extrapolate with particular mechanisms between different systems.

Macrophages

These cells, along with eosinophils, are perhaps the most important accessory cells in the response against parasites and seem to be equally effective at phagocytosing protozoan parasites as degranulating on to the surface of larger organisms. It should be remembered that macrophages are a highly heterogeneous population of cells, even if the various subclasses are not easily distinguished other than by function; it is reasonable to expect that phagocytosing macrophages may belong to a different lineage from degranulating macrophages.

Giardia lamblia trophozoites were observed to be ingested and destroyed by activated macrophages in association with immune serum (Hill and Pearson, 1987), while in *Toxoplasma gondii* infections, macrophage-mediated ADCC appeared to play a role in *in vivo* protection, as mice injected with activated macrophages and specific antibody developed fewer *T. gondii* cysts in the brain than controls (Eisenhauer *et al.*, 1988).

Rat macrophages, in association with IgE, have been shown to damage *Acanthocheilonema viteae* microfilariae (Haque *et al.*, 1980, Ouaissi *et al.*, 1981) and *S. mansoni* schistosomula (Capron *et al.*, 1977) *in vitro*, while in *N. brasiliensis* infections, ADCC involving IgA may be more important, as the number of alveolar macrophages expressing IgA-specific Fc receptors increased from 14% in control mice to 30% in infected mice (Gauldie *et al.*, 1983). Although the relevance of this latter finding to parasite killing was not demonstrated, it is not surprizing to find IgA playing a more prominent role in a mucosal site, such as the lungs, since IgA is usually the predominant antibody isotype in mucosal secretions (Chapter 10).

Neutrophils, platelets, basophils, and mast cells

Neutrophils have been demonstrated to have a cytotoxic function against parasites *in vitro* in a few systems. For example, neutrophil-mediated ADCC has been shown to be effective against protozoan trichmonads (Rein *et al.*, 1980) and against *S. mansoni* schistosomula (Incani and McLaren, 1981). In addition, they have also been observed surrounding degenerating exoerythrocytic forms in *Plasmodium berghei* infections, ingesting parasite material (Meis *et al.*, 1987). Figure 5.9 shows a *Trypanosoma theileri* epimastigote being seriously damaged as it is phagocytosed by a bovine neutrophil (from Townsend *et al.*, 1982).

In the last few years it has become apparent that platelets, in addition to mediating blood clot formation, may also be able to act as cytotoxic effectors in association with antibody. This was first observed by Joseph *et al.* (1983) who showed that the killing of schistosomes (see Table 5.3) was dependent on IgE binding to the Fc-epsilon receptor on platelets.

The role of basophils in the response against endoparasites is rather ill-defined, and evidence for their participation in ADCC depends on circumstan-

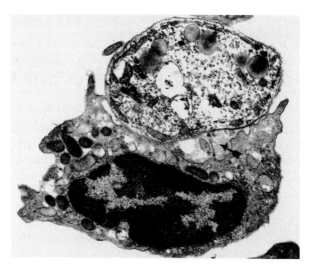

Figure 5.9. A *Trypanosoma theileri* epimastigote in the process of being phagocytosed by a bovine neutrophil. The epimastigote is already showing signs of serious damage. The arrowed area between the neutrophil and the epimastigote may contain secreted products of the neutrophil. This photograph was taken after one minute of incubation (taken from Townsend *et al.*, 1982; used with permission of the authors and the Company of Biologists Ltd).

Table 5.3.   Transfer of platelets from immune animals can protect against *S. mansoni* infection (data adapted from Joseph *et al.*, 1983).

Group	Total worms	% Protection
Normal platelets	164	—
Day 46 immune[a] platelets	60	63
Day 67 immune[a] platelets	106	36

[a] Immune platelets were taken from rats infected with *S. mansoni* either 46 or 67 days previously.

tial findings such as those of Vuitton *et al.* (1988) who described *Echinococcus multilocularis*-specific IgE bound to circulating basophils. A cytotoxic action was not described.

Mast cells have been described in many parasitic infections, although their function has never been clearly identified (see Chapters 3 and 10). In recent work importance has often been attached to the mast cell protease which is a characteristic of these cells. However, Tambourgi *et al.* (1989) have recently demonstrated mast cell-mediated ADCC of *Trypanosoma cruzi* blood trypomastigotes, with subclasses of IgG leading to optimal killing. Interestingly, degranulation of the mast cell did not appear to have taken place, and the precise cytotoxic mechanism was not elucidated. In addition, these cells have been shown to act in a co-operative capacity with eosinophils in killing *S. mansoni* schistosomula (Capron *et al.*, 1978).

5.9. PROTECTIVE EFFECTS *IN VIVO*—THE REAL WORLD

The impression to be gained from the preceding sections is that there is a highly effective immune response against most of the major parasitic infections, with many different mechanisms being able to damage the parasite. However, this is extremely misleading as most parasites are able to survive in the host for at least long enough to reproduce and in many cases for many months or even years. The fact is that most of the above work was carried out in *in vitro* systems with the conditions optimized to give rise to maximal responses. This, of course, permits various individual environmental components to be precisely controlled and examined in detail. But the inevitable corollary to this is that the *in vitro* system becomes greatly simplified and may no longer bear any useful similarities to *in vivo* conditions. In addition, any suppressive mechanisms employed by the parasite may be reduced under these artificial conditions.

A second problem with the *in vitro* approach, and the more critical one in relation to the prospects for the production of successful vaccines, is that animals with different genetic backgrounds often respond with widely differing abilities to the same infection. For example, in the section on T cells above, use was made of strains of mice resistant or susceptible to *Leishmania major* to understand better the T-cell phenotype involved in protection. The problem arises when high-responding strains of animals are used to show best the effects of the different components of the immune system. A vaccine developed from this kind of work may protect the majority of the outbred population, but it is almost certain that a significant proportion will not be protected, simply because the ability to respond well is not there (see Chapter 6).

This section will therefore bring together and summarize a few examples of experiments from the above which do show some kind of *in vivo* protection (although not taking into account the difference between high and low responders).

T cells

In vivo depletions of CD4 and CD8 T-cell subpopulations has shown that the CD4 Th cells are important for protection in leishmaniasis (Sadick *et al.*, 1987; Stern *et al.*, 1988), in schistosomiasis (Phillips *et al.*, 1987a, 1987b; Kelly and Colley, 1988), and in *Nippostrongylus brasiliensis* infection (Katona *et al.*, 1988). This type of technique has also demonstrated that CD8 Tc cells protect mice from secondary infection by malaria (Schofield *et al.*, 1987b, Weiss *et al.*, 1988).

In vivo transfer of protection by injecting parasite-specific T-cell lines or clones before infection (Brake *et al.*, 1986, 1988, in *Plasmodium chabaudi* infection, Lammas *et al.*, 1987, in *Mesocestoides corti* infection, and Riedlinger *et al.*, 1986, in *Trichinella spiralis* infection) has reinforced the importance of Th cells in initiating the immune response.

Lymphokines

Enhancement of immunity to *P. chabaudi* (Clark *et al.*, 1987) and to *Trypanosoma cruzi* (Reed, 1988) by *in vivo* administration of IFN-gamma (and TNF as well, in Clark *et al.*, 1987) demonstrates the role it has to play in these infections, probably via its action on macrophages. Immunity to *Eimeria vermiformis* is reduced if mice are first treated with an IFN-gamma-specific monoclonal antibody (Rose *et al.*, 1989).

Accessory cells

Macrophages have been observed *in vivo* phagocytosing *Giardia muris* (Owen *et al.*, 1981), and administration of activated macrophages to *T. gondii* reduced the brain pathology seen in controls (Eisenhauer *et al.*, 1988).

Meis *et al.* (1987) demonstrated neutrophils ingesting fragments of *P. berghei* exoerythrocytic forms, while administration of a monoclonal antibody to deplete eosinophils *in vivo* led to an increase in the number of *T. spiralis* larvae encysting in muscle (Grove *et al.*, 1977). The importance of platelets in schistosome infections has been demonstrated by transferring platelets from immune animals to naive recipients and showing that these animals are protected to a significant degree from subsequent infection (Joseph *et al.*, 1983).

Antibody

Transfer of parasite-specific immunoglobulins to naive animals can also lead to an enhanced immunity in the recipients, presumably through ADCC: IgE in *S. japonicum* (Kojima *et al.*, 1987) and *S. mansoni* (Verwaerde *et al.*, 1987); IgG in *S. mansoni* (Mangold and Dean, 1986); IgA and IgG in *Taenia taeniaeformis* (Lloyd and Soulsby, 1978). Transfer of protection with immune serum has also been observed with *N. brasiliensis* (Miller, 1980) and with *Trichuris muris* (Selby and Wakelin, 1973).

5.10. GRANULOMA FORMATION AND ENCAPSULATION

In many parasitic infections, the immune system may be unable to damage the organism itself, and this situation may result in the formation of a tightly packed collection of various cell types around the parasite. This response is known as a granuloma and is seen in infections by many different parasites (e.g. tapeworms, filarial worms, hookworms), but the best-studied examples are those caused by *Schistosoma* spp. eggs.

Ultrastructural analysis of the granulomata induced by *Brugia malayi* (Vincent *et al.*, 1980), *S. mansoni* (Bentley *et al.*, 1982), and *S. japonicum* (Warren *et al.*, 1978) have revealed the main cell types found in the granuloma to be macrophages, lymphocytes, eosinophils, and fibroblasts. It is this last cell type which lays down the collagen and fibrinogen leading to fibrosis of the granuloma (Grimaud *et al.*, 1987). Overwhelming evidence now indicates that granuloma formation is caused by the immune system, usually T cells.

The lymphatic lesions in infection by *Brugia pahangi* are much greater in animals which have been presensitized to the parasite (Klei *et al.*, 1982). In *S. mansoni* infections, Phillips *et al.* (1980) demonstrated that athymic mice had much smaller granulomata than euthymic controls, and that thymic reconstitution reversed this. They concluded from *in vitro* experiments that granuloma formation was dependent on Th cells and macrophages. Similarly, Cheever *et al.* (1985) found that nude mice, which are congenitally athymic, failed to develop as large granulomata as normal mice following *S. japonica* infection. Schistosome egg granulomata also appear to undergo a modulation after a few weeks. In the case of *S. mansoni* (Chensue and Boros, 1979; Weinstock and Boros, 1981) this effect was mediated by T cells, while in the case of *S. japonica* this modulation can be mediated by either IgG1 (Olds *et al.*, 1982) or T cells (Boros, 1986).

This immune induction of granulomata followed by immunomodulation leading to a reduction in their size has led many people to hypothesize whether the role of the granuloma is of benefit to the host or to the parasite. Perhaps the most satisfactory explanation is one which sees the immune system walling off the egg in an effort to minimize toxic effects of possible factors secreted by the egg or to arrest its further development [as appears to be the case with the granuloma formed around *Heligmosomoides polygyrus* in mice (Behnke and Parish, 1979)]. The subsequent immunosuppression seen in schistosomiasis may be an attempt by the host to limit the non-specific damage to surrounding tissue by the inflammatory infiltrate. McKenzie (1984) suggests that sequestration is of benefit to the parasite as it ensures prolonged survival of the host by preventing it succumbing to an overwhelming infection. Similarly Doenhoff *et al.* (1986) present data supporting the view that the immunopathology caused by granuloma formation supports parasite survival (see Chapter 1).

5.11. CONCLUSIONS

In this chapter, I have attempted to explain how the mammalian host tries to defend itself following invasion by eukaryotic parasites. It is clear that the immune system is well organized and has a wide range of different mechanisms, both cellular and humoral, on which to call in the ensuing battle. This host/ parasite relationship can be viewed as a battle since, as is explained in Chapter 13, the parasite itself is equally well equipped to evade the immune system, either by rendering itself effectively invisible to the host or by using other means to compromize the optimum functioning of the immune response. It is therefore clear that the host does not have things all its own way, and this is an important point to bear in mind if the overall impression from this chapter is one of an invincible immune system.

There is still much work to be done in fully understanding the protective immune response to a parasite, but progress is being made especially more recently with the use of T-cell clones and the purification of lymphokines. As the questions about the mechanisms involved are intrinsic to the regulation and functioning of immunity in general, the information gained from such studies will be of interest not only to parasitologists but to researchers in the field of pure immunology as well. Similarly, for exactly the same reasons, fundamental immunology research will provide many answers and clues for parasitologists.

REFERENCES

Albright, J.W., Munger, W.E., Henkart, P.A. and Albright, J.F. (1988), The toxicity of rat large granular lymphocyte tumor cells and their cytoplasmic granules for rodent and African trypanosomiasis, *Journal of Immunology*, **140**, 2774–8.

Auriault, C., Balloul, J.-M., Pierce, R.J., Damonneville, M., Sondermeijer, P. and Capron, A. (1987), Helper T cells induced by a purified 28-kilodalton antigen of *Schistosoma mansoni* protect rats against infection, *Infection and Immunity*, **55**, 1163–9.

Behnke, J.M. and Parish, H.A. (1979), *Nematospiroides dubius*: Arrested development of larvae in immune mice, *Experimental Parasitology*, **47**, 116–27.

Bentley, A.G., Doughty, B.L. and Phillips, S.M. (1982), Ultrastructural analysis of the cellular response to *Schistosoma mansoni*. III. The *in vitro* granuloma, *American Journal of Tropical Medicine and Hygiene*, **31**, 1168–80.

Boros, D.L. (1986), Immunoregulation of granuloma formation in murine schistosomiasis mansoni, *Annals of the New York Academy of Science*, **465**, 313–23.

Brake, D.A., Weidanz, W.P. and Long, C.A. (1986), Antigen-specific, interleukin 2-propagated T lymphocytes confer resistance to a murine malarial parasite, *Plasmodium chabaudi adami*, *Journal of Immunology*, **137**, 347–52.

Brake, D.A., Long, C.A. and Weidanz, W.P. (1988), Adoptive protection against *Plasmodium chabaudi adami* malaria in athymic nude mice by a cloned T cell line, *Journal of Immunology*, **140**, 1989–93.

Butterworth, A.E., Wassom, D.L., Gleich, G.J., Loegering, D.A. and David, J.R. (1979), Damage to schistosomula of *Schistosoma mansoni* induced directly by eosinophil major basic protein, *Journal of Immunology*, **122**, 221–9.

Butterworth, A.E., David, J.R., Francis, D., Mahmoud, A.A.F., David, P.H., Sturrock, R.F. and Houba, V. (1977), Antibody-dependent eosinophil-mediated damage to ^{51}Cr-labelled schistosomula of *Schistosoma mansoni*: Damage by purified eosinophils, *Journal of Experimental Medicine*, **145**, 136–47.

Canessa, A., Pistoia, V., Roncella, S., Merli, A., Melioli, G., Terragna, A. and Ferrarini, M. (1988), An *in vitro* model for *Toxoplasma* infection. Interaction between CD4$^+$ monoclonal T cells and macrophages results in killing of trophozoites, *Journal of Immunology*, 140, 3580–8.

Capron, A., Dessaint, J-P., Capron, M., Joseph, M. and Pestel, J. (1980), Role of anaphylactic antibodies in immunity to schistosomes, *American Journal of Tropical Medicine and Hygiene*, 29, 849–57.

Capron, A., Dessaint, J-P., Joseph, M., Rousseaux, R., Capron, M. and Bazin, H. (1977), Interaction between IgE complexes and macrophages in the rat: a new mechanism of macrophage activation, *European Journal of Immunology*. 7, 315–22.

Capron, M., Capron, A., Torpier, G., Bazin, H., Bout, D. and Joseph, M. (1978), Eosinophil-dependent cytotoxicity in rat schistosomiasis: Involvement of rat IgG2a antibody and role of mast cells, *European Journal of Immunology*, 8, 127–33.

Cesbron, J-Y., Hayaski, M., Joseph, M., Lutsch, C., Grzych, J-M. and Capron, A. (1988), *Onchocerca volvulus*. Monoclonal anti-idiotype antibody as antigen signal for the microfilaricidal cytotoxicity of diethylcarbamazine-treated platelets, *Journal of Immunology*, 141, 279–85.

Cheever, A.W., Byram, J.E. and von Lichtenberg, F. (1985), Immunopathology of *Schistosoma japonicum* infection in athymic mice, *Parasite Immunology*, 7, 387–98.

Chensue, S.W. and Boros, D.L. (1979), Modulation of granulomatous hypersensitivity. I. Characterization of T lymphocytes involved in the adoptive suppression of granuloma formation in *Schistosoma mansoni*-infected mice, *Journal of Immunology*, 123, 1409–14.

Clark, I.A., Virelizier, J-L., Carswell, E.A. and Wood, P.R. (1981), Possible importance of macrophage-derived mediators in acute malaria, *Infection and Immunity*, 31, 1058–66.

Clark, I.A., Hunt, N.H., Butcher, G.A. and Cowden, W.B. (1987), Inhibition of murine malaria (*Plasmodium chabaudi*) *in vivo* by recombinant interferon-gamma of tumor necrosis factor, and its enhancement by butylated hydroxyanisole, *Journal of Immunology*, 139, 3493–6.

Cook, R.M., Ashworth, R.F., and Chernin, J. (1988), Cytotoxic activity of granulocytes against *Mesocestoides corti*, *Parasite Immunology*, 10, 97–109.

Damonneville, M., Wietzerbin, J., Pancre, V., Joseph, M., Delanoye, A., Capron, A. and Auriault, C. (1988), Recombinant tumor necrosis factors mediate platelet cytotoxicity to *Schistosoma mansoni* larvae, *Journal of Immunology*, 140, 3962–5.

David, J.R., Butterworth, A.E. and Vadas, M.A. (1980), Mechanism of the interaction mediating killing of *Schistosoma mansoni* by human eosinophils, *American Journal of Tropical Medicine and Hygiene*, 29, 842–·8.

Davies, C. and Goose, J. (1981), Killing of newly encysted juveniles of *Fasciola hepatica* in sensitised rats, *Parasite Immunology*, 3, 81–96.

Doenhoff, M.J., Hassounah, O., Murare, H., Bain, J. and Lucas, S. (1986), The schistosome egg granuloma: immunopathology in the cause of host protection or parasite survival, *Transactions of the Royal Society of Tropical Medicine and Hygiene*, 80. 503–16.

Doria, G., Goodman, J.W., Gengozian, N. and Congdon, C.C. (1962), Immunologic study of antibody-forming cells in mouse radiation chimaeras, *Journal of Immunology*, 88, 20–30.

Eisenhauer, P., Mack, D.G. and McLeod, R. (1988), Prevention of peroral and congenital acquisition of *Toxoplasma gondii* by antibody and activated macrophages, *Infection and Immunity*, 56, 83–7.

Esparza, I., Mannel, D., Ruppel, A., Falk, W. and Krammer, P.H. (1987), Interferon gamma and lymphotoxin or tumor necrosis factor act synergistically to induce macrophage killing of tumor cells and schistosomula of *Schistosoma mansoni*, *Journal of Experimental Medicine*, 166, 589–94.

Gauldie, J., Richards, C. and Lamontagne, L. (1983), Fc receptors for IgA and other immunoglobulins on resident and activated alveolar macrophages, *Molecular Immunology*, 20, 1029–37.

Goddeeris, B.M., Morrison, W.I., Teale, A.J., Bensaid, A. and Baldwin, C.L. (1986), Bovine cytotoxic T cell clones specific for cells infected with the protozoan parasite *Theileria parva*: parasite strain specificity and Class I major histocompatibility complex restriction, *Proceedings of the National Academy of Sciences (USA)*, 83, 5238–42.

Gorczynski, R.M. (1987), Immunization with *Leishmania*-specific T cell not B cell lines or hybridomas can modulate the response of susceptible mice infected with viable parasites, *Journal of Immunology*, 139, 3070–5.

Grau, G.E., Fajardo, L.F., Piguet, P-F., Allet, B., Lambert, P-H. and Vassalli, P. (1987), Tumor necrosis factor (cachectin) as an essential mediator in murine cerebral malaria, *Science*, 237, 1210–12.

Grimaud, J.A., Boros, D.L., Takiya, C., Mathew, R.C. and Emonard, H. (1987), Collagen isotypes, laminin, and fibronectin in granulomas of the liver and intestines of *Schistosoma mansoni*-infected mice, *American Journal of Tropical Medicine and Hygiene*, 37, 335–44.

Grove, D.I., Mahmoud, A.A.F. and Warren, K.S. (1977), Eosinophils and resistance to *Trichinella spiralis*, *Journal of Experimental Medicine*, 145, 755–9.

Haque, A., Joseph, M., Ouaissi, M.A., Capron, M. and Capron, A. (1980), IgE antibody-mediated cytotoxicity of rat macrophages against microfilariae of *Dipetalonema viteae in vitro*, *Clinical and Experimental Immunology*, 40, 487–95.

Heinzel, F.P., Sadick, M.D., Holaday, B.J., Coffman, R.L. and Locksley, R.M. (1989), Reciprocal expression of interferon gamma or interleukin 4 during the resolution or progression of murine leishmaniasis, *Journal of Experimental Medicine*, 169, 59–72.

Herrington, D.A., Clyde, D.F., Losonsky, G., Cortesia, M., Murphy, J.R., Davis, J., Baqar, S., Felix, A.M., Heimer, E.P., Gillesen, D., Nardin, E., Nussenzweig, R.S., Nussenzweig, V., Hollindale, M.R. and Levine, M.M. (1987), Safety and immunogenicity in man of a synthetic malaria vaccine against *Plasmodium falciparum* sporozoites, *Nature*, 328, 257–9.

Hill, D.R. and Pearson, R.D. (1987), Ingestion of *Giardia lamblia* trophozoites by human mononuclear phagocytes, *Infection and Immunity*, 55, 3155–61.

Hoffman, S.L., Oster, C.N., Plowe, C.V., Woollett, G.R., Beier, J.C., Chulay, J.D., Wirtz, R.A., Hollingdale, M.R. and Mugambi, M. (1987), Naturally acquired antibodies to sporozoites do not prevent malaria: vaccine development implications, *Science*, 237, 639–42.

Hughes, H.P.A., Kasper, L.H., Little, J. and Dubey, J.P. (1987), Absence of a role for natural killer cells in the control of acute infection by *Toxoplasma gondii* oocysts, *Clinical and Experimental Immunology*, 72, 394–9.

Incani, R.N. and McLaren, D.J. (1981), Neutrophil-mediated cytotoxicity to schistosomula of *Schistosoma mansoni in vitro*: studies on the kinetics of complement and/or antibody-dependent adherence and killing, *Parasite Immunology*, 3, 107–26.

Jerne, N.K. (1974), Towards a network theory of the immune system, *Annals of Immunology* (Paris), 125C, 373–89.

Joseph, M., Dessaint, J.P. and Capron, A. (1977), Characteristics of macrophage cytotoxicity induced by IgE immune complexes, *Cellular Immunology*, 34, 247–58.

Joseph, M., Auriault, C., Capron, A., Vorng, H. and Viens, P. (1983), A new function for platelets: IgE-dependent killing of schistosomes, *Nature* 303, 810–12.

Kassis, A.I., Aikawa, M. and Mahmoud, A.A.F. (1979), Mouse antibody-dependent eosinophil and macrophage adherence and damage to schistosomula of *Schistosoma mansoni*, *Journal of Immunology*, **122**, 398–405.

Katona, I.M., Urban, Jr., J.F. and Finkelman, F. (1988), The role of L3T4$^+$ and Lyt2$^+$ T cells in the IgE response and immunity to *Nippostrongylus brasiliensis*, *Journal of Immunology*, **140**, 3206–11.

Kazura, J.W. and Aikawa, M. (1980), Host defense mechanisms against *Trichinella spiralis* infection in the mouse: eosinophil-mediated destruction of newborn larvae *in vitro*, *Journal of Immunology*, **124**, 355–61.

Kelly, E.A.B. and Colley, D.G. (1988), *In vivo* effects of monoclonal anti-L3T4 antibody on immune responsiveness of mice infected with *Schistosoma mansoni*. Reduction of irradiated cercariae-induced resistance, *Journal of Immunology*, **140**, 2737–45.

Kelso, A. and Gough, N.M. (1988), Co-expression of granulocyte-macrophage colony-stimulating factor, gamma interferon, and interleukins 3 and 4 is random in murine alloreactive T-lymphocyte clones, *Proceedings of the National Academy of Sciences (USA)*, **85**, 9189–93.

Klei, T.R., Enright, F.M., Blanchara, D.P. and Uhl, S.A. (1982), Effects of presensitisation on the development of lymphatic lesions in *Brugia pahangi*-infected jirds, *American Journal of Tropical Medicine and Hygiene*, **31**, 280–91.

Kojima, S., Niimura, M. and Kanazawa, T. (1987), Production and properties of a mouse monoclonal IgE antibody to *Schistosoma mansoni*, *Journal of Immunology*, **139**, 2044–9.

Kumar, S., Miller, L.H., Quakyi, I.A., Keister, D.B., Houghton, R.A., Maloy, W.L., Moss, B., Berzofsky, J.A. and Good, M.F. (1988), Cytotoxic T cells specific for the circumsporozoite protein of *Plasmodium falciparum*, *Nature*, **334**, 258–60.

Kunkel, H.G., Mannik, M. and William, R.C. (1963), Individual antigenic specificities of isolated antibodies, *Science*, **140**, 1218–19.

Lammas, D.A., Mitchell, L.A. and Wakelin, D. (1987), Adoptive transfer of enhanced eosinophilia and resistance to infection in mice by an *in vitro* generated T-cell line specific for *Mesocestoides corti* larval antigen, *Parasite Immunology*, **9**, 591–601.

Liew, F.W., Hodson, K. and Lelchuk, R. (1987), Prophylactic immunization against experimental leishmaniasis. VI. Comparison of protective and disease-promoting T cells, *Journal of Immunology*, **139**, 3112–17.

Lloyd, S. and Soulsby, E.J.L. (1978), The role of IgA immunoglobulin in the passive transfer of protection to *Taenia taeniaeformis* in the mouse, *Immunology*, **34**, 939–45.

Male, D., Champion, B. and Cooke, A. (1987), *Advanced Immunology*, London: Gower Medical Publishing.

Mangold, B.L. and Dean, D.A. (1986), Passive transfer with serum and IgG antibodies of irradiated cercariae-induced resistance against *Schistosoma mansoni* in mice, *Journal of Immunology*, **136**, 2644–8.

Manson-Smith, D.F., Bruce, R.G. and Parrott, D.M.V. (1979), Villous atrophy and expulsion of intestinal *Trichinella spiralis* are mediated by T cells, *Cellular Immunology*, **47**, 285–92.

McKenzie, C.D. (1984), Sequestration—beneficial to both host and parasite, *Parasitology*, **88**, 593–5.

McLaren, D.J. and Terry, R.J. (1982), The protective role of acquired host antigens during schistosome maturation, *Parasite Immunology*, **4**, 129–48.

McLaren, D.J., McKenzie, C.D. and Ramalho-Pinto, F.J. (1977), Ultrastructural observations on the *in vitro* interaction between rat eosinophils and some parasitic helminths (*Schistosoma mansoni, Trichinella spiralis, and Nippostrongylus brasiliensis*), *Clinical and Experimental Immunology*, **30**, 105–18.

McLaren, D.J., Ramalho-Pinto, F.J. and Smithers, S.R. (1978), Ultrastructural evidence for complement and antibody-dependent damage to schistosomula of *Schistosoma mansoni* by rat eosinophils *in vitro*, *Parasitology*, **77**, 313–24.

Mehta, K., Sindhu, R.K., Subrahmanyam, D. and Nelson, D.S. (1980), IgE-dependent adherence and cytotoxicity of rat spleen and peritoneal cells to *Litomosoides carinii* microfilariae, *Clinical Experimental Immunology*, **41**, 107–14.

Meis, J.F.G.M., Jap, P.H.K., Hollingdale, M.R. and Verhave, J-P. (1987), Cellular response against exoerythrocytic forms of *Plasmodium berghei* in rats, *American Journal of Tropical Medicine and Hygiene*, **37**, 506–10.

Mellouk, S., Maheshwari, R.H., Rhodes-Feuillette, A., Beaudoin, R.L., Berbiguier, N., Matile, H., Miltgen, F., Landau, I., Pied, S., Chigot, J.P., Friedman, R.M. and Mazier, D. (1987), Inhibitory activity of interferons and interleukin 1 on the development of *Plasmodium falciparum* in human hepatocyte cultures, *Journal of Immunology*, **139**, 4192–5.

Miller, H.R.P. (1980), Expulsion of *Nippostrongylus brasiliensis* from rats protected with serum. I. The efficacy of sera from singly and multiply infected donors related to time of administration and volume of serum injected, *Immunology*, **40**, 325–34.

Mosmann, T.R. and Coffman, R.L. (1987), Two types of mouse helper T-cell clone: implications for immune regulation, *Immunology Today*, **8**, 223–7.

Mosmann, T.R., Schumacher, J.H., Street, N. and Coffman, R.L. (1988), Possible role of two types of helper T cell in immune regulation, *FASEB Journal*, **2**, A8882.

Niederkorn, J.Y., Stewart, G.L., Ghazizadeh, S., Mayhew, E., Ross, J. and Fischer, B. (1988), *Trichinella pseudospiralis* larvae express natural killer (NK) cell-associated asialo-GM1 antigen and stimulate pulmonary activity, *Infection and Immunology*, **56**, 1011–16.

Olabuenaga, S.E., Cardoni, R.L., Segura, E.L., Riera, N.E. and de Bracco, M.M.E. (1979), Antibody-dependent cytolysis of *Trypanoosoma cruzi* by human polymorphonuclear leukocytes, *Cellular Immunology*, **45**, 85–93.

Olds, G.R., Olveda, R., Tracy, J.W. and Mahmoud, A.A.F. (1982), Adoptive transfer of modulation of granuloma formation and hepatosplenic disease in murine schistosomiasis japonica by serum from chronically infected animals, *Journal of Immunology*, **128**, 1391–3.

Ouaissi, M.A., Haque, A. and Capron, A. (1981), *Dipetalonema viteae*: ultrastructural study on the *in vitro* interaction between rat macrophages and microfilariae in the presence of IgE antibody, *Parasitology*, **82**, 55–62.

Oudin, J. and Michel, M. (1963), Une nouvelle forme d'allotypie des globulines gamma du sérum de lapin apparemment liée à la fonction et à la spécificité anticorps, *Comptes Rendus Hebdomadaires des Seances de l'Académie des Sciences* (Paris), **257**, 805–8.

Owen, R.L., Allen, C.L. and Stevens, D.P. (1981), Phagocytosis of *Giardia muris* by macrophages in Peyer's patch epithelium in mice, *Infection and Immunity*, **33**, 591–601.

Phillips, S.M., Reid, W.A., Doughty, B.L. and Bentley, A.G. (1980), The immunologic modulation of morbidity in schistosomiasis. Studies in athymic mice and *in vitro* granuloma formation, *American Journal of Tropical Medicine and Hygiene*, **29**, 820–31.

Phillips, S.M., Linette, G.P., Doughty, B.L., Byram, J.E., von Lichtenberg, F. (1987a), *In vivo* T cell depletion regulates resistance and morbidity in murine schistosomiasis, *Journal of Immunology*, **139**, 919–26.

Phillips, S.M., Walker, D., Abdel–Hafez, S.K., Linette, G.P., Doughty, B.L., Perrin, P.J., El Fathelbab, N. (1987b), The immune response to *Schistosoma mansoni* infection in inbred rats. VI. Regulation by T cell subpopulations, *Journal of Immunology*, **139**, 2781–7.

Potocnjak, P., Yoshida, N., Nussenzweig, R.S. and Nussenzweig, V. (1980), Monovalent fragments (Fab) of monoclonal antibodies to a sporozoite surface antigen (Pb44) protect mice against malarial infection, *Journal of Experimental Medicine*, **151**, 1504–13.

Reed, S.G. (1988), *In vivo* administration of recombinant IFN-gamma induces macrophage activation, and prevents acute disease, immune suppression, and death in experimental *Trypanosoma cruzi* infections, *Journal of Immunology*, **140**, 4342–7.

Reed, S.G., Nathan, C.F., Pihl, D.L., Rodricks, P., Shanebeck, K., Conlon, P.J. and Grabstein, K.H. (1987), Recombinant granulocyte/macrophage colony-stimulating factor activates macrophages to inhibit *Trypanosoma cruzi* and releases hydrogen peroxide: Comparison with interferon gamma, *Journal of Experimental Medicine*, 1734–6.

Rein, M.F., Sullivan, J.A. and Mandell, G.L. (1980), Trichomonacidal activity of human polymorphonuclear neutrophils: killing by disruption and fragmentation, *Journal of Infectious Diseases*, **142**, 575–85.

Riedlinger, J., Grencis, R.K. and Wakelin, D (1986), Antigen-specific T-cell lines transfer protective immunity against *Trichinella spiralis in vivo*, *Immunology*, **58**, 57–61.

Rose, M.E., Wakelin, D. and Hesketh, P. (1989), Gamma interferon controls *Eimeria vermiformis* primary infection in BALB/c mice, *Infection and Immunity*, **57**, 1599–1603.

Sadick, M.D., Heinzel, F.P., Shigekane, V.M., Fisher, W.L. and Locksley, R.M. (1987), Cellular and humoral immunity to *Leishmania major* in genetically susceptible mice after *in vivo* depletion of L3T4 cells, *Journal of Immunology*, **139**, 1303–9.

Salata, R.A., Murray, H.W., Rubin, B.Y. and Ravdin, J.I. (1987), The role of gamma interferon in the generation of human macrophages cytotoxic for *Entamoeba histolytica* trophozoites, *American Journal of Tropical Medicine and Hygiene*, **37**, 72–8.

Schofield, L., Ferreira, A., Altszuler, R., Nussenzweig, V. and Nussenzweig, R.S. (1987a), Interferon-gamma inhibits the intrahepatocytic development of malaria parasites *in vitro*, *Journal of Immunology*, **139**, 2020–5.

Schofield, L., Villaquiran, J., Ferreira, A., Schellekens, H., Nussenzweig, R.S. and Nussenzweig, V. (1987b), Gamma interferon, CD8[+] T cells, and antibodies required for immunity to malaria sporozoites, *Nature*, **330**, 664–6.

Scott, P., Natovitz, P., Coffman, R.L., Pearce, E. and Sher, A. (1988), Immunoregulation of cutaneous leishmaniasis. T cell lines that transfer protective immunity or exacerbation belong to different T helper subsets and respond to distinct parasite antigens, *Journal of Experimental Medicine*, **168**, 1675–84.

Selby, G.R. and Wakelin, D. (1973), Transfer of immunity against *Trichuris muris* in the mouse by serum and cells, *International Journal of Parasitology*, **3**, 717–22.

Stern, J.J., Oca, M.J., Rubin, B.Y., Anderson, S.L. and Murray, H.W. (1988), Role of L3T4[+] and Lyt-2[+] cells in experimental visceral leishmaniasis, *Journal of Immunology*, **140**, 3971–7.

Tambourgi, D.V., Kipnis, T.L. and da Silva, W.D. (1989), *Trypanosoma cruzi*: antibody-dependent killing of bloodstream trypomastigotes by mouse bone marrow-derived mast cells and by mastocytoma cells, *Experimental Parasitology*, **68**, 192–201.

Taverne, J., Tavernier, J., Fiers, W. and Playfair, J.H.L. (1987), Recombinant tumor necrosis factor inhibits malaria parasites *in vivo* but not *in vitro*, *Clinical and Experimental Immunology*, **67**, 1–4.

Thorne, K.J.I., Glauert, A.M., Svvennsen, R.J. and Franks, D. (1979), Phagocytosis and killing of *Trypanosoma dionisii* by human neutrophils, eosinophils and monocytes, *Parasitology*, **79**, 367–79.

Thorson, R.E. (1956), Proteolytic activity in extracts of the oesophagus of adults of *Ancylostoma caninum* and the effect of immune serum on this activity, *Journal of Parasitology*, **46**, 21–5.

Townsend, J., Duffus, W.P.H. and Glauert, A.M. (1982), An ultrastructural study of the interaction *in vitro* between *Trypanosoma theileri* and bovine leucocytes, *Journal of Cell Science*, **56**, 389–407.

Tyan, M.L. (1968), Studies on the ontogeny of the mouse immune system. I. Cell-bound immunity, *Journal of Immunology*, **100**, 535–42.

Verwaerde, C., Joseph, M., Capron, M., Pierce, R.J., Damonneville, M., Velge, F., Auriault, C. and Capron, A. (1987), Functional properties of a rat monoclonal IgE antibody specific for *Schistosoma mansoni*, *Journal of Immunology*, **138**, 4441–6.

Vincent, A.L., Ash, L.R., Rodrick, G.E. and Sodeman, Jr., W.A. (1980), The lymphatic pathology of *Brugia malayi* in the Mongolian jird, *Journal of Parasitology*, **66**, 613–20.

Vuitton, D-A., Bresson–Hadni, S., Lenys, D., Flausse, F., Liance, M., Wattre, P., Miguet, J.P. and Capron, A. (1988), IgE-dependent humoral immune response in *Echinococcus multilocularis* infection: circulating and basophil-bound specific IgE against *Echinococcus* antigens in patients with alveolar echinococcosis, *Clinical and Experimental Immunology*, **71**, 247–52.

Warren, K.S., Grove, D.I. and Pelley, R.P. (1978), The *Schistosoma japonicum* egg granuloma. II. Cellular composition, granuloma size, and immunologic concomitants, *American Journal of Tropical Medicine and Hygiene*, **27**, 271–5.

Weinstock, J.V. and Boros, D.L. (1981), Heterogeneity of the granulomatous response in the liver, colon, ileum, and ileal Peyer's patches to schistosome eggs in murine *Schistosoma mansoni*, *Journal of Immunology*, **127**, 1906–9.

Weiser, W.Y., van Niel, A., Clark, S.C., David, J.R. and Remold, H.G. (1987), Recombinant human granulocyte/macrophage colony-stimulating factor activates intracellular killing of *Leishmania donovani* by human monocyte-derived macrophages, *Journal of Experimental Medicine*, **166**, 1436–46.

Weiss, W.R., Sedegah, M., Beaudoin, R.L., Miller, L.H. and Good, M.F. (1988), CD8[+] T cells (cytotoxic/suppressors) are required for protection in mice immunised with malaria sporozoites, *Proceedings of the National Academy of Sciences (USA)*, **85**, 573–6.

Wirth, J.J. and Kierszenbaum, F. (1988), Recombinant tumor necrosis factor enhances macrophage destruction of *Trypanosoma cruzi* in the presence of bacterial endotoxin, *Journal of Immunology*, **141**, 286–8.

Yoshida, N., Nussenzweig, R.S., Potocjnak, P., Nussenzweig, V. and Aikawa, M. (1980), Hybridoma produces protective antibodies directed against the sporozoite stage of malaria parasite, *Science*, **207**, 71–3.

6. Genetically determined variation in host response and susceptibility to pathological damage

R.K. Grencis

6.1. INTRODUCTION

Infection by parasitic organisms is widespread throughout the animal kingdom including man. However, there is little consistency and uniformity in the patterns observed, with populations affected by a variety of invasive organisms including protozoan and metazoan parasites exhibiting a spectrum of disease states. For example, the protozoan parasites of the genus *Leishmania* can give rise to an array of conditions in infected humans. Although the basic differences between the cutaneous and visceral forms of the disease are related to parasite species, within species disease states differ markedly. In the cutaneous form, disease manifestations range from self-healing to disseminated lesions and in the visceral form vary from subclinical to fatal forms. Although *Leishmania* is transmitted by a vector, the variation observed in host response is not thought to be vector related.

A metazoan parasitic disease, again vector transmitted, which exhibits a wide spectrum of disease states is that caused by filarial nematodes, such as *Wuchereria* and *Brugia* spp., which are responsible for lymphatic filariasis. In endemic regions where transmission is intense, at least 50% of the population may appear free of infection although a proportion may be infected subclinically. The remaining population can show a variety of disease conditions ranging from filarial fever, tropical eosinophilia through to chronic lymphatic obstruction (elephantiasis; see Chapter 1). Again the variation in

120

host response is not thought to be vector related as in endemic regions all individuals of the population are likely to have been bitten by infected insects.

Variation in host response is not restricted to vector-transmitted parasites. It has been known for many years that infections by gastrointestinal nematodes show a marked variation between individuals within a population. Studies carried out some 50 to 60 years ago established that there were variations in hookworm infection levels between peoples of different races. Recent work has confirmed that under natural conditions helminths are aggregated within hosts, with only a few individuals in any population being infected by high numbers of worms. This has led to the concept of 'wormy people' (Croll and Ghadirian, 1981) who will make a significant contribution to the transmission of the infection. Indeed, a number of intensive studies carried out in endemic regions upon the gastrointestinal-dwelling nematodes *Ascaris*, hookworm and *Trichuris* have shown that within populations certain individuals are predisposed to heavy worm infections. Predisposition has also been demonstrated in human populations infected with the blood flukes *Schistosoma haematobium* and *S. mansoni* (see reviews by Anderson, 1986; Butterworth and Hagan, 1987; Chapter 12).

Variation in host responses in domestic animals was first discussed in detail by Whitlock during the 1950s (Whitlock, 1958). He commented upon the observations that shepherds could pick out sheep within the flock which, relative to others, were resistant to infection by gastrointestinal nematodes. Variation in resistance to infection in domestic animals has also been observed for parasitic protozoa. For example, the N'Dama cattle of west and central Africa are termed 'trypanotolerant', i.e. they can contain infections of trypanosomes which, in other breeds of cattle, cause debilitating or fatal disease.

Taken together, these and many other observations suggest a genetic basis for the variation in host response to parasitic infection observed in natural populations. Of course, data drawn from naturally outbred host populations are often difficult to interpret. However, with the introduction of strains of laboratory animals with defined genetic histories, it became possible to examine, in detail, whether the spectrum of responses seen in the wild could be reproduced experimentally in populations of animals with strictly controlled genetic constitutions. The present chapter will attempt to highlight current theories and interests relating to the investigation of genetic variation in host responses to parasitic infections in laboratory models and will attempt to relate the findings to field observations. Finally, the problems of vaccinating a genetically heterogeneous host population against parasitic disease will be considered.

6.2. LABORATORY MODELS

Whilst for a number of species of laboratory animal different strains are evident, the most widely available, diverse, and therefore most intensively studied species is the mouse. The present discussion will concentrate upon investigations of parasitic infection utilizing the mouse as a model host. This species has been particularly useful in the study of genetic control of the immune response to both infectious and non-infectious agents.

The most widely used strains of mice for such studies are known as inbred. An inbred strain is one in which all the individuals are genetically identical to one another. Such strains are produced by the successive mating of related individuals, usually brother and sister. In this way controlled parasite infections of inbred strains can be compared and their responses monitored. In essence, differences between such strains will correspond to differences between distinct individuals in an outbred population.

With the discovery of a group of genes which exert a profound influence upon the immune response, the major histocompatibility complex (MHC, HLA in man, H-2 in the mouse; Figure 6.1), selective breeding of particular mouse strains allowed the production of strains which differ from one another only at the H-2 region. These strains are called H-2 congenic strains. The H-2 region consists of a number of genes which exhibit a high degree of polymorphism. A particular allele at each locus within the H-2 region is designated by a small letter; a particular combination of H-2 alleles is referred to as the H-2 haplotype and is designated by a small letter superscript in association with H–2, e.g. H–2^b. Thus, H–2 congenic mice have the same background genes but differ only in their H–2 haplotype. Different inbred strains can have the same haplotype but different background genes. Further selective breeding between H–2 congenic mouse strains has also allowed the development of mouse strains which differ only within regions of the H–2

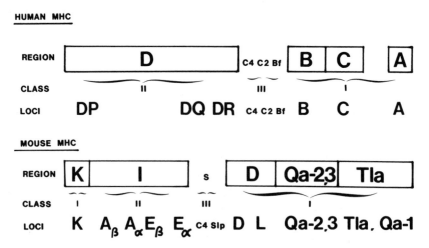

Figure 6.1. Diagrammatic presentation of the major histocompatibility complex (MHC) in man and mouse. The major loci of the different class regions are shown. The murine MHC is found on chromosome 17 in the mouse; the human MHC (HLA) is found on chromosome 6. The Class III genes, which primarily encode for complement components, should not be considered as MHC genes even though they are located within the MHC complex.

Table 6.1. H–2 congenic and H–2 recombinant mouse strains showing haplotype and alleles expressed at different H–2 loci.

					H–2 Loci			
	Strain	Haplotype	K	A_β	A_α	E_β	E_α	D
H–2 Congenic	B10	*b*	*b*	*b*	*b*	*b*	*b*	*b*
	B10S	*s*	*s*	*s*	*s*	*s*	*s*	*s*
	B10BR	*k*	*k*	*k*	*k*	*k*	*k*	*k*
	B10G	*q*	*q*	*q*	*q*	*q*	*q*	*q*
H–2 Recombinant	B10.T(6R)	y^2	*q*	*q*	*q*	*q*	*q*	*d*
	B10.S(8R)	as^1	*k*	*k*	*k*	*k*	*s*	*s*

complex, and these are known as H–2 recombinant mouse strains (see Table 6.1). As can be deduced, the use of such strains has greatly facilitated studies of genetically controlled immune responses to infection by both protozoan and metazoan parasites.

Helminths

Genetically determined variation in the host response to more than 20 species of parasitic helminths, covering all the major taxonomic groups, has been studied in mice. In order to illustrate current trends, only a selection will be discussed here, and readers are referred to Wakelin (1985, 1988) for further details.

Trichinella spiralis

This gastrointestinal parasite represents arguably one of the most intensively studied models with respect to genetic control of host resistance (Figure 6.2). Immunity is T-cell dependent and can affect the parasite in both the enteral and parenteral sites. Variation between mouse strains has been noted for a number of parameters. One enteral parameter which has been extensively studied is the expulsion of the worm from the gut.

Expulsion is thought to be the result of a T-cell mediated inflammatory response (see Chapter 8). A series of experiments by Wakelin and colleagues have shown that T helper cells within the draining lymph node of the small intestine (mesenteric lymph node, MLN) are involved in this process. Such helper cells are thought to exert their effects by the release of cytokines which control the growth and differentiation of inflammatory cell types (see Chapter 3). Further experiments demonstrated that the overall rapidity of the response was determined not by the T-cell component but by the ability of the host

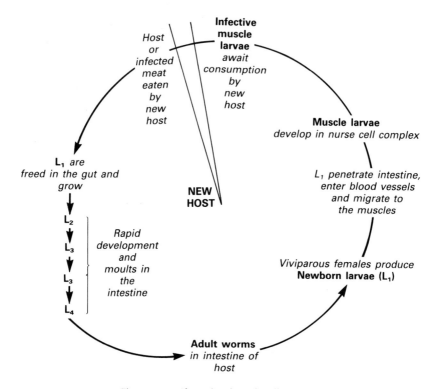

Figure 6.2. Life cycle of *Trichinella spiralis*.

to generate inflammatory cells in the bone marrow (see Wakelin and Denham, 1983).

Different inbred strains of mice vary considerably in the time taken to expel a primary infection from the small intestine. However, there has been some disagreement between different laboratories on the performances of strains tested independently and this may be attributable in part to the experimental techniques employed (e.g. the intensity of infection is now recognized as particularly important in determining the onset of expulsion). Nevertheless, certain key strains have been recognized as good or rapid responders and others as poor or slow responders. It was evident that mouse strains of some haplotypes, e.g. H–2^k were more resistant to infection than mice of other haplotypes, e.g. H–2^f. However the first convincing evidence demonstrating a role for H–2-linked genes came from the work of Wassom and colleagues (1983). They compared resistance to infection using a number of different H–2 congenic mouse strains with C57BL/10 background genes. Again the evidence suggested that the H–2^k haplotype was associated with susceptibility, relative to H–2^q and H–2^s which were associated with resistance or good responder phenotype. Using H–2 recombinant mice, mapping studies showed that at least two I-region alleles were involved. One important allele was present at the A locus and was designated *Ts–1* and one was present at a new locus between S and D regions and was termed *Ts–2*.

Expression of the *s*, *q*, *f* or *b* alleles at the *Ts–1* locus was associated with resistance, but this was modulated if the *d* allele was present at the *Ts–2* locus (Figure 6.3). This conclusion was verified through experiments exploiting B10.T(6R) mice which have resistant alleles at all loci except *Ts–2*, at which they express the *d* alleles. These mice were more susceptible to infection than mice which expressed resistant alleles at all the H–2 loci. It was also shown that B10.T(6R) mice contained, within their MLN, immune T cells capable of accelerating worm expulsion following adoptive transfer to resistant strains of mice challenged with larvae. However, B10.T(6R) mice could not themselves achieve accelerated worm expulsion when provided with immune MLN T cells known to be effective in accelerating worm expulsion in resistant strains (Figure 6.4).

On the basis of these findings, but taken within the context of many other studies analysing the mechanisms controlling worm expulsion, Wassom and colleagues (1984) proposed two hypotheses. The first states that the *Ts–2 d* allele exerts its effect by regulating the responses of bone marrow-derived cells. The second suggests that there may be two distinct T-cell populations, one within the MLN and one outside the MLN. Genes at the *Ts–1* locus

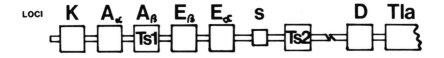

Figure 6.3. Diagram showing the position of the two genes which influence resistance to infection by *T. spiralis* (data from Wassom *et al.*, 1983).

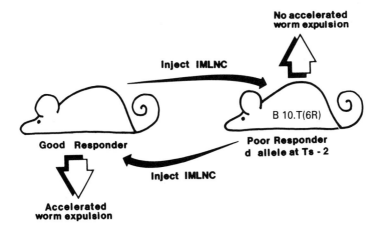

Figure 6.4. Effect of the *d* allele at the *Ts–2* locus upon the capacity to generate protective immunity against *T. spiralis* in the mouse (see text for details).

regulate T helper cells responding to worm antigens; the *Ts–2* genes influence another T-cell population outside the MLN which, in turn, exerts its effect on the bone marrow-derived cells involved in intestinal inflammation (Wassom *et al.*, 1984). Of course such hypotheses do not explain all the results, and bearing in mind the complexity of parasite antigens, they need not be mutually exclusive. However, they do form a sound basis for future investigations of the mechanisms by which the regulation of parasitic infection is controlled at the individual gene level.

Although H–2 genes clearly influence worm expulsion, comparison of the expulsion kinetics of a number of different inbred strains sharing the same haplotype but differing in background genes, shows that non-H–2 genes exert an equally marked effect upon the rapidity of the response (Figure 6.5). These and other observations suggest that H–2 genes exert their effect only within the constraints imposed by background genes. The latter genes remain to be located and described, but already some progress has been made. Bell and co-workers have identified an autosomal dominant gene outside the MHC which seems to exert a strong effect upon the rapid elimination of *T. spiralis* following challenge infection of immune animals (Bell *et al.*, 1984).

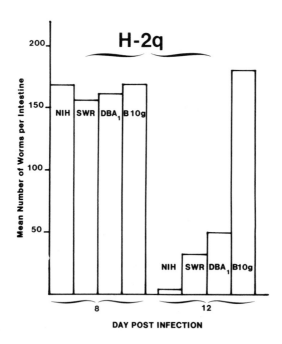

Figure 6.5. Effect of non-H-2 (background) genes upon the course of intestinal infection with *T. spiralis* in different strains of inbred mice sharing the same haplotype (H–2q). Levels of infection were assessed on days 8 and 12 post-infection.

Heligmosomoides polygyrus

Whilst the details of the genetic control of resistance may differ between *T. spiralis* and other helminth infections, the broad concepts established with this species may apply to others. Another rodent model of gastrointestinal nematode infection which has been studied with respect to genetic control is *H. polygyrus*. In contrast to other laboratory models in which infection tends to be acute, *H. polygyrus* causes a chronic infection, a characteristic of many intestinal dwelling nematodes affecting man. Again immunity to this worm appears to be strongly T-cell dependent although detailed mechanisms have yet to be defined. Different inbred strains of mice show variation in pathology, infection and response to immunization. The ability to expel primary infections can be inherited as a dominant trait, and gene complementation has been reported. This latter phenomenon was observed in experiments in which two parental strains which did not expel primary infection adults within ten weeks of the administration of larvae (NIH and C57BL/10) produced F1 progeny when crossed which cleared worms by week seven (Behnke and Robinson, 1985; Robinson *et al.*, 1989; Wahid *et al.*, 1989). Detailed analysis using congenic strains of mice has demonstrated that resistance to challenge infection is influenced by MHC genes (Enriquez *et al.*, 1988) although background genes can also exert strong influences.

Whilst inbred strains of mice have been utilized extensively in studies of the genetic control of parasitic infection, the use of outbred populations can also provide interesting data, and this latter approach has been examined in the *H. polygyrus* system. Brindley and Dobson (1981) have demonstrated that selection of wild or outbred populations of mice on the basis of liability to infection can lead to the establishment of liable or refractory lines. Mice were selected on the basis of the number of parasite eggs passed in the faeces at 3–4 weeks post infection, and by the end of the period of selection it was found that mice in the liable line passed 58% more parasite eggs per gram of faeces than a control random bred line, with the refractory group passing 31% fewer than the random line. When the numbers of adult worms present in the small intestine were examined, however, there was little to distinguish the groups. Thus the main effect was on parasite fecundity, the criterion used in selection. The differences observed were considered to be the result of innate immunity although later experiments suggested that liable and refractory lines also differed with respect to specific acquired immunity.

The results of challenge infections in refractory and liable mice established that liable mice required more intense immunization than refractory mice to stimulate equivalent levels of protection. Specific antibody responses to *H. polygyrus* were higher in mice of the refractory line and were effective in transferring immunity passively when given to naive recipient mice. Indeed it is known that anti-worm IgG antibodies play an important role in host protective immunity to *H. polygyrus* (Pritchard *et al.*, 1983; Williams and Behnke, 1983) and that there are marked differences in the antigens recognized by serum taken from different inbred strains of mice following infection.

The overall situation is likely to be complicated by the fact that in *H. polygyrus* infections, there is a marked degree of immunosuppression associated with the adult stages of the parasite. The mechanisms responsible for the generation of the immunosuppression are as yet undefined but are almost certain to be under genetic control (see Behnke, 1987).

Protozoans

Many protozoan infections, including species of *Leishmania*, *Eimera*, *Plasmodia*, *Trypanosoma*, *Giardia*, *Trichomonas* and *Toxoplasma*, have been subject to the investigation of variation in host response. To illustrate some of the characteristics of variation in host response, two infections will be highlighted here, namely leishmaniasis and malaria (*Plasmodium*). However, readers are directed to Skamene (1985) and Blackwell (1988) for more detailed information.

Leishmania

The *Leishmania* are parasitic protozoa which are transmitted to vertebrate hosts by sandfly vectors. The infective promastigote stage invades host macrophages and transforms into the replicative amastigote stage. Genetic control of the response to infection in mouse models has been the subject of intense research over the last few years. The two models most studied are *L. donovani*, which causes the visceral form of the disease, and *L. major* which causes the cutaneous form. The current understanding of immune aspects of the host/parasite relationship suggests two stages during control of infection: 1) An early stage at which components of the innate defence system eliminate or control parasite replication and 2) a mechanism operating later in which acquired immunity regulates parasite survival in the macrophage through T-cell dependent mechanisms.

Experiments on *L. donovani* carried out by Bradley and colleagues during the 1970s (Bradley, 1977) demonstrated a clear separation of inbred mouse strains in terms of their ability to control early replication of parasites within the macrophage. Analysis by selective breeding demonstrated that this control was exerted by a single dominant gene found on mouse chromosome 1. This autosomal gene exists in two allelic forms termed Lsh^r (resistant) and Lsh^s (susceptible). Interestingly, traditional genetic mapping analysis has suggested that the *Lsh* gene is indistinguishable from genes which control early resistance to a number of bacterial pathogens including *Salmonella typhimurium* and *Mycobacterium lepraemurium* (see Blackwell, 1988).

The mechanism by which the *Lsh* gene operates within the macrophage is unknown, although it functions post invasion and affects parasite replication. However, in order to study this problem in detail, a congenic mouse strain has been produced in which the resistant allele from one inbred strain of mice was transposed onto the genetic background of another susceptible strain.

This was accomplished by repeated backcross mating of the heterozygous resistant progeny onto the homozygous *Lsh*[s] susceptible strain (Figure 6.6) and selecting at each stage mice with the resistant phenotype. Analysis of macrophage populations from this congenic strain should provide some clues as to how the products of the *Lsh* gene modify macrophage function to produce resistance to *L. donovani*.

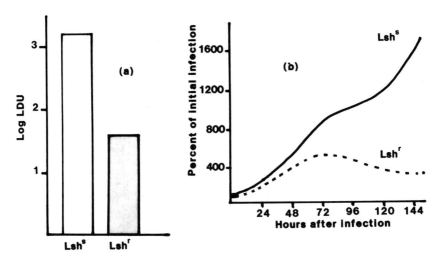

Figure 6.6. Influence of the *Lsh* gene upon resistance to *L. donovani* in mice. (a) *L. donovani* infections (LDU; *L. donovani* unit, i.e. the number of parasites/1000 host cell nuclei × weight of the organ) in the livers of *Lsh*[s] or *Lsh*[r] mice on day 15 after infection with 10^7 amastigotes (data from Crocker *et al.*, 1984). (b) Growth of *L. donovani* in liver macrophages isolated from *Lsh*[s] or *Lsh*[r] mice. Liver macrophages were infected for 2 h with 2×10^5 amastigotes. The graph shows % of the initial infection developing at different time intervals (data from Crocker *et al.*, 1987).

The separation of inbred mouse strains into susceptible or resistant strains with regard to early phase responses, provided a sound basis for the analysis of the later operating, T-cell dependent responses (Blackwell *et al.*, 1980). Experiments with homozygous recessive (*Lsh*[s]) strains of inbred mice demonstrated two response phenotypes, cure or non-cure, and use of H–2 congenic strains established differences between mice bearing different alleles at loci coding for the MHC Class I and Class II molecules. Attempts to identify important alleles within the H–2 region using H–2 recombinant mice gave rise to complex results. A close examination showed that strains of mice which possessed the *k* allele at the IE region of H–2 exhibited a highly variable response, some individuals self-cured, others tolerated high parasite loads for extended periods. Such variation within inbred mouse strains was unexpected and difficult to reconcile with contemporary understanding of the mechanisms controlling resistance to *Leishmania*. Since then it has become apparent that several additional factors may influence the outcome of infection, notably the intensity of parasitaemia, and the generation of suppressor T cells by the host (see Section 6.3 for discussion). In mouse strains expressing the *k* allele at IE, the course an infection will follow is determined by the interaction of these factors. A shift in the balance one way or the other in individual mice determines whether self-cure or chronic infection will occur.

Background (non-H–2) genes have also been shown to exert effects upon the rate of recovery from an infection. One non-H–2 linked gene which has been shown to modify the cure response is the H–11 linked gene. Possession of such a gene causes a more pronounced non-cure response than normal. In addition it is unique as being the only gene identified so far which exerts the same influence on both visceral and cutaneous forms of the disease.

When studying *L. major*, a comparison of different inbred mouse strains resulted in a variety of disease profiles being recognized. Mendelian inheritance analysis, by selective breeding employing key strains, enabled the identification of a single gene-controlling susceptibility to infection. This gene maps to a different chromosomal location from the *Lsh* gene and has been termed *ScL–1*, but at present, little is known about the gene products and their influence on processes which determine resistance and susceptibility.

H–2 differences between strains of mice give rise to only minor variations in the disease progression following subcutaneous inoculation of the parasite, and the *Lsh*ʳ allele does not appear to exert any influence over lesion growth. However, as mentioned earlier, H–11 differences can have a profound effect upon lesion growth.

It is possible that *ScL–1* and H–11 exert their effects at the level of inhibiting parasite replication within the macrophage, and this view is supported by observations made both *in vivo* and *in vitro*. Subsequent control may be via T-cell influence on macrophage activation, with disease progression dependent upon initial antigenic load (controlled by *ScL–1*, perhaps). It is believed that the intensity of infection is a critical factor in determining whether helper or suppressor T cells are generated. Certainly T cells of the same phenotype [expressing the CD4 (L3T4) molecules] have been found to either cure or exacerbate the disease (Liew, 1986).

Plasmodium spp.

Human malaria species do not infect non-primates, but a number of rodent malarias (*P. berghei*, *P. yoelii*, and *P. chabaudi*) have been studied with regard to variation in host response (see review by Blackwell, 1989). Examination of *P. berghei* infections following injection of sporozoites or infected blood in different inbred strains of mice enabled segregation of strains into susceptible or relatively resistant groups. From blood infection studies it was concluded that parasitaemia and host survival were only partially interdependent variables. Breeding analysis suggested that polygenic control was involved. Examination of Biozzi mice (strains of mice bred with selection for their overall ability to produce high or low levels of antibody) after immunization with irradiated parasitized red blood cells suggested that low antibody-producing mice did not respond to vaccination, unlike the high antibody-producing mice. Backcross breeding experiments again suggested polygenic control.

In the case of *P. yoelii* infections, qualitative and quantitative differences between different inbred strains of mice in the IgG antibody response to a varied array of malaria antigens were demonstrated although the relationship to genes determining host protective resistance is still unknown. The Lsh^r congenic mouse strains developed for genetic studies of resistance to *Leishmania* (see above) have been used for studies with *P. yeolii*, and recent experimental evidence suggests that the *Lsh* gene may exert some early control over sporozoite invasion of liver macrophages (Kupffer cells). However this still remains controversial and further work in this area is required.

Infection of inbred strains of mice with *P. chabaudi* has again enabled identification of resistant and susceptible strains, and the available data is consistent with control by a single dominant autosomal gene termed *mcp* located on chromosome 2. However, in common with many parasite host systems, the gene products of *mcp* have not been identified and their precise role in controlling infection remains speculative.

6.3. GENETIC CONTROL OF REGULATORY MECHANISMS—A COMMON THEME?

The examples discussed above may be taken to imply that non-responsiveness or poor responsiveness to a parasite is simply a reflection of a quantitative deficiency in one or other component of the immune response. Whilst this may be true in some situations, recent evidence has given rise to the suggestion that resistance and susceptibility may also be determined by the genes which affect the regulatory components involved in homeostatic control of the immune system. Evidence in support of this hypothesis comes from several studies utilizing the protozoan and helminth/mouse models discussed in Section 6.2.

The experimental data are centred around the pivotal role that T cells play in the generation of acquired immunity and particularly upon the mechanisms of recognition of parasite antigens by T helper cells (see Chapter 4 for details). T helper cells recognize antigen only in association with self molecules present on the surface of antigen-presenting cells. In the mouse these molecules are controlled by genes located at the Class II loci of the H–2 complex (see Figure 6.7). The molecules encoded by this region are known as Class II or Ia molecules. Most inbred strains of mice express two Class II molecules termed IA and IE. However, not all strains express the IE product, e.g. mice with haplotypes *s*, *b*, *q* or *f*. The reason is that these mice have defects in the genes or their products which are involved in the production and

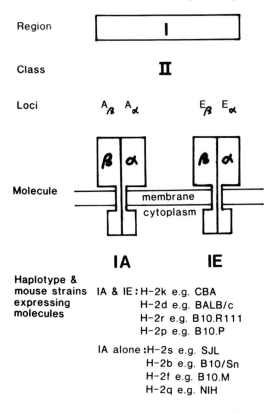

Figure 6.7. MHC Class II molecules in the mouse. Examples of different inbred strains of mice which exhibit variation in expression of the IA and IE molecules on the cell surface.

expression of the IE molecule on the cell surface. However, mice which fail to express the IE molecule are, in most other respects, normal.

In connection with parasitic infection, especially with infections of *T. spiralis*, there appears to be a correlation between the responsiveness of the strain to infection and the expression of the IE molecule. Mice which do not express IE (*s*, *b*, *q* and *f* haplotypes) are relatively more resistant than those which do express the molecule. The F_1 progeny of matings between IE expressing and non-expressing strains are more susceptible than the resistant parent. Wassom *et al.* (1987) hypothesized that the presentation of parasite antigens, when seen in association with IE molecules, preferentially induced the generation of CD4$^+$ T cells which down regulate or suppress the proliferation of CD4$^+$ T cells which see parasite antigen in association with IA molecules (Figure 6.8). This hypothesis may be particularly pertinent to the *T. spiralis* model as the *Ts–1* gene has been mapped to the IA region of the H–2 complex. Moreover, the T cells which are thought to be responsible for the protective immune response against intestinal stages of infection bear the

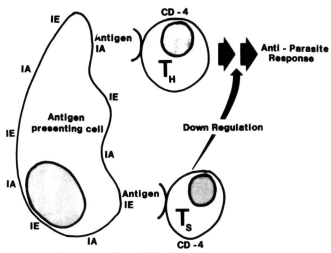

Figure 6.8. Diagrammatic representation of the hypothesis to explain the influence of antigen presentation when seen in conjunction with IE molecules upon anti-parasite responses. Antigen presented in association with IA molecules generate T helper (T_H) cells which lead to the generation of anti-parasite responses. Antigen presented in association with IE molecules generate T suppression (T_s) cells which down regulate responses generated by T_H cells (redrawn from Wassom *et al.* 1987).

marker L3T4 (CD4 marker equivalent). Such cells are known to recognize antigen in association with Class II MHC molecules.

In the *H. polygyrus* model, Enriquez *et al.* (1988) have mapped two H–2 linked genes which influence levels of resistance to infection, one of which probably encodes a Class II molecule. Using a series of H–2 congenic mouse strains, Wassom *et al.* (1987) found that, by and large, strains which did not express the IE product were relatively resistant to a challenge infection.

Further support for this hypothesis may be found in experiments with *Leishmania*. Mouse strains which express IE molecules are known to have highly variable responses to infection with *L. donovani*, the causative agent of the visceral form of the disease. Again in this infection CD4$^+$ bearing T cells are thought to play a major role in regulating macrophage activity and therefore controlling the infection. Recently experiments have been carried out in which a non-curing strain of mice was infected with *L. donovani* and treated with monoclonal antibodies specific for the IA or IE molecules (Blackwell and Roberts, 1987). The results demonstrated that while anti-IE treatment was associated with an enhanced clearance of parasites from the liver and spleen, anti-IA treatment caused an exacerbation of the disease. Whether the mechanisms which account for these observations reflect an interference with antigen presentation or are due to suppression remains unresolved. Nevertheless the data do lend support to the idea that infection and therefore disease may be genetically controlled at the level of the antigen-presenting cell.

The presence or absence of a particular Class II molecule on host cells involved in antigen presentation and their affinity to parasite antigens determine which T-cell population is stimulated and hence may be expected to exert a crucial influence upon the subsequent course of disease. Indeed, experiments in mice with a cloned malaria protein antigen, a candidate sporozoite vaccine against *Plasmodium falciparum*, have shown that only mice expressing the *b* allele at the IA locus were able to respond to the vaccine by the production of specific antibody (Good *et al.*, 1986).

It is important to remember, however, that despite the accumulating evidence supporting the importance of IE in the control of responses to parasitic infection, the association is not always so clear cut; for example, Behnke and Robinson (1985) demonstrated that C57BL/10 mice (H–2b, non–IE expressing) were relatively poor responders in their ability to express primary and acquired immunity to *H. polygyrus*. This is not surprising bearing in mind the importance of other genes within and outside the MHC which can exert profound effects upon the outcome of infection.

6.4. GENETIC CONTROL OF PARASITE-INDUCED PATHOLOGY

Hitherto the discussion has focused on host resistance and control of protective immunity, with little attention being given to the pathology accompanying infection. The interrelationships between host genes and the immunopathological consequence of infection have been studied in several parasite models but notably in schistosome infections in rodents. Again, mouse models have provided much of the information because the major human parasites, *S. mansoni* and *S. japonicum*, can be maintained in this host in the laboratory. Variation in resistance to infection is seen in infections of mice with both *S. mansoni* and *S. japonicum* and operates at a number of distinct sites around the body. Differences in antibody production and T-cell-mediated inflammation have been implicated and variation observed following primary and secondary infection or vaccination with attenuated parasites. Gene control appears complex with both MHC and non-MHC genes playing important roles.

The pathology associated with schistosome infection in mice is primarily caused by the egg stage. For *S. mansoni* and *S. japonicum* the eggs are released from female worms after migration into the small venules carrying blood from the small intestine. The eggs pass into the gut lumen after moving through the intestinal tissue by a combination of enzymatic digestion, movement induced by intestinal peristalsis and mucosal epithelial turnover. The eggs subsequently leave the body with the faeces. However, not all the eggs enter the intestinal lumen, some (up to 70%) are carried away in the bloodstream and become lodged in the lungs or the liver. Pathology arises from a T-cell-mediated hypersensitivity reaction against egg antigens derived from the miracidium contained within the egg. As a consequence granulomata form around the eggs which are trapped in the tissues. One consequence of extensive granuloma formation in the liver (see Chapter 5) is the restriction of blood flow through the hepatic sinusoids. This in turn causes portal hypertension and gives rise to a number of pathological changes including hepatomegaly, splenomegaly, oesophageal varices and periportal fibrosis. As a consequence of these pathological changes, schistosome larvae from a challenge infection may fail to complete their migration from the lungs to the mesenteric veins via the liver. This is due to the formation of portal caval anastomoses into which blood and hence migrating larvae are diverted. Thus it is possible that the variation observed in resistance to challenge infections could be partly accounted for by variation in the degree of pathological responsiveness.

An interesting feature of granuloma development is that as infection progresses the size of the granulomata which form decreases (Warren, 1982). Greatest size is attained shortly after the female worm begins to lay eggs, as the host becomes hypersensitive. The exact mechanisms involved in modulation of granuloma size are unresolved, although involvement of suppressor T cells bearing the CD8 marker has been demonstrated for both *S. mansoni* and *S. japonicum* infections. An additional factor has been shown to be involved following *S. japonicum* infections, namely IgG_1 antibody specific for parasite egg antigen. Moreover, IgG_1 anti-idiotypic antibodies (antibodies directed against the combining site of the IgG_1 anti-egg antigen antibodies; Chapter 14) were observed to develop as the infection progressed. Such anti-idiotypic antibodies were found to modulate granulomatous inflammation although the precise mechanisms involved remain to be described (see review by Stavitsky, 1987).

In both infections granuloma modulation exhibits genetically determined variation. This has serious consequences as the degree of modulation is a contributing factor to host mortality. In *S. mansoni* infections both granuloma size and host mortality have been shown to be strain dependent, although the data concerning the influence of H–2 haplotypes upon pathology is conflicting and remains to be clarified. In the *S. japonicum* model, strains of mice which exhibited a strong acute lung granulomatous response following injection of preparations of egg antigens also showed the highest antibody responses to an egg antigen thought to be involved in the immunopathological processes associated with infection (Mitchell *et al.*, 1983). This suggests that variation in pathology may be related to anti-egg antibody production. The genetic basis for the variation in modulation due to anti-egg antibody and anti-idiotypic antibodies remains to be determined but is likely to be influenced by both H–2 and non-H–2 genes.

6.5.   THE SITUATION IN THE FIELD

Although variation in disease profiles in wild host populations is evident, analysis of the mechanisms underlying the variation in host response to parasitic infection is extremely difficult under field conditions. Unlike the laboratory animal-based studies, all observations of individuals are upon those exposed to natural (and therefore varying) infections. Environmental factors exert strong influences, and more often than not individual hosts are infected by more than one species of parasite. Thus, interpretation of field data is complicated unless resistance or susceptibility can be correlated with single gene defects, the products of such genes or complexes of genes, notably the MHC. However a number of important studies have been carried out and interesting observations have been made (see reviews by Anderson, 1986; Hagan, 1987; Bundy and Cooper, 1989), a sample of which will be discussed below.

Helminths

An important observation made over the years concerning helminth infection in humans is that parasites show an aggregated distribution, i.e. a few individuals possess high levels of infection (many parasites) whereas the majority of the population possess fewer parasites. This has been found in areas of uniform parasite exposure and is referred to as an overdispersed distribution of parasites (see Chapters 12 and 13 for details). Examples can be found in infections by filarial worms such as *Onchocerca*, the schistosomes and a number of intestinal nematodes including *Ascaris*, *Trichuris* and hookworms. An overdispersed distribution of parasites in a population mirrors the experimental situation in which outbred strains of laboratory animals, or panels comprizing inbred strains of animals are viewed together. The variation in human infection levels corresponds to the variation observed in the pathology of the disease. Such observations when viewed in the light of laboratory experiments, strongly suggest that genetic factors controlling resistance to parasites explain some of the inter-host variation which typifies epidemiological studies of human communities.

  A number of detailed epidemiological studies of infection by *S. mansoni* and *S. haematobium* have been carried out in Africa. Results from these studies suggest that individuals acquire immunity to schistosomes over many years. Long-term studies have presented evidence of predisposition to infection, i.e. certain individuals who harbour large numbers of worms are predisposed to reacquiring high levels of infection again after the initial worm burden has been removed by chemotherapy (wormy people). It has also been noted that individuals showing severe pathology and disease symptoms tend to be restricted to a small proportion of the population. Furthermore, studies of infection levels within families and between non-identical twins have shown that individuals can exhibit pronounced differences in parasite loads despite similar exposure levels. The variation observed also holds true for parasite burdens developing after naturally acquired reinfection. The mechanisms underlying the differences remain to be resolved, although the capacity to generate eosinophils (which can kill invading schistosome larvae in conjunction with antibody, see Sections 5.8 and 5.9) has been implicated (see Hagan, 1987).

The relationship between the degree of pathology and blood group or HLA markers has been studied by a number of workers. A positive correlation between severe hepatosplenic disease following *S. mansoni* infection and the possession of blood group A has been established in Brazil. This has been corroborated in Egypt (Abdel Salem *et al.*, 1979) where an increased risk of developing hepatomegaly was found in individuals possessing the A1 and B5 specificities at the HLA A and B region (Class I molecules). The mechanisms by which these markers influence disease is still unknown.

It has been shown in *S. japonicum* that a very strong correlation exists between HLA markers and the response to infection (Sasazuki *et al.*, 1980). Individuals were examined for their lymphocyte proliferation response to parasite antigens and typed for HLA markers. Low response to antigen was associated with the haplotype HLA – BW52 – DW12. DR2. DQW1 (Figure 6.9). Individuals who possessed this haplotype also showed a decrease in the pathological condition known as post-schistosomal liver cirrhosis. None of the low responders were homozygous, a finding taken to imply that low

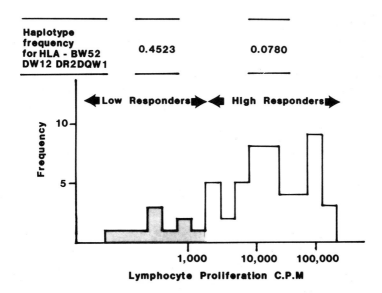

Figure 6.9. Immune responsiveness to *S. japonicum* adult worm antigen in a population (n = 57) with prior schistosomal infection. Responsiveness was measured by antigen-specific proliferation of peripheral blood T lymphocytes *in vitro*. Frequency for the 'susceptible' haplotype is shown for the 'low' and 'high' responders (data from Sasazuki *et al.*, 1980).

responsiveness was caused by a single dominant immunosuppression gene. The mechanisms underlying the immunosuppression are thought to involve antigen-specific T suppressor cells. Further examination of this regulatory mechanism has utilized T-cell lines grown *in vitro* from high or low responder individuals (Hirayama *et al.*, 1987). The data establish that MHC Class II molecules play a vital role in determining the outcome of infection. HLA–DR molecules influence the helper response whereas HLA–DQ gene products are involved in suppression. These observations are particularly interesting in the light of the experimental evidence from laboratory models of helminth infection and our understanding of the role of Class II MHC molecules in determining immune response to parasites (see Section 6.3).

It has been known for a long time that there are distinct differences between breeds and within breeds of domestic animals, in regard to helminth infection resistance. Although genetic analysis of the basis for this variation still suffers from many of the problems encountered in human studies, controlled experiments have been undertaken using animals of relatively defined genetic status. The most intensively studied helminths in this regard are the gastrointestinal nematodes, e.g. *Haemonchus contortus* and *Trichostrongylus colubriformis*. Both species are economically important and exert most effect upon young animals. Heavy infections cause considerable intestinal pathology, although even light infections are associated with depressed growth. Resistance to *H. contortus* is thought to be mediated through a combination of immunological responses, which expel the worm from the intestine, together with a physiological component, which controls the capacity of the animal to regulate the pathophysiological changes that accompany infection. A genetic element is known to underly both components. Certain breeds of sheep (e.g. Scottish Blackface) developed lower worm burdens, suffered less pathology and expelled a challenge infection more rapidly than other breeds (e.g. Finn Dorset; Figure 6.10; Altaif and Dargie, 1978). Furthermore, a correlation has been identified between susceptibility and haemoglobin type. Breeds possessing haemoglobin type A are generally more susceptible than breeds with haemoglobin type B; heterozygotes expressed an intermediate response status. Considerable attention has also been devoted to understanding the inheritance of resistance to *H. contortus*. Long-term studies involving over 600 lambs from 40 sire groups were examined for their inherited resistance. One group was of particular interest and showed an exceptional degree of resistance. It was postulated that the ram which sired the lambs from this group possessed a rare, dominant, major resistance gene (see Gray, 1987).

The other intestinal-dwelling nematode which has received much attention is *T. colubriformis*. Studies of this parasite have benefited from the availability of a laboratory host (guinea pig) which has enabled closer examination of certain complex aspects of the host/parasite relationship.

In Australia, Dineen and colleagues carried out a number of experiments in which sheep were selected for breeding based upon their relative response to experimental immunization with an irradiated parasite vaccine. The lambs, sired by responder rams or derived from responder rams and responder ewes,

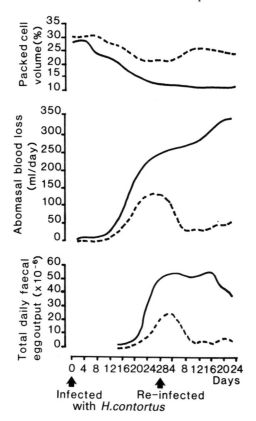

Figure 6.10. Genetic resistance to *H. contortus* in sheep. Development of primary and secondary infections in Finn Dorset (————) and Scottish Blackface (– – – –) sheep. Marked differences in packed cell volume and abomasal blood loss can be seen between the relatively resistant Scottish Blackface and the more susceptible Finn Dorset as evidenced by faecal egg output (redrawn from Altaif and Dargie, 1978).

were found to be much more resistant to a challenge infection than lambs from susceptible parents (Dineen and Windon, 1980). Attempts have been made to relate responder status with levels of specific lymphocyte stimulation. Greater levels of lymphocyte proliferation were observed with cells from responder individuals than with cells from non-responders. Recent work has also shown a degree of correlation between resistance and the possession of a sheep lymphocyte marker. Sheep with lymphocytes bearing the ovine Sy–1 marker tended to be more resistant than sheep bearing any of the other five markers (Sy–2–6; Outteridge *et al.*, 1985).

Protozoan infections

The difficulties emphasized in studies of human variation in response to helminth parasites apply equally to the study of protozoan diseases. Little is known of the genetic control of resistance to *Leishmania* in man, although as mentioned earlier, the spectrum of disease profiles parallels the variation observed in resistance between inbred mouse strains and may therefore reflect similar immunogenetic mechanisms.

Variation in host susceptibility to malaria probably represents one of the most intensely studied aspects of the human disease. Certainly genetic traits which influence red blood cell physiology and structure can affect the capacity of malaria parasites to enter and survive in erythrocytes. Such traits include haemoglobin S, glucose-6-phosphate dehydrogenase deficiency, thalasaemia and Duffy blood group determinants. Haemoglobin S produces sickle cell anaemia. Individuals heterozygous for this trait are resistant to the lethal consequences of *Plasmodium falciparum* disease, and such an advantage explains the persistence of this trait in endemic regions despite its debilitating effects on subjects homozygous for the gene. Associations between HLA and susceptibility have also been noted with certain haplotypes being correlated with resistance and levels of anti-parasite antibody (see review by Blackwell, 1989).

Among domestic animals, interbreed variation in resistance to protozoal infections is best illustrated by resistance to bovine trypanosomiasis. Variation between different cattle breeds to tsetse fly challenge has been recorded from the beginning of the century and has been confirmed in recent studies (see review by Dolan, 1987). In western Africa the humpless N'Dama and West African shorthorn are superior breeds to the zebu in terms of trypanotolerance. Again, little is known of the mechanisms involved, but based on laboratory models it seems likely that control is complex and unlikely to be the result of one major gene. Trypanotolerance is the result of two factors, both of which may be under complex genetic control: 1) Trypanotolerant cattle may become infected less frequently, often have lower parasitaemias and 2) following infection trypanotolerant cattle may survive better. Correlation between markers (such as MHC) and resistance have yet to be established but will no doubt progress alongside advances in ruminant immunology.

6.6. CONCLUSIONS

The preceding discussion has emphasized the intense interest which surrounds the study of the genetic basis of variation in resistance to parasitic infection. It has also stressed the complexity of the genetic control of both innate and acquired immunity to parasites. Defined laboratory models have and will continue to generate a wealth of useful information concerning the subtle genetic mechanisms by which the host influences the aetiology of parasitic disease. However, there is a clear urgency to learn more about immune mechanisms which control parasites both in laboratory hosts and during natural infection. Only with a detailed knowledge of host immunity, will it be possible to understand the functional significance of particular gene expression and the consequences for host/parasite survival. Considerable advances have already been made in this respect, as for example leishmaniasis. The isolation and localization of the murine-resistance gene, *Lsh*, may provide a convenient probe for use in molecular genetic studies of human leishmaniasis. The murine model eminently illustrates the complex mechanisms involved in the generation of acquired immunity, particularly at the level of the T cell and antigen-presenting cell.

Data from several laboratory models, including both protozoan and metazoan species, are beginning to show similarities in genetic control at the level of antigen presentation, especially in relation to the MHC molecules involved (IA *versus* IE molecules). Further experimentation with other species is important and must not be seen simply as duplication of existing data. It is vital that the scope of application of the hypotheses which have been formulated is rigorously examined on a far broader base than that existing to date. Investigation of human interrelationships with parasites at the level of genetic control is a priority. Will individuals expressing certain alleles at the DQ and DR loci (human IA and IE equivalents) tend to be particularly resistant or susceptible to parasitic infection? Of course the difficulties in analysing data from the field cannot be understated. Nevertheless, previous and ongoing long-term studies of parasitic infection in endemic areas are providing essential information concerning the development of acquired immunity to parasites and about individual predisposition to infection, e.g. schistosomes. Again, the investigation of basic immune mechanisms is crucial, e.g. difficulties arise in explaining the basis for predisposition to GI dwelling nematodes in man because there is little evidence for acquired immunity to these parasites in humans (see Behnke, 1987). Reinfection studies in the field, however, have shown correlation between worm burdens before and after chemotherapy, with individuals reacquiring worm burdens similar to pre-treatment levels, and this may be suggestive of genetically determined variation in resistance (Schad and Anderson, 1985).

Detailed biochemical and structural information concerning parasite antigens and important epitopes is still scarce. The definition of functional parasite molecules will aid analysis of their interaction with host MHC molecules and may provide information which will enable accurate prediction of the type of immune response expected in individuals of particular genotypes. Again, such studies have already been initiated, and in the case of malaria, still generally considered to be the most important of the parasite species affecting man, progress is being made. Whilst analysis of the genetic variation to medically important parasites will aid the rational design of control strategies, an understanding of the genetic basis for variation in resistance to parasites of veterinary importance may have additional benefits. Parasitic diseases are responsible for immense economic losses from mortality as well as from impaired productivity. Knowledge of the genetic basis for resistance may enable accurate and suitable selection of breeding programmes for the improvement of domestic breeds of cattle, sheep and pigs. Moreover, with the advent of molecular techniques for insertion and expression of DNA coding for resistance (e.g. the genes controlling trypanotolerance), it may be possible to enhance or even circumvent selective breeding programmes.

In summary, the information available to date strongly implies that the variation observed in resistance and disease resulting from parasitic infection has a considerable genetic basis, aside from variation arising from environmental causes. The genetic mechanisms operating are undoubtedly complex, but the functional consequences of particular gene expression are still largely unknown. With the recent advances in molecular genetics and immunology, it may be possible in the foreseeable future to predict which individuals are likely to be susceptible or resistant to particular parasitic diseases and to attempt appropriate vaccination. However, a note of caution; the segregation of a population into relatively resistant or susceptible individuals carries a number of implications with regard to vaccination strategies. Given uniform exposure to parasite transmission stages, individuals with low numbers of parasites may be considered to be resistant and are either innately refractive to infection or mount a strong, protective acquired immune response. These subjects will not make a large contribution to parasite transmission in nature, will not show marked pathology and should respond well to vaccination because they are genetically equipped to respond to parasite antigens in a host-protective manner. However, such individuals are not those who would benefit most from successful vaccination. It is the susceptible proportion in the population, who have high parasite burdens and show severe disease symptoms, who more urgently require assistance in coping with their parasite burdens. They do not mount a suitable protective anti-parasite response, and it is plausible to suggest that a vaccine composed of parasite antigens which should stimulate a good protective response (by definition assessed in terms of a resistant individual's immune response) may not be effective.

Indeed, laboratory studies have shown that certain strains of mice fail to respond to vaccination against various parasites including schistosomes (James et al., 1984), *Trichinella* (Wakelin et al., 1986) and *Plasmodium* (Good et al., 1986). It is possible that in vector-transmitted infections vaccination of 'intermediates' (neither totally susceptible nor totally resistant) may contribute to a reduction in parasite transmission and subsequent pathology and disease. This is unlikely to happen with non-vector transmitted, direct life-cycle parasites and with infections showing overdispersion, e.g. gastrointestinal nematodes. As a consequence vaccine design needs to be directed towards the non-responder individuals, perhaps more so than towards the responders. Hence understanding the genetic control of parasitic infections is crucial when evaluating the usefulness of candidate parasite molecules for inclusion in vaccines.

ACKNOWLEDGEMENTS

I would like to thank Professor D. Wakelin for reviewing the manuscript, Dr. T. I. A. Roach for helpful discussion, Mrs J. Crosby for photography and Mrs. J. Jolley for excellent secretarial assistance.

REFERENCES

Abdel Salem, E., Ishaac, S. and Mahmoud, A.F.F. (1979), Histocompatibility-linked susceptibility for hepatosplenomegaly in human Schistosomiasis mansoni, *Journal of Immunology*, **123**, 1829–32.

Altaif, K.I. and Dargie, J.D. (1978), Genetic resistance to helminths. The influence of breed and haemoglobin type or the response of sheep to re-infection with *Haemonchus contortus*, *Parasitology*, **77**, 177–87.

Anderson, R.M. (1986), The population dynamics and epidemiology of intestinal nematode infections, *Transactions of the Royal Society of Tropical Medicine and Hygiene*, **80**, 686–96.

Behnke, J.M. (1987a), Evasion of immunity by nematode parasites causing chronic infections, *Advances in Parasitology*, **26**, 2–71.

Behnke, J.M. (1987b), Do hookworms elicit protective immunity in man? *Parasitology Today*, **3**, 200–6.

Behnke, J.M. and Robinson, M. (1985), Genetic control of immunity to *Nematospiroides dubius*: a 9-day anthelmintic abbreviated immunizing regime which separates weak and strong responder strains of mice, *Parasite Immunology*, **7**, 235–53.

Bell, R.G., Adams, L.S. and Ogden, R.W. (1984), A single gene determines rapid expulsion of *Trichinella spiralis* in mice, *Infection and Immunity*, **45**, 273–5.

Blackwell, J.M. (1989), Protozoal infections, in Wakelin, D. and Blackwell, J.M. (Eds.), *Genetics of Resistance to Bacterial and Parasitic Infection*, pp. London: Taylor & Francis.

Blackwell, J.M. and Roberts, M.B. (1987), Immunomodulation of murine visceral leishmaniasis by administration of monoclonal anti-Ia antibodies: differential effects of anti-I-A *vs* anti-I-E antibodies, *European Journal for Immunology*, **17**, 1669–72.

Blackwell, J.M., Freeman, J.C. and Bradley, D.J. (1980), Influence of H–2 complex on acquired resistance to *Leishmania donovani* infection in mice, *Nature*, **283**, 72–4.

Bradley, D.J. (1977), Regulation of Leishmania populations within the host. II. Genetic control of acute susceptibility of mice to *Leishmania donovani* infection, *Clinical and Experimental Immunology*, **30**, 130–40.

Brindley, P.J. and Dobson, C. (1981), Genetic control of liability to infection with *Nematospiroides dubius* in mice. Selection of refractory and liable populations of mice, *Parasitology*, **83**, 51–65.

Bundy, D.A.P. and Cooper, E.S. (1989), *Trichuris* and trichuriasis in humans, *Advances in Parasitology*, 107–73.

Butterworth, A.E. and Hagan, P. (1987), Immunity in human schistosomiasis, *Parasitology Today*, **3**, 11–16.

Crocker, P.R., Blackwell, J.M. and Bradley, D.J. (1984), Transfer of innate resistance and susceptibility to *Leishmania donovani* infection in mouse radiation bone marrow chimaeras, *Immunology*, **52**, 417–22.

Crocker, P.R., Davies, E.V. and Blackwell, J.M. (1987), Variable expression of the murine natural resistance gene *Lsh* in different macrophage populations infected *in vitro* with *Leishmania donovani*, *Parasite Immunology*, **9**, 705–19.

Croll, N.A. and Ghadirian, E. (1981), Wormy persons: Contributions to the nature and patterns of overdispersion with *Ascaris lumbricoides*, *Ancylostoma duodenale*, *Necator americanus* and *Trichuris trichiura*, *Tropical and Geographical Medicine*, **33**, 241–8.

Dineen, J.K. and Windon, R.G. (1980), The effect of sire selection on the response of lambs to vaccination with irradiated *Trychostrongylus colubriformis* larvae, *International Journal for Parasitology*, **10**, 189–96.

Dolan, R.B. (1987), Genetics and trypanotolerance, *Parasitology Today*, **3**, 137–43.

Enriquez, F.J., Brooks, B.O., Cypess, R.M., David, C.S. and Wassom, D.L. (1988), *Nematospiroides dubius*: Two H–2-linked genes influence levels of resistance to infection in mice, *Experimental Parasitology*, **67**, 221–6.

Good, M.F., Berzofsky, J.A., Maloy, W.L., Hayashi, Y., Fujii, N., Hockmeyer, W.T. and Miller, L.H. (1986), Genetic control of the immune response in mice to a *Plasmodium falciparum* sporozoite vaccine. Widespread nonresponsiveness to single malaria T epitope in highly repetitive vaccine, *Journal of Experimental Medicine*, **164**, 655–60.

Gray, G.D. (1987), Genetic resistance to Haemonchosis in sheep, *Parasitology Today*, **3**, 253–5.

Hagan, P. (1987), The human immune response to schistosome infection. In Rollinson, D. and Simpson, A.J.G. (Eds.), *The Biology of Schistosomes: From Genes to Latrines*, pp. 295–320, London: Academic Press.

Hirayama, K., Matsushita, S., Kikuchi, I., Iuchi, M., Ohta, N. and Sasazuki, T. (1987), HLA–DQ is epistatic to HLA–DR in controlling the immune response to schistosomal antigen in humans, *Nature (London)*, **327**, 426–30.

James, S.L., Correa-Oliviera, R. and Leonard, E.J. (1984), Defective vaccine-induced immunity to *Schistosoma mansoni* in P strain mice, *Journal of Immunology*, **133**, 1587–93.

Liew, F.Y. (1986), Cell mediated immunity in experimental cutaneous Leishmaniasis, *Parasitology Today*, 264–70.

Mitchell, G.F., Cruise, K.M., Garcia, E.G., Vadas, M.A. and Munoz, J.J. (1983), Attempts to modify lung granulomatous responses to *Schistosoma japonicum* eggs in low and high responder mouse strains, *Australian Journal of Experimental Biology and Medical Science*, **61**, 411–24.

Outteridge, P.M., Windon, R.G. and Duneen, J.K. (1985), An association between a lymphocyte antigen in sheep and the response to vaccination against the parasite *Trychostrongylus colubriformis*, *International Journal for Parasitology*, **15**, 121–7.

Pritchard, D.I., Williams, D.J.L., Behnke, J.M. and Lee, T.D.G. (1983), The role of IgG$_1$ hypergammaglobulinaemia in immunity to the gastrointestinal nematode *Nematospiroides dubius*. The immunochemical purification, antigen specificity and *in vivo* antiparasitic effect of IgG$_1$ from immune serum, *Immunology*, **49**, 353–65.

Robinson, M., Wahid, F., Behnke, J.M. and Gilbert, F.S. (1989), Immunological relationships during primary infection with *Heligmosomoides polygyrus* (*Nematospiroides dubius*): dose-dependent expulsion of adult worms, *Parasitology*, **98**, 1–15.

Sasazuki, T., Ohta, N., Kaneoka, R. and Kojima, S. (1980), Association between an HLA haplotype and low responsiveness to schistosomal worm antigen in man, *Journal of Experimental Medicine*, **152**, 314–18.

Schad, G.A. and Anderson, R.M. (1985), Predisposition to hookworm infection in humans, *Science*, **228**, 1537–40.

Skamene, E. (Ed.) (1985), Genetic control of host resistance to infection and malignancy, New York: Alan R. Liss.

Stavitsky, A.B. (1987), Immune regulation in *Schistomiasis japonica*, *Immunology Today*, **8**, 228–33.

Wahid, F.N., Robinson, M. and Behnke, J.M. (1989), Immunological relationships during primary infection with *Heligmosomoides polygyrus* (*Nematospiroides dubius*): expulsion of adult worms from fast responder syngeneic and hybrid strains of mice, *Parasitology*, in press.

Wakelin, D. (1985), Genetic control of immunity to helminth infections, *Parasitology Today*, **1**, 17–23.

Wakelin, D. (1988), in Wakelin, D.R. and Blackwell, J.M. (Eds.), *Genetics of Resistance to Bacterial and Parasitic Infection*, London: Taylor & Francis.

Wakelin, D. and Denham, D.A. (1983), *The Immune Response*, in Campbell, W.C. (Ed.), Trichinella *and* Trichinosis, pp. 265–308. New York: Plenum Press.

Wakelin, D., Donachie, A.M., Mitchell, L.A. and Grencis, R.K. (1986), Genetic control of immunity to *Trichinella spiralis* in mice. Response of rapid and slow responder strains to immunization with parasite antigens, *Parasite Immunology*, **8**, 159–70.

Warren, K.S. (1982), Mechanisms of immunopathology in parasitic infections, in Cohen, S. and Warren, K.S. (Eds.), *Immunology of Parasitic Infections*, pp. 116–37, Oxford: Blackwell Scientific Publications.

Wassom, D.L., Krco, C.J. and David, C.S. (1987), I–E expression and susceptibility to parasite infection, *Immunology Today*, **8**, 39–43.

Wassom, D.L., Brooks, B.O., Babish, J.G. and David, C.S. (1983), A gene mapping between the S and D regions of the H–2 complex influences resistance to *Trichinella spiralis* infections of mice, *Journal of Immunogenetics*, **10**, 371–8.

Wassom, D.L., Wakelin, D., Brooks, B.O., Krco, C.J. and David, C.S. (1984), Genetic control of immunity to *Trichinella spiralis* infections of mice. Hypothesis to explain the role of H–2 genes in primary and challenge infections, *Immunology*, **51**, 625–31.

Whitlock, J.H. (1958), The inheritance of resistance to Trichostrongyloidosis in sheep. I. Demonstration of the validity of the phenomena, *Cornell Veterinarian*, **48**, 127–33.

Williams, D.L.J. and Behnke, J.M. (1983), Host protective antibodies and serum immunoglobulin isotypes in mice chronically infected or repeatedly immunized with the nematode parasite *Nematosporoides dubius*, *Immunology*, **348**, 34–47.

7. Protozoan parasites of erythrocytes and macrophages

I.A. Clark and M.J. Howell

7.1. INTRODUCTION

Both erythrocytes and macrophages have been successfully parasitized by protozoans. This is not usually innocuous to the rest of the host, and various debilitating diseases, of practical importance, ensue. In many cases surprisingly little information is available on the pathogenesis of the resultant systemic diseases, and the mechanisms of protective immunity are also uncertain. Both aspects of the more important diseases caused by intracellular parasites are summarized in this chapter.

7.2. INTRA-ERYTHROCYTIC PROTOZOAN PARASITES

Protozoa belonging to several different genera have evolved to exploit circulating red blood cells in a wide array of vertebrate hosts, ranging from reptiles and birds to mankind. This section discusses examples that cause important human and bovine diseases, predominantly in tropical countries. We have focused mainly on how these parasites are thought to cause illness and pathology and also summarized current ideas on which of the many ways that the host responds to their presence might contribute to their destruction. Most of the following section is taken up with parasites that cause malaria since this group of protozoans has attracted by far the most research attention.

Plasmodium sp.

Although now eradicated from first-world countries, human malaria is as big a problem in most of the tropical world as it was a century ago. Currently it infects more than 200 million people a year and threatens 800 million more. The disease is caused by four members of the genus *Plasmodium*, *P. falciparum* (the chief concern and often fatal), *P. vivax*, *P. malariae* and *P. ovale*. Many different *Anopheles* mosquitoes can spread human malaria, and their effectiveness as carriers varies widely. *Anopheles gambiae*, from central Africa, is generally considered the most efficient vector for this disease.

As shown in Figure 7.1, malaria parasites have complex life cycles in both the mosquito and vertebrate hosts. In brief, both male and female gametocytes are ingested along with the blood meal of the female mosquito and sexual reproduction occurs in the stomach. The zygote migrates through the stomach wall and while attached to its outer surface forms a cyst, inside which it

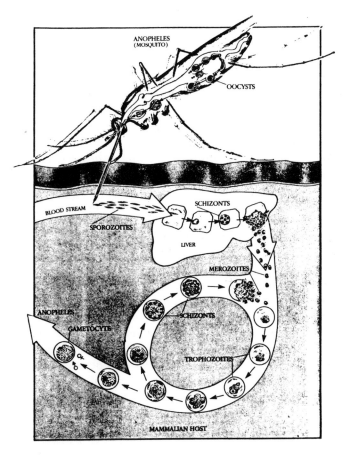

Figure 7.1. Life cycle of the malaria parasite (reproduced, with permission, from *The NYU Physician*/Carol Ann Morley).

undergoes asexual division to form large numbers of sporozoites. These migrate to the salivary glands and are injected when a blood meal is taken. Asexual multiplication occurs twice in the vertebrate host, the first time in the liver, without causing illness, and then, in repeated cycles, in circulating erythrocytes. As discussed below, this stage of the life cycle initiates illness and host pathology.

Malarial illness and pathology

There is as yet no consensus on how the presence of small numbers of intra-erythrocytic protozoa, typically 50–100/mm^3 blood in a previously unexposed person (Kitchen, 1949), causes systemic illness and pathology. It had been observed in the last century that, in synchronous infections, when all of the infected red cells burst (schizogony) within a short space of time, illness reliably follows 2–3 h later. Not surprisingly, this led to the assumption that parasite material (malarial toxin) released into the circulation at schizogony was directly responsible for the illness of malaria. No direct toxin has, however, been identified despite decades of effort (reviewed by Fletcher, 1987). The past ten years have seen the emergence of the concept that the host is instead harmed by soluble products of its own leukocytes. These products are all recognized as useful immunomodulators when produced in lesser quantities. Before summarizing the work that led to this proposal and saw its consolidation, it is necessary to give a synopsis of the pathology of malaria.

Typically the illness of human malaria starts as headache, myalgia, nausea with vomiting, and then fever and rigors, rather like an acute viral or bacterial infection. Anaemia may develop, and foetal death may occur. Nothing distinguishes falciparum malaria, the kind that can be life-threatening, at this stage. As has been reviewed recently (Phillips and Warrell, 1986; Warrell, 1987), untreated falciparum malaria can further develop into any of a series of life-threatening complications, including cerebral malaria, hypoglycaemia, circulatory collapse, severe anaemia, pulmonary oedema and acute renal tubular necrosis. What sets *P. falciparum* apart is its more rapid multiplication rate and the propensity of red cells containing mature parasites to adhere to vascular endothelium.

Bird, rodent and monkey malarias have been used experimentally to gain insight into the nature of the human disease. The most striking contrast with human malaria is that illness in these model infections does not occur until the parasites have attained a much higher density than the concentration required for severe pathology in man. When patients and monkeys infected with *P. falciparum* are compared, this parasite so toxic to humans, is much less harmful for a given parasite load to those monkey species it can infect (Jervis *et al.*, 1972). The same principle holds for *Babesia microti* which causes a malaria-like disease in man at very low parasite densities, yet no illness in mice at less than 60% parasitaemia. Hence sensitivity to low densities of malaria or babesia parasites is a host characteristic, not a reflection of some innate parasite pathogenicity. In addition, *P. falciparum* can sometimes cause the red cells it inhabits to adhere to certain endothelial cells in man but not, evidently, in *Aotus* monkeys (Miller, 1969). Any model for the pathogenesis of malarial illness and pathology should take these observations into account.

In this context, our research group first observed that the susceptibility of various host species to bacterial endotoxin correlated with the parasite density at which onset of illness occurred in malaria or babesiosis (Clark, 1982). This strengthened our proposal (Clark, 1978; Clark *et al.*, 1981) that tumour necrosis factor (TNF) and other monokines might mediate illness in malarial illness, since it was emerging that the cells that produce these soluble factors were probably important in endotoxin-induced illness (Michalek *et al.*, 1980). TNF was first described by Carswell *et al.* (1975) as an undefined functional entity with anti-tumour activity. Like many other such mediators it was given little credence in many quarters until it was available in recombinant form a decade later. This development has enabled rapid advances, and the importance of TNF in many fields, ranging from immunoregulation to inflammation and sepsis, is now acknowledged (reviewed by Beutler and Cerami, 1987; Nathan, 1987).

Several groups have detected TNF in serum from patients ill with malaria (Scuderi *et al.*, 1986; Teppo and Maury, 1987; van der Meer *et al.*, 1988), and it is also present in at least one of the rodent models (Clark and Chaudhri, 1988a). Furthermore, there is now evidence that malaria parasites can trigger *in vitro* release of TNF (Bate *et al.*, 1988). Seen alongside the capacity of exogenous TNF to reproduce the foetal death (Clark and Chaudhri, 1988b), lung (Clark *et al.*, 1987a) and bone marrow (Clark and Chaudhri, 1988a) pathology that occur in experimental and human malaria, this greatly strengthens the argument that the onset of malarial pathology involves TNF. An important principle behind these series of experiments is that mice with sub-clinical malarial infections are much more sensitive to TNF toxicity than are normal mice (Clark *et al.*, 1987a). Recent reviews (Clark, 1987a, 1987b) give more detailed descriptions of the congruency of TNF-induced illness and pathology and that seen in malaria. These include biochemical changes that have led this monokine to be also termed 'cachectin' (reviewed by Beutler and Cerami, 1987).

Once other lymphokines and monokines were available in recombinant form, it was soon realized that none of them could be understood in isolation, since overlapping activity, synergy, and enhanced release of one in the presence of another were commonplace (reviewed by Le and Vilcek, 1987; Old, 1987, Chapters 3 and 6). For instance, release of TNF and interleukin-1 is greater when interferon-gamma (IFN-γ; a T-cell product which also upregulates TNF receptors) is present. TNF, interleukin-1 and lymphotoxin overlap in many of their functions, and TNF induces release of the colony-stimulating factor (GM-CSF) that induces development of both granulocyte and macrophage precursors in bone marrow (Chapter 2). In addition, TNF enhances the capacity of both neutrophils (Klebanoff *et al.*, 1986) and macrophages (Hoffman and Weinberg, 1987) to release oxygen-derived free radicals, molecules that can damage the structural molecules, particularly lipids and proteins, of host cells. These processes were reviewed, in a parasite context, several years ago (Clark *et al.*, 1986). In brief, the essential property of free radicals (atoms or molecules with one or more unpaired electrons in their external orbit) that makes them destructive to parasites is their capacity to remove electrons from adjacent molecules. Typically this reaction reduces the stability of these molecules and alters their structure and function. The structure of certain molecules, such as vitamin E, enables them to undergo this encounter without loss of stability. Such molecules are termed 'free radical scavengers' because they can absorb radicals without harm, protecting adjacent molecules in the process.

The single-step reduction of oxygen to water involves the formation of, in sequence, hydrogen peroxide, superoxide (O_2^-) and hydroxyl radical ($\cdot$OH). Aerobic organisms have evolved biochemical pathways, involving the antioxidant enzymes superoxide dismutase and catalase, and also the glutathione system (which facilitates a 2-electron reduction of hydrogen peroxide to water, avoiding $\cdot$OH formation) to minimize the generation of these harmful intermediate forms.

Normal erythrocytes are amply stocked with free radical scavengers, antioxidant enzymes and reduced glutathione. In combination, these are usually sufficient to protect malarial parasites from background levels of oxygen-derived free radicals. If the protective mechanisms are imperfect, such as in glucose-6-phosphate dehydrogenase deficiency (this enzyme being necessary to re-charge the glutathione system), or the local concentration of free radicals is increased (for example, a nearby TNF-primed leukocyte secreting O_2^-), the parasite suffers accordingly.

Radicals are very short-lived, so can act only over an extremely short range. In practice, therefore, most of the damage radicals inflict on parasites is likely to be mediated by long-lived products of radical-induced lipid peroxidation. These have proved toxic, in low concentrations, to *P. falciparum* (Clark *et al.*, 1987c). This proposal is consistent with the recent finding that serum lipid peroxides are increased when TNF is generated *in vivo*, and that more than 70% of the anti-malarial activity of such serum is in the lipoprotein fraction (Rockett *et al.*, 1988).

Although *P. falciparum* causes a more severe disease than does *P. vivax*, the feature that distinguishes it most clearly is its capacity to cause cerebral malaria, a condition in which parasitized red cells preferentially adhere to cerebral vascular endothelium (MacPherson *et al.*, 1985). The mechanism of this cytoadherence is still obscure, but it has been argued that it depends on thrombospondin (Roberts *et al.*, 1985; Rock *et al.*, 1988), a glycoprotein whose sources include endothelium cells. Thrombospondin can be induced by platelet-derived growth factor (PDGF; Majack *et al.*, 1985), and TNF can instruct cultured endothelium to produce PDGF (Hajjar *et al.*, 1987). Thus it seems reasonable that, as reported recently, TNF would induce endothelial cells to generate messenger RNA for thrombospondin (Marks *et al.*, 1988). It seems clear, therefore, that unravelling cerebral malaria could depend, like much of the rest of the pathology of malaria, on a closer understanding of TNF and its functional relatives.

The first published experimental attempt to use this approach reported that the administration of polyclonal antibodies to recombinant mouse TNF completely suppressed the development of cerebral pathology in mice infected with *P. berghei*. This result is consistent with TNF having a central role in the pathogenesis of malaria (Grau *et al.*, 1987).

The destruction of malaria parasites by the host

This topic has generated a large literature which can be only briefly summarized in the space available. Historically, the phagocytic activity of macrophages was thought responsible. References to antibody began to emerge, first as opsonins aiding phagocytosis and then acting alone, largely against merozoites (Cohen *et al.*, 1969). In a recent review of this area, Anders (1986) summarized current attempts to utilize these ideas to develop a vaccine against malaria. Macrophages again were invoked, but this time as a source of soluble mediators, when it was found that pretreating mice with macrophage activators led to intra-erythrocytic death of rodent malaria and babesia parasites (reviewed in Clark, 1987a; see Figure 7.2). As this area developed in the hands of several groups, both oxygen-derived free radicals and TNF were shown to cause *in vivo* intra-erythrocytic death in various experimental models (reviewed by Clark *et al.*, 1987b; Clark, 1987a). These concepts,

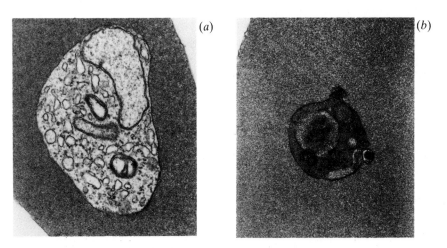

(a)   *(b)*

Figure 7.2. a. Healthy intra-erythrocytic *Babesia microti* trophozoite. b. Degenerate intra-erythrocytic *B. microti* 24 h after injection into BCG-pretreated mouse (X3500; Clark, I. (1976), U. of London PhD Thesis).

which have become embodied in the term 'cell-mediated immunity' (CMI), carry with them the obvious implication that if the mediators are not precisely focused, or their production tightly controlled, the host may also be harmed. One possibility often canvassed is that CMI may limit the infection in its early stages, when the host is sick, and be subsequently focused by the action of antibody that brings the macrophage and infected cell in close contact (Figure 7.3). These processes would, it is reasoned, allow the host time in which to develop an immunity based entirely on antibody. This is no sterile academic argument; scientists trying to improve the outcome of vaccination experiments using defined *P. falciparum* antigens, such as those being undertaken on *Aotus* monkeys and human subjects, are operating in the dark unless they know which component of the immune response needs to be enhanced to generate host-protective immunity.

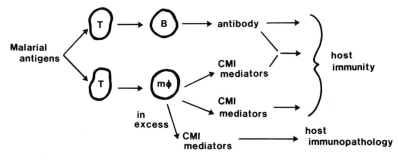

Figure 7.3. The interrelated roles of antibody and CMI in malaria (reproduced with permission, from *Parasitology Today*).

The same picture holds true for anti-sporozoite immunity, which was originally thought to operate through antibody in the period before the arrival of sporozoites in the liver (Nussenzweig and Nussenzweig, 1986). A prospective field study has, however, cast serious doubt on the protective value of naturally acquired antisporozoite antibodies (Hoffman *et al.*, 1986). Moreover, it has recently been shown that inserting the gene for the relevant sporozoite antigen into *Salmonella typhimurium* and using this to infect mice orally, generates protection without detectable antibody (Sadoff *et al.*, 1988). This is consistent with the studies of Ferreira *et al.* (1986), who reported that interferon-gamma (IFN-γ) could kill sporozoites in liver cells without the presence of antibody.

Babesia sp.

Life cycle and pathology

The other two parasites discussed in this section are termed piroplasms, so-named because of the pear shape of their intra-erythrocytic stages. The first to be considered, *Babesia* sp., has a host range that spans rodents, dogs, horses and cattle; sometimes human infections occur. The life cycle of *Babesia* sp. is recognizably similar to that of malaria parasites. The disease is spread by ticks rather than mosquitoes, and several tick genera may carry each of the species of *Babesia*. Like the mosquito, ticks are host to a developmental phase that involves sexual fusion of ingested gametes and also an asexual reproduction stage that produces sporozoites. Unlike malaria-infected mosquitoes or ticks infected with other piroplasms, *Babesia*-infected ticks may acquire ovarian infection and thus pass the parasite on to their offspring. As in malaria, sporozoites, along with saliva, are injected into the new host but in this case enter erythrocytes directly without a prior liver stage. A further asexual stage matures in circulating erythrocytes. As in malaria, it is multiplication and release of this stage (schizogony) that heralds the onset of illness. A detailed account of the life cycle is given by Mehlhorn and Schein (1984). A recent review by Young and Morzaria (1986) also covers the general biology of the parasite and is more oriented to the disease and its control.

As originally documented by Maegraith *et al.* (1957), the disease caused by *Babesia* sp. (babesiosis) can closely resemble malaria. This is most marked when *B. bovis* infections in cattle are compared to severe human falciparum malaria; in each case fever, severe anaemia, pulmonary oedema, cerebral symptoms and disseminated intravascular coagulation may be present when peripheral parasitaemias are low (reviewed by Clark *et al.*, 1986). Even more compelling are the striking clinical similarities between malaria and human babesiosis, a disease that is not common but is well documented, particularly along the north eastern seaboard of the United States (reviewed by Ruebush, 1980). The similarities suggest a common pathogenesis. Furthermore, the host macrophage response that leads to intra-erythrocytic death of parasites is, from rodent studies, common to both malaria and babesiosis (Cox, 1978; Clark, 1978). In short, the reasoning that led to the development of the monokine approach to malaria pathology (reviewed by Clark, 1987a, 1987b) suggests that monokines and lymphokines might also cause the illness and pathology seen in babesiosis. However, this has not yet been conclusively tested.

The destruction of Babesia *sp. by the host*

There is little precise knowledge of how cattle acquire immunity to babesiosis. Antibody is often assumed to be important but on no firm evidence; transferred serum from recovered calves will protect against *B. argentina* (= *bovis*) but only if repeated daily (Mahoney, 1967). A single large dose was of no value, so the active component of the serum may not have been antibody. Evidence from rodent models is consistent with this interpretation since serum transfer will not protect mice against *B. rodhaini* (Mitchell *et al.*, 1978). Attempts to develop *in vitro* inhibition assays using purified antibody, such as those in use for *P. falciparum*, have not been successful (I.G. Wright, pers. comm.).

Macrophage activators will, in the absence of antibody, very readily protect mice against *B. microti* and *B. rodhaini* causing intra-erythrocytic death of the parasites as occurs in undisturbed infections (reviewed by Clark *et al.*, 1986). This implies that some soluble factor of macrophage origin is important in this process. No one has as yet developed this approach in cattle.

Theileria sp.

Life cycle and immunity

Various members of the genus *Theileria*, another piroplasm, cause illness in cattle, sheep and goats. We will restrict our comments to the diseases caused in cattle by *Theileria parva* (east coast fever) and *Theileria annulata* (tropical theileriosis). These are serious illnesses, characterized by lymphoproliferation and acute febrile illness, which is often fatal. *Theileria parva*, which is the more pathogenic, is restricted to central, eastern and southern Africa, whereas *T. annulata* occurs in north Africa, Asia Minor and tropical and sub-tropical Asia. As with babesiosis, theileriosis is spread by ticks, but transovarian transmission has not been recorded in these species.

Details of the life cycle and vectors involved are reviewed by Mehlhorn and Schein (1984). Sporozoites are injected when the tick feeds and enter the cytoplasm of circulating leukocytes. There is a suggestion, quoted in Dyer and Tait (1987), that *T. parva* preferentially invades T lymphocytes and *T. annulata* prefers B lymphocytes. In some way not yet understood, the parasite then orchestrates lymphocyte proliferation so that there is a clonal expansion of infected cells with parasites and leukocytes dividing synchronously (reviewed in Dyer and Tait, 1987). Fever, depression, weight loss and pulmonary oedema with dyspnoea accompany this stage of parasite division. The merozoites produced at these divisions can enter erythrocytes as well as further lymphocytes and here give rise to the characteristic small piroplasm forms. Further schizogony occurs in red cells.

The pathogenesis of illness and pathology in theileriosis, beyond the effects on lymphoid cells, has not been studied. The close congruency of the changes seen in this disease and malaria and babesiosis leads us to suggest that the possible involvement of lymphokines and monokines should be investigated.

Knowledge of the mechanism of immunity theileriosis is hampered by the lack of a laboratory model. Nevertheless, *in vitro* studies using contemporary cell culture techniques and biotechnology have provided answers to some of these questions and to what is perhaps the most intriguing aspect of any haemoprotozoan disease, the lymphoproliferation of theileriosis.

As reviewed by Irvin (1985), cattle produce *T. parva* antisporozoite antibodies that neutralize the *in vitro* infectivity of sporozoites to lymphocytes. Subsequent proliferation is thus prevented. The relevant antigen is common to various strains of *T. parva*, including those that do not cross-protect *in vivo*. Antisera directed towards *T. annulata* also suppress lymphocyte transformation (the step that precedes proliferation) and in addition inhibit *in vitro* invasion of lymphocytes (Preston and Brown, 1985). Furthermore, antibodies to *T. parva* macroschizonts, the forms found in infected lymphoblasts, are also elicited by infection but do not seem to be protective *in vivo* (reviewed by Irvin, 1985). Instead, exposure to macroschizonts generates cytotoxic T cells that will kill parasitized lymphocytes provided they share the same Class I major histocompatibility complex (MHC) antigens as the immune cell donor (Emery and Morrison, 1980; Emery *et al.*, 1981). Thus, in the absence of evidence implicating other systems, it is currently believed that the major effector mechanism controlling lymphoproliferative theileriosis operates through genetically-restricted cytotoxic T lymphocytes. There is as yet no evidence for protective mechanisms involving merozoites or piroplasms.

Lymphoproliferation in theileriosis

In some way not yet fully understood, *Theileria* parasites can take over control of multiplication of the host lymphocyte that they invade. Current theories have been summarized by Dyer and Tait (1987). Treating the infection stops blastogenesis, so it seems unlikely that the parasite may insert DNA into or near a host gene responsible for cell transformation since this would be irreversible once underway. These authors suggest that the most likely mechanism involves the *yes* oncogene (a retroviral transforming gene encoding a tyrosine kinase) since *T. annulata* DNA contains sequences homologous to its structure. Dyer and Tait (1987) report the presence of at least four kinase substrates (two phosphorylated at tyrosine) specific to *T. annulata*-infected lymphocytes. These, they argue, could alter protein kinase activity in host cells and thus stimulate them to divide.

7.3. MACROPHAGE PARASITES

Several protozoans have the ability to parasitize macrophages—an important cell concerned with both afferent (antigen presentation Chapter 4) and efferent (effector mechanisms) elements of the immune response. These parasites are *Toxoplasma gondii*, *Leishmania* spp. and *Trypanosoma cruzi* whose lifestyles raise interesting questions about survival in a particularly hostile environment. Only brief details can be considered here; further information is available from major reviews (Sethi, 1982; Thorne and Blackwell, 1983; Mauel, 1984). The relationship of each of these parasites with the macrophage is shown in Figure 7.4.

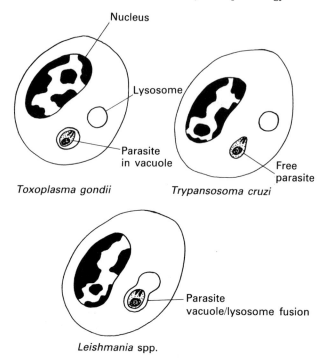

Figure 7.4. Relationships between protozoan parasites and the macrophage lysosome.

Toxoplasma gondii

Life cycle and immunity

The biology of immunology of *T. gondii* has been reviewed in detail by Krahenbuhl and Remington (1982) and Sethi and Piekarski (1987). It is a very widespread parasite occurring in three distinct forms during its life cycle—the tachyzoite and tissue cyst in the intermediate host and the oocyst in the cat. As a tachyzoite, it can invade almost any mammalian cell type; entry into macrophages is achieved by phagocytosis, which is presumably initiated by some form of binding between receptors on the macrophage and ligands in the parasite plasma membrane or glycocalyx. It is known that neither Ig nor complement receptors are involved (cf. *Leishmania*) but a 'penetration enhancing factor' is apparently produced by the parasite. Other cells are infected by active penetration, but precise details are not available. Intracellular multiplication occurs with generation time depending on the virulence of the particular strain since virulent strains multiply more rapidly than avirulent strains. A host response (involving antibody, T cells and macrophages) develops that is partially effective in regulating parasite numbers, killing all parasites other than those in intracellular sites in tissue (almost exclusively tissue cysts). Resistance to reinfection is established, and the relationship between host and parasite is one of concomitant immunity, since tissue cysts (and perhaps tachyzoites) derived from the primary infection persist. Interestingly, resistance is non-specific as infected hosts are protected against infection with some phylogenetically unrelated organisms.

A principal effector mechanism in resistance to *T. gondii* appears to be the activated macrophage (perhaps induced by IFN-γ), and both oxidative and non-oxidative properties of the cell have been implicated in killing the parasite. *In vitro* studies have shown that antibodies can opsonize parasites and earmark them for destruction by normal macrophages. In these circumstances, phagosomes containing antibody-coated parasites fuse with lysosomes and the parasites are digested.

Pathology

Toxoplasma gondii rarely causes severe clinical disease in humans and only 10–20% of infected adults exhibit symptoms—usually prolonged lymphadenopathy, sometimes confused with infectious mononucleosis (Howell and Rowell, 1976). Once symptoms have abated, there are no apparent inflammatory reactions elicited by surviving tissue cysts. Immunosuppression of chronically infected humans (with persistent tissue cysts) induced either by chemotherapy or acquired immunodeficiency disease syndrome (AIDS) can lead to widespread dissemination of the parasite in the body, culminating in death often as a result of invasion of the central nervous system. Infection of the human foetus from an acutely-infected mother can lead to a spectrum of serious problems for the neonate (e.g. hydrocephaly, chorioretinitis) but apparently not abortion. There is no agreement on whether parasite multiplication is directly responsible for the ocular lesions or whether these occur as a result of immune responses directed at *T. gondii* antigens released from cysts. The importance of the parasite as a human pathogen is also debated (Hughes, 1985; Apt, 1985).

Studies of *T. gondii* in experimental animals indicate that parasite strains vary in their degree of virulence and that chronically-infected female mice can transmit infection to their offspring. Moreover, a toxic factor has been identified in peritoneal cell exudate in infected mice, and this may be related to 'toxofactor', a substance isolated from *in vitro* cultures of *T. gondii* that has teratogenic effects in mice.

Unexplained aspects of the biology of T. gondii

Some important questions raised by the biology of *T. gondii* include the following:

i) How does the parasite survive in macrophages in the early stages of infection? First, normal macrophages produce only low levels of peroxide and free radicals, and *T. gondii* inhibits the oxidative burst usually accompanying phagocytosis. The parasite also contains high levels of catalase that contribute to the destruction of hydrogen peroxide and prevent its interaction with superoxide. Second, the vacuole enclosing the phagocytosed parasite (the phagosome) does not fuse with lysosomes, but the reasons for this are not established. In this respect, the phagosome behaves rather like other membranous cell organelles, such as rough endoplasmic reticulum and mitochondria. As noted above, if the parasite's surface is coated with antibody the phagosome fuses with the lysosome and the parasite is killed. Thus, it would seem that a naked parasite plasma membrane and glycocalyx is crucial for survival.

ii) How does the parasite survive in the immune host? It has been suggested that survival in cells (other than in activated macrophages that kill the parasite) and especially the persistence of tissue cysts in the brain reflect the immunologically-privileged nature of these habitats. Since the outer membrane of the tissue cyst is of host origin, parasite antigens are presumably weakly expressed, if at all, on its surface; otherwise some sort of reaction would be expected. Clearly, the exacerbation of disease in immunosuppressed hosts bears testimony to the effectiveness of immunological control mechanisms in limiting parasite multiplication in competent hosts.

Trypanosoma cruzi

The course of infection

Trypanosoma cruzi gives rise to a chronic infection in humans (Chagas' disease). It can also infect many other hosts (including mammals and some reptiles—the latter particularly so after immunosuppressive treatment) but not birds. In humans, three phases of infection can be clinically recognized: an acute phase occurring almost exclusively in children under the age of 15 in which the parasite multiplies in macrophages and local tissue cells as an amastigote. Symptoms may be lacking in about two-thirds of individuals. The remainder show either skin lesions (chagoma) or oedema, enlarged lymph nodes and conjunctivitis (Romana's signs), these changes being legacies of where the triatomid vector took its blood meal and infected the host. Individuals who recover from acute *T. cruzi* infection carry the parasite for life, at low levels, in the blood and tissues. In some, there may be a period of asymptomatic infection (latent phase) but this, if present, is usually replaced by an inevitable progression of infection over periods as long as twenty years. The long-lasting chronic phase frequently presents with extensive cardiac and digestive tract pathology, and the prognosis is poor. Clearly, the parasite seems to be held in check by immunological mechanisms that also prevent reinfection (i.e. a state of concomitant immunity exists), but the degenerative consequences of infection eventually assert themselves.

Immunity

The immunology of *T. cruzi* infection is comprehensively reviewed by Scott and Snary (1982) and Teixeira (1987). Factors that appear to be important in regulating infection are activated macrophages, antibodies and complement, antibody-dependent cell-mediated cytotoxicity (ADCC) and cytotoxic T cells. These mechanisms are influenced by the genetic background of the host, and a broad spectrum of susceptibilities is seen both between and within species.

   T. cruzi faces similar problems to other macrophage-inhabiting protozoans in avoiding destruction by these aggressive cells. In contrast to *Toxoplasma gondii* and *Leishmania* spp., phagocytosed *T. cruzi* trypomastigotes transform to amastigotes and escape from the phagosome into the surrounding cytoplasm where they apparently are out of the reach of lysosomes and can divide freely. Neither Fc nor C3b receptors on macrophages are involved in the phagocytic uptake of *T. cruzi*, but apparently pronase-sensitive molecules on the macrophage, together with parasite surface glycoproteins, are involved. When macrophages become activated (probably as a result of induction by IFN-γ, whose concentration in serum increases following infection), phagosomes fuse with lysosomes and the parasites are killed by reactive oxygen species and then digested. There is also evidence that in mice, α/β interferon-induced Natural Killer (NK) cells play a role in resistance to *T. cruzi*, but few details are available.

   The bloodstream trypomastigotes are at risk from attack by antibodies that can prepare them for complement lysis, ADCC and enhanced destruction by activated macrophages. There are several suggestions as to how these immune effector mechanisms are either avoided or their potential to kill the parasite is diminished. These include: i) Non-specific immunosuppression by parasite-derived substances; ii) the protective nature of the parasite's surface (surface molecules with anti-phagocytic and anti-complement, especially anti-C3, activities); iii) the ability of parasite enzymes to cleave Ig at the cell surface leaving only Fab (antigen-binding) bearing fragments attached, separated from the Fc (complement activating) fragments; this diminishes the prospect of both antibody-complement-mediated lysis as well as ADCC; iv) capping and dislodgement of antibody-antigen complexes that form on the cell surface; and v) antigen sharing by the parasite and host.

   Like *Toxoplasma gondii*, *T. cruzi* can invade many types of cells other than macrophages, in particular cardiac and smooth muscle cells; the mode of penetration is not known but a parasite lectin-like protein and an inducible parasite protease may be involved. In one study a high proportion (>90%) of patients with chronic disease had parasites only inside smooth muscle cells of venous vessels in the adrenal gland. Thus, it was proposed that high local concentrations of adrenal corticosteroids impair local immune mechanisms, conferring immunological privilege on that site.

Pathology

The pathology of Chagas' disease was once thought to be due entirely to the destruction of host cells by parasites and the release of toxic substances that caused localized inflammation. However, this view now seems to be incorrect. Indeed, the pronounced inflammatory lesions in the heart, skeletal muscle and digestive tract which are characteristic of chronic Chagas' disease usually arise in the absence of parasites. These observations have led to much debate (Hudson, 1985; Kierszenbaum, 1985) as to whether autoimmunity develops during the course of *T. cruzi* infection. Argument has centred mainly on whether inflammatory lesions are elicited by host tissue components, perhaps modified and rendered immunogenic by parasite products, or whether the host and parasite share epitopes and tissue damage is due to the production of autoantibodies and the stimulation of self-reactive lymphocytes. Another possibility, which does not appear to have been canvassed, is that TNF is involved in the pathology of the disease, as in malaria.

Leishmania spp.

Immunity

Immunity to the many species of *Leishmania* has been intensively studied, and for a detailed consideration of the topic the reader should refer to recent reviews (Mauel and Behin, 1982; Wakelin, 1984; Chang and Bray, 1985; Liew, 1986; Preston, 1987).

Protective immunity in cutaneous leishmaniasis is mediated by T cells that produce IFN-γ and MAF; macrophages become activated and intracellular parasites are killed. Other mechanisms (e.g. antibody, complement, phago-cytosis by eosinophils) may be involved, but the essential role of the activated macrophage seems undisputed. Studies on *L. major* (Mitchell and Handman, 1985) indicate that T cells in infected mice can have resistance-promoting or disease-promoting effects. The relative balance of each type of T cell is determined by host genotype (via the MHC-complex) and thus the pattern of disease in different strains of mice can be remarkably different (Chapter 6). It is argued that a lipid-containing glycoconjugated antigen of the parasite plays a central role in the disease. When this antigen is anchored in the macrophage and oriented in relation to certain Class II MHC molecules, it is able to induce a set of Lyt^{1+2-} T cells, T_{MA}, which in turn activate macrophages. Reactive oxygen intermediates are then produced which have deleterious consequences for the parasite. The number of T_{MA} cells is apparently regulated by the number of suppressor T cells (Lyt^{1+2}) induced by the delipidized glycoconjugated antigen attached to a receptor on the macrophage surface—the more suppressor cells, the less effective resistance (i.e. fewer activated macrophages) and the host dies from disseminated disease (BALB/c mouse); with fewer suppressor cells, the disease resolves (CBA/H mouse). The genetics of both host and parasite have a bearing on the outcome of infection and non-genetic factors such as antigenic load, impaired lymphatic drainage and route of infection also play a role (Chapter 6).

The tissue damage seen in cutaneous leishmaniasis in humans may result from the multiplication of parasites and destruction of surrounding cells, but it may also, as with *T. cruzi*, proceed in the almost total absence of parasites. This is also a parallel with malaria, so the possible involvement of TNF and other cytokines should be considered in future research. Question marks hang over whether hypersensitivity reactions are responsible for both the resolution of lesions in some cases and the persistence of lesions in others. In the latter, autoimmune phenomena may be responsible and again parallels with *T. cruzi* are evident.

Visceral leishmaniasis is less well understood than the cutaneous disease. Clearly there are differences between populations of parasites in the degree to which they induce immunosuppression in humans; equally, there is a broad spectrum of immune responsiveness to the parasite among humans and other hosts (Chapter 6). There is some evidence that human neutrophils are more effective than monocytes in digesting amastigotes, but little is known of the immune effector mechanisms responsible in cases where disease has resolved. Hamsters provide the most appropriate laboratory model of *L. donovani* infection in humans because they exhibit a number of similar pathological changes. These include hepatosplenomegaly, depressed T-cell responses but elevated gammaglobulin production, and immune complex glomerulonephritis (Sartori *et al.*, 1987). In general, it seems that, as with cutaneous species of *Leishmania*, activated macrophages are responsible for the expression of protective immunity. Some elegant genetic studies with mice have demonstrated the importance of certain host genes (both MHC and non-MHC linked) in both innate resistance and the long-term outcome of infection (Chapter 6).

Survival in macrophages

Entry of *Leishmania* spp. into macrophages is achieved by phagocytosis, and the mechanisms involved are fairly well understood. Phagocytosis of promastigotes (the infective forms inoculated by the vector) involves interaction between complement (especially the C3 component) and a surface glycoprotein of the parasite (GP63), which in turn interacts with inactivated C3b and mannose-fucose receptors on the macrophage surface (Blackwell *et al.*, 1986; Figure 7.5). GP63 is rich in mannose, has protease activity and has been designated parasite surface protease by Bordier (1987; Chapter 2.2). The role of C3 is clearly to facilitate close contact between the promastigote and the macrophage. In the case of amastigotes (the replicating form in the mammalian host) the mechanism of entry into the macrophage is not so clear, and a role for the surface protease has not been established.

As with *T. gondii* and *T. cruzi*, a number of adaptive strategies are deployed

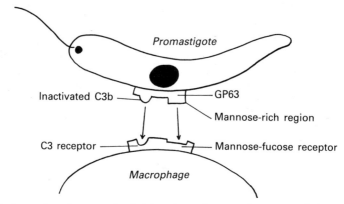

Figure 7.5. Interactions between the *Leishmania* surface protein GP63 and the macrophage receptor as a prelude to phagocytosis (adapted from Blackwell *et al.*, 1986).

by the parasite to secure its continued survival and replication within macrophages before they become activated (Chang, 1983). Amastigotes do not trigger a respiratory burst in macrophages when phagocytosis is initiated. It has been suggested that this may be due to either the effects of parasite surface acid phosphatase/lipase activity on the macrophage membrane or through more generalized interference with macrophage physiology (Blackwell *et al.*, 1986). The parasites also contain superoxide dismutase and trypanothione (a glutathione-spermidine conjugate) which protect against oxidant damage. The cutaneous-infecting species may achieve respite from the leishmanicidal activities of macrophages because of the impaired activity of these cells at temperatures below 37°C in the skin (Scott, 1985).

Unlike *T. gondii* and *T. cruzi*, phagocytosed *Leishmania* fuse with lysosomal vacuoles to form phagolysosomes, so the parasites would seem to be at special risk of destruction by digestive enzymes of the lysosomes. Suggested possibilities on how this is avoided include (i) the parasite possesses a proton pump that lowers the pH of the lysosomal vacuole and leads to sub-optimal functioning (or inhibition) of lysosomal enzymes; (ii) megasomes—unusual cytoplasmic organelles present in some species—in some way alter the lysosomal environment in the parasite's favour perhaps by changing the pH or inactivating lysosomal enzymes (Figure 7.6); (iii) the parasite produces and secretes factors—so-called 'excreted factors' (including an acid phosphatase) with effects as described for megasomes; (iv) the surface of the parasite may bear structural resemblances to lysosome membrane proteins.

7.4.   CONCLUSIONS

As can be seen from this short summary, the diseases that result from parasitism by protozoa of erythrocytes and macrophages are not yet well

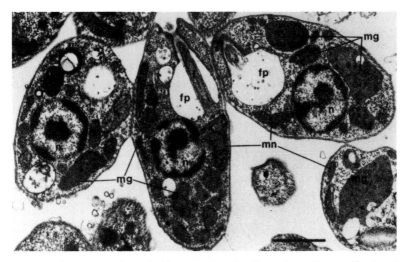

Figure 7.6. *Leishmania* amastigotes inside macrophage showing various cell organelles. Longitudinal sections of *Leishmania m. mexicana* amastigotes showing lysosome-like organelles known as megasomes (mg) which contain high levels of cysteine proteinase activity. k = kinetoplast, n = nucleus, f = flagellum, fp = flagellar pocket, mn = mitochondrial network. Bar = 1 μm (reproduced, with permission, from *Parasitology Today*).

understood and leave more questions posed than have been resolved. The antigens that prime the immune response to these parasites are distinctive to each organism, and the disease states their cause can be superficially dissimilar. Nevertheless, as experimental evidence accumulates, it is becoming clear that similar host responses are involved throughout the range of parasites. These responses, leading to both host immunity and host pathology, are being increasingly recognized to involve cytokines, a broad category of non-antibody proteins that includes interferons, interleukins, colony-stimulating factors and tumour-necrosis factor. The mediators are secreted by leukocytes and possess many overlapping and synergistic functions. New functions for those already described are regularly appearing in the immunological literature. As protozoan parasites of erythrocytes and macrophages are examined in the light of this new information, much that is now obscure should become clear.

REFERENCES

Anders, R.F. (1986), Multiple cross-reactivities amongst antigens of *Plasmodium falciparum* impair the development of protective immunity against malaria, *Parasite Immunology*, **8**, 529–39.

Apt, W. (1985), Toxoplasmosis in developing countries, *Parasitology Today*, **1**, 44–6.

Bate, C.A.W., Taverne, J. and Playfair, J.H.L. (1988), Malarial parasites induce TNF production by macrophages, *Immunology*, in press.

Beutler, B. and Cerami, A. (1987), Cachectin: more than a tumor necrosis factor, *Science*, **316**, 379–85.

Blackwell, J., McMahon-Pratt, D. and Shaw, J. (1986), Molecular biology of *Leishmania, Parasitology Today*, **2**, 45–53.

Bordier, C. (1987), The promastigote surface protease of *Leishmania, Parasitology Today*, **3**, 151–3.

Carswell, E.A., Old, L.J., Kassel, R.L., Green, S., Fiore, N. and Williamson, B. (1975), An endotoxin-induced serum factor that causes necrosis of tumors, *Proceedings of the National Academy of Sciences (USA)*, **72**, 3666–70.

Chang, K.-P. (1983), Cellular and molecular mechanisms of intracellular symbiosis in Leishmaniasis, *International Reviews of Cytology (supplement)*, **14**, 267–305.

Chang, K.P. and Bray, R.S. (Eds.) (1985), *Leishmaniasis*, **Volume 1** Amsterdam: Elsevier Biomedical Press, 503 pp.

Clark, I.A. (1978), Does endotoxin cause both the disease and parasite death in acute malaria and babesiosis? *Lancet*, **2**, 75–7.

Clark, I.A. (1982), Correlation between susceptibility to malaria and *Babesia* parasites and to endotoxicity, *Transactions of the Royal Society of Tropical Medicine and Hygiene*, **76**, 4–7.

Clark, I.A. (1987a), Cell-mediated immunity in protection and pathology of malaria, *Parasitology Today*, **3**, 300–5.

Clark, I.A. (1987b), Monokines and lymphokines in malarial pathology, *Annals of Tropical Medicine and Parasitology*, **81**, 577–85.

Clark, I.A. and Chaudhri, G. (1988a), Tumour necrosis factor may contribute to the anaemia of malaria by causing dyserythropoiesis and erythrophagocytosis, *British Journal of Haemotology*, in press.

Clark, I.A. and Chaudhri, G. (1988b), Tumour necrosis factor and abortion in malaria, *American Journal of Tropical Medicine and Hygiene*, in press.

Clark, I.A., Hunt, N.H. and Cowden, W.B. (1986), Oxygen-derived free radicals in the pathogenesis of parasitic disease, *Advances in Parasitology*, **26**, 1–44.

Clark, I.A., Hunt, N.H. and Cowden, W.B. (1987b), Immunopathology of malaria, in Soulsby, E.J.L. (Ed.), *Immune Responses in Parasitic Infections*, **Vol. 4**, pp. 1–34, Boca Raton: CRC Press.

Clark, I.A., Virelizier, J.-L., Carswell, E.A. and Wood, P.R. (1981), The possible importance of macrophage-derived mediators in acute malaria, *Infection and Immunity*, **32**, 1058–66.

Clark, I.A., Cowden, W.B., Butcher, G.A. and Hunt, N.H. (1987a), Possible roles of tumour necrosis factor in the pathology of malaria, *American Journal of Pathology*, **129**, 192–9.

Clark, I.A., Butcher, G.A., Buffinton, G.D., Hunt, N.H. and Cowden, W.B. (1987c), Toxicity of certain products of lipid peroxidation to the human malarial parasite *Plasmodium falciparum, Parasitology*, **94**, 553–76.

Cohen, S., McGregor, I.A. and Carrington, S. (1969), Gamma-globulin and acquired immunity to human malaria, *Nature*, **192**, 733–7.

Cox, F.E.G. (1978), Heterologous immunity between piroplasms and malaria parasites: the simultaneous elimination of *Plasmodium vinckei* and *Babesia microti* from the blood of doubly infected mice, *Parasitology*, **77**, 55–60.

Dyer, M. and Tait, A. (1987), Control of lymphoproliferation by *Theileria annulata, Parasitology Today*, **3**, 3309–11.

Emery, D.L. and Morrison, W.I. (1980), Generation of autologous mixed leukocyte reactions during the course of infection with *Theileria parva* (East Coast Fever) in cattle, *Immunology*, **40**, 229–38.

Emery, D.L., Eugui, E.M., Nelson, R.T. and Tenywa, T. (1981), Cell-mediated immune responses to *Theileria parva* (East Coast Fever) during immunization and lethal infections in cattle, *Immunology*, **43**, 323–36.

Ferreira, A., Schofield, L., Enea, V., Schellekens, H., van der Meide, P., Collins, W.E., Nussenzweig, R.S. and Nussenzweig, V. (1986), Inhibition of development of exoerythrocytic forms of malaria parasites by -interferon, *Science*, **232**, 881–4.

Fletcher, K.A. (1987), Biochemical approaches to research in malaria, *Annals of Tropical Medicine and Parasitology*, **81**, 587–98.

Grau, G.E., Fajardo, L.F., Piquet, P.-F., Allet, B.; Lambert, P.-H. and Vassali, P. (1987), Tumor necrosis factor (cachectin) as an essential mediator in murine cerebral malaria, *Science*, **237**, 1210–12.

Hajjar, K.A., Hajjar, D.P., Silverstein, R.L. and Nochman, R.L. (1987), Tumor necrosis factor-mediate release of platelet-derived growth factor from cultured endothelial cells, *Journal of Experimental Medicine*, **166**, 235–45.

Hoffman, S.L., Wistar, R., Ballou, W.R., Hollingdale, M.R., Wirtz, R.A., Schneider, I., Marwoto, H.A. and Hockmeyer, W.T. (1986), Immunity to malaria and naturally acquired antibodies to the circumsporozoite protein of *Plasmodium falciparum*, *New England Journal of Medicine*, **315**, 601.

Hoffman, M. and Weinberg, B.J. (1987), Tumor necrosis factor α induces increased hydrogen peroxide and Fc receptor expression, but not increased Ia antigen expression by peritoneal macrophages, *Journal of Leukocyte Biology*, **42**, 704–7.

Howell, M.J. and Rowell, A.M. (1976), Toxoplasmosis in Canberra, *Medical Journal of Australia*, **2**, 465.

Hudson, L. (1985), Autoimmune phenomena in chronic chagasic cardiopathy, *Parasitology Today*, **1**, 6–9.

Hughes, H.P.A. (1985), Toxoplasmosis—a neglected disease, *Parasitology Today*, **1**, 41–4.

Irvin, A.D. (1985), Immunity to Theileriosis, *Parasitology Today*, **1**, 124–8.

Jervis, H.R., Spritz, H., Johnson, A.J. and Wellde, B.T. (1972), Experimental infection with *Plasmodium falciparum* in *Aotus* monkeys. II. Observations on host pathology, *American Journal of Tropical Medicine and Hygiene*, **21**, 272–81.

Kierszenbaum, F. (1985), Is there autoimmunity in Chagas disease? *Parasitology Today*, **1**, 4–6.

Kitchen, S.F. (1949), Falciparum malaria, in Boyd, M.F. (Ed.), *Malariology*, pp. 995–1016, Philadelphia: Saunders.

Klebanoff, S.J., Vadas, M.A., Harlan, J.M., Sparks, L.H., Gamble, J.R., Agosti, J.M., Waltersdorph, A.M. (1986), Stimulation of neutrophils by tumor necrosis factor, *Journal of Immunology*, **136**, 4220–5.

Krahenbuhl, J.L. and Remington, J.S. (1982), The immunology of toxoplasma and toxoplasmosis, in Cohen, S. and Warren, K.S. (Eds.), *Immunology of Parasitic Infections*, pp. 356–421, Oxford: Blackwell Scientific Publications.

Le, J. and Vilcek, J. (1987), Tumor necrosis factor and interleukin-1: cytokines with multiple overlapping biological activities, *Laboratory Investigation*, **56**, 234–8.

Liew, F.Y. (1986), Cell mediated immunity in experimental cutaneous leishmaniasis, *Parasitology Today*, **2**, 264–70.

MacPherson, G.G., Warrell, M.J., White, N.J., Looareesuwan, S. and Warrell, D.A. (1985), Human cerebral malaria. A quantitative ultrastructural analysis of parasitized erythrocyte sequestration, *American Journal of Pathology*, **119**, 385–401.

Maegraith, B., Gilles, H.M. and Devakul, K. (1957), Pathological processes in *Babesia canis* infections, *Zeitschrift für Tropenmedizin und Parasitologie*, **4**, 485–514.

Mahoney, D.F. (1967), Bovine babesiosis: the passive immunization of calves against *Babesia argentina*, with special reference to the role of complement-fixing antibodies, *Experimental Parasitology*, **20**, 119–24.

Majack, R.A., Cook, S.C. and Bornstein, P. (1985), Platelet-derived growth factor and heparin-like glycoasaminoglycans regulate thrombospondin synthesis and deposition in the matrix by smooth muscle cells, *Journal of Cellular Biology*, **101**, 1059–70.

Marks, R.M., Ward, P.A., Kunkel, S.L. and Dixit, V.M. (1988), Tumor necrosis factor induces mRNA for thrombospondin in human endothelial cells, *FASEB Journal*, **2**, A1599.

Mauel, J. (1984), Mechanisms of survival of protozoan parasites in mononuclear phagocytes, *Parasitology*, **88**, 579–92.

Mauel, J. and Behin, R. (1982), Leishmaniasis, in Cohen, S. and Warren, K.S. (Eds.), *Immunology of Parasitic Infections*, pp. 299–355, Oxford: Blackwell Scientific Publications.

Mehlhorn, H. and Schein, E. (1984), The piroplasms: life cycle and sexual stages, *Advances in Parasitology*, **23**, 37–103.

Michalek, S.M., Moore, R.M., McGhee, J.R., Rosenstreich, D.L. and Merghen, S.E. (1980), The primary role of lymphoreticular cells in the mediation of host responses to bacterial endotoxin, *Journal of Infectious Disease*, **141**, 55–63.

Miller, L.H. (1969), Distribution of mature trophozoites and schizonts of *Plasmodium falciparum* in Aotus monkeys. II. Observations on host pathology, *American Journal of Tropical Medicine and Hygiene*, **21**, 272–81.

Mitchell, G.F. and Handman, E. (1985), T-lymphocytes recognise *Leishmania* glycoconjugates, *Parasitology Today*, **1**, 61–73.

Mitchell, G.F., Handman, E. and Howard, R.J. (1978), Protection of mice against *plasmodium* and *babesia* infections: Attempts to raise host-protective sera, *Australian Journal of Experimental Science*, **56**, 553–9.

Nathan, C.F. (1987), Secretory products of macrophages, *Journal of Clinical Investigation*, **79**, 319–26.

Nussenzweig, V. and Nussenzweig, R.S. (1986), Development of a sporozoite malaria vaccine, *American Journal of Tropical Medicine and Hygiene*, **35**, 678–88.

Old, L.J. (1987), Polypeptide mediator network, *Nature*, **326**, 330–1.

Phillips, R.E. and Warrell, D.A. (1986), The pathophysiology of severe falciparum malaria, *Parasitology Today*, **2**, 271–82.

Preston, P.M. (1987), The immunology, immunopathology, and immunoprophylaxis of *Leishmania* infections, in Soulsby, E.J.L. (Ed.), *Immune Responses in Parasitic Infections: Immunology, Immunopathology and Immuonoprophylaxis*, **Vol. 3**, pp. 119–81, Florida: CRC Press.

Preston, P.M. and Brown, C.G.D. (1985), Inhibition of lymphocyte invasion by sporozoites and the transformation of trophozoite infected lymphocytes *in vitro* by serum from *Theileria annulata* immune cattle, *Parasite Immunology*, **7**, 301–14.

Roberts, D.D., Sherwood, J.A., Spitalnik, S.L., Paton, L.J., Howard, R.J., Dixit, V.M., Frazier, W.A., Miller, L.H. and Ginsberg, V. (1985), Thrombospondin binds falciparum malaria parasitized erythrocytes and may mediate cytoadherence, *Nature*, **318**, 64–6.

Rock, E.P., Roth, E.F., Rohas-Corona, R.R., Sherwood, J.A., Nagel, R.L., Howard, R.J. and Kaul, D.K. (1988), Thrombospondin mediates the cytoadherence of *Plasmodium falciparum*-infected red cells to vascular endothelium in shear flow conditions, *Blood*, **71**, 71–5.

Rockett, K.A., Targett, G.A.T. and Playfair, J.H.L. (1988), Killing of blood-stage *Plasmodium falciparum* by lipid peroxides from tumor necrosis factor, *Infection and Immunity*, **56**, 3180–3.

Ruebush, T.K. (1980), Human babesiosis in North America, *Transactions of the Royal Society of Tropical Medicine and Hygiene*, **74**, 149–53.

Sadoff, J.G., Ballou, W.R., Baron, L.S., Marjarian, W.R., Brey, R.M., Hockmeyer, W.T., Young, J.F., Cryz, S.J., Ou, J., Lowell, G.H. and Chulay, J.D. (1988), Oral *Salmonella typhimurium* vaccine expressing circumsporozite protein protects against malaria, *Science*, **240**, 336–8.

Sartori, A., de Oliveira, A.V., Roque-Barreira, M.C., Rossi, M.A. and Campos-Neto, A. (1987), Immune complex glomerulonephritis in experimental Kala-azar, *Parasite Immunology*, **9**, 93–103.

Scott, P. (1985), Impaired macrophage leishmanicidal activity at cutaneous temperature, *Parasite Immunology*, **7**, 277–88.

Scott, M.T. and Snary, D. (1982), American trypanosomiasis (Chaga's disease), in Cohen, S. and Warren, K.S. (Eds.), *Immunology of Parasite Infections*, pp. 261–98, Oxford: Blackwell Scientific Publications.

Scuderi, P., Sterling, K.E., Lam, K.S., Finley, P.R., Ryan, K.J., Ray, C.G., Petersen, E., Shymen, D.J. and Salmon, S.E. (1986), Raised serum levels of tumour necrosis factor in parasitic infections, *Lancet*, **ii**, 1364–5.

Sethi, K.K. (1982), Intracellular killing of parasites by macrophages, *Clinics in Immunology and Allergy*, **2**, 541–65.

Sethi, K.K. and Piekarski, G. (1987), Immunological aspects of toxoplasmosis, in Soulsby, E.J.L. (Ed.), *Immune Responses in Parasitic Infections: Immunology, Immunopathology and Immunoprophylaxis*, **Vol. 3**, pp. 313–36, Florida: CRC Press.

Teixeira, A.R.L. (1987), The stercorarian trypanosomes, in Soulsby E.J.L. (Ed.), *Immune Responses in Parasitic Infections: Immunology, Immunopathology and Immunoprophylaxis*, **Vol. 3**, pp. 25–118, Florida: CRC Press.

Teppo, A.-M. and Maury, C.P. (1987), Radioimmunoassay of tumor necrosis factor in serum, *Clinical Chemistry*, **33**, 2024–7.

Thorne, K.J.I. and Blackwell, J.M. (1983), Cell mediated killing of protozoa, *Advances in Parasitology*, **22**, 43–151.

van der Meer, J.W.M., Endres, S., Lonnemann, G., Cannon, J.G., Ikejima, T., Okusawa, S., Gelfand, J.A. and Dinarello, G.A. (1988), Concentrations of immunoreactive human tumor necrosis factor alpha produced by human mononuclear cells *in vitro*, *Journal of Leukocyte Biology*, **43**, 216–23.

Wakelin, D. (1984), *Immunity to Parasites*, London: Edward Arnold, p. 165.

Warrell, D.A. (1987), Pathophysiology of severe falciparum malaria in man, *Parasitology*, **94**, 553–76.

Young, A.S. and Morzaria, S.P. (1986), Biology of Babesia, *Parasitology Today*, **2**, 211–19.

8. The cutaneous inflammatory response to parasite infestation

Diane J. McLaren

8.1. INTRODUCTION

The skin is the largest organ of the human body and it functions not only as a physico-mechanical barrier to infection, but also for the synthesis of many and varied biologically active substances; it may thus be considered a first line of defense which in turn develops crucial interactions with the immune system. It seems appropriate to begin this chapter with a brief description of the structure and organization of the skin and to summarize the functions of its many and varied constituent parts in order to form a basis against which the role of the skin in the immunology and pathology of selected parasitic diseases may be reviewed and evaluated.

The layers of the skin

The skin essentially comprizes three main layers, an epidermis or outer layer, an underlying dermis, and a subcutaneous layer that serves to attach the skin to deeper structures (Figure 8.1). The overall thickness of the skin varies in different parts of the body; the epidermis is composed of stratified squamous epithelial tissue, while the dermis contains dense, irregularly arranged connective tissue and various cell types (to be mentioned in more detail below), as well as blood and lymphatic vessels, nerves, hair follicles, sweat glands, sebaceous glands and some muscle. The subcutaneous layer comprizes fat, blood vessels, nerves, hair follicles, and sebaceous and sweat glands.

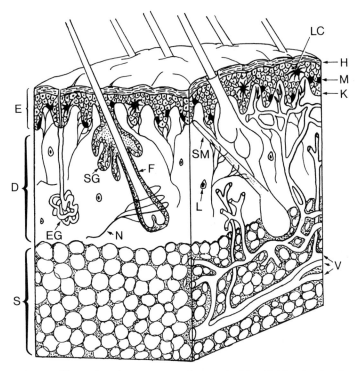

Figure 8.1. Diagram illustrating the structure of normal skin. Epidermis, E; Dermis, D; Subcutaneous layer, S; eccrine gland, EG: follicle, F; horny layer, H; keratinocyte, K; lymphocyte, L; Langerhan's cell, LC; melanocyte, M; nerve fibre, N; sebaceous gland, SG: smooth muscle, SM; blood and lymphatic vessels, V.

Skin cells and their functions

Interestingly, it is now considered that the number of cell types exerting immunological functions in skin [keratinocytes, dendritic antigen-presenting cells (Langerhan's cells, indeterminate cells, veiled cells), tissue macrophages, granulocytes, vascular endothelial cells, lymphatic endothelial cells and homing T cells] may equal the number of cells which, at this time, have no obvious immunological function (melanocytes, Merkel cells, fibroblasts, eccrine gland cells, duct cells, myoepithelial cells, apocrine secretory cells, sebocytes, smooth muscle cells, neural receptor cells and pericytes). This organ is therefore well equipped with immunologically active cells which play a vital role in inflammatory processes that often determine the outcome of an infection. The following summary gives only a flavour of the functional capabilities of such cells and the interested reader should consult more comprehensive and specialized reviews (e.g. Roitt *et al.*, 1985; Bos and Kapsenberg, 1986).

Langerhan's cells, indeterminate cells and veiled cells are accessory cells inhabiting respectively the epidermis, the epidermis/dermis and the skin-draining lymph vessels; they trap antigens that have penetrated skin and present them to paracortical T cells in draining lymph nodes. This results in T-cell activation, proliferation and differentiation, which in turn leads to the proliferation and recruitment of effector cells. Epidermal keratinocytes have also been suggested to play a role in T-cell differentiation. Monocytes and macrophages (sometimes called tissue histiocytes) are scattered throughout the dermis in large numbers; in normal skin they function principally as scavenger cells, although, as we shall see later, they also serve as crucial effector cells in some parasitic infections. Neutrophils and eosinophils are infrequent constituents of normal skin, but they do function to clear immune complexes under normal conditions; both however, participate in the inflammatory reactions that accompany parasitic infestation. Mast cells, which occur in quite high numbers in normal skin, regulate vasculature, participate in hypersensitivity reactions and are thought to play an important role in T-cell trafficking; their mediators are known to recruit and arm other granulocytes for parasite destruction. The blood and lymphatic vessels form networks that basically connect the cutaneous tissues with the rest of the immune system. Postcapillary venules in the skin permit the migration of inflammatory cells and are sites of histamine-induced permeability and immune complex formation; their lining endothelial cells have also been shown to function as antigen-presenting cells *in vitro*. The skin lymphatic system permits the unidirectional flow of lymph and in turn facilitates the movement of antigen-presenting cells towards the regional skin-related lymph nodes. A scattering of T lymphocytes can always be identified in normal healthy skin and it has been suggested, therefore, that a proportion of circulating T cells, called homing T cells, reside in this organ. The functional capabilities of the skin are therefore multiple and diverse, but as far as immune responses are concerned, the antigen-presenting dendritic cells, together with the circulating and recruited T cells, are probably the most important components.

The primary skin response to infection

Unbroken skin serves as an excellent mechanical barrier which is impenetrable to most micro-organisms. However, certain parasites possess abrasive armature, enzymes and other secretions which enable them to breach this first line of defense, and it is then that the body's second line of defense, inflammation, comes into play. The characteristic signs of inflammation are heat, pain, redness and swelling, followed by loss of function. The redness results from an increased blood supply to the infected area, while increased capillary permeability caused by retraction of the endothelial cells, allows soluble mediators of immunity to reach the area. Oedema follows as tissue fluid accumulates in the intercellular spaces. Subsequently neutrophils, and to a lesser extent macrophages, migrate from the capillaries into the tissues and move, through chemotaxis, to the site of infection. These phagocytic cells have receptors on their surface membranes that enable them to recognize and attach non-specifically to a variety of micro-organisms, but attachment is much enhanced if the invader has been opsonized by the C3b component of complement, generated as a consequence of complement activation at the site of injury. Following attachment the cells generally try to engulf the invader and internalize it within a phagosome where it is destroyed by toxic oxygen products generated through the respiratory burst and/or lysosomal enzymes. Although this is the normal course of events as far as micro-organisms, protozoa and small helminth larvae are concerned, larger parasites present a formidable target and the phagocytes must instead release their lysosomal enzymes extracellularly.

Key characteristics of the secondary skin response to infection

The cutaneous inflammatory reactions induced by a second encounter with the invading organism are accelerated and enhanced. In essence, the specificity of the adaptive immune response is based upon the specificity of the antibodies and lymphocytes. Since each lymphocyte is capable of recognizing only one antigen, an adequate response is generated by clonal selection (i.e. antigen binds to those lymphocytes which recognize it, and this event in turn induces the lymphocytes to proliferate). This applies to both B lymphocytes, which proliferate and mature into antibody-producing cells, and T lymphocytes, which are involved with the recognition and destruction of infected cells. Following clonal activation by antigen, the T lymphocytes produce lymphokines which in turn stimulate phagocytes to destroy infectious agents more effectively. Specificity and memory are thus key elements in the adaptive immune response. Complement components generated at sites of inflammation act directly on local vasculature and in addition activate mast cells to release mediators which cause vasodilation and increased vascular permeability. Anaphylactic antibodies also trigger the release of mast-cell mediators that in turn attract and arm other effector cells, like macrophages and eosinophils, which then function as potent cytotoxic effector cells.

The following account describes and evaluates the histopathological skin responses which accompany primary and secondary infestation of the host by a variety of parasitic organisms and includes, wherever possible, complementary evidence from experimental studies that these responses actively contribute to the control and/or elimination of the foreign invader.

8.2. PROTOZOAN PARASITES

Since their discovery by Anton van Leeuwenhoek in the seventeenth century, unicellular protozoa have been extensively studied in relation to many public health problems and diseases. The parasitic protozoa relevant to this account belong essentially to the Mastigophora, which are believed to be the oldest eukaryotic organisms and the ancestors of many other forms of life; such parasites possess at least one flagellum at some stage of their life cycle.

Patterns of skin disease associated with *Leishmania* species

The disease, leishmaniasis, occurs throughout Africa, Asia, the Americas and parts of Europe and is caused by flagellates belonging to the genus *Leishmania*. The parasites are transmitted by *Phlebotomus* sandfly vectors, and they have been isolated from a number of reservoir hosts including wild animals and the domestic dog. There are three principal disease manifestations: visceral, mucocutaneous and cutaneous leishmaniasis. The former does not concern us in this discussion, while the latter is probably the most widespread. The self-healing form of cutaneous disease is known as oriental sore or skin ulcer and is caused by *L. tropica*, and *L. mexicana*; it is characterized by localized as opposed to metastatic lesions and is accompanied by strong, delayed-type, skin-test reactions. Interestingly, the associated immunopathology has certain analogies to tuberculoid leprosy.

Chronic cutaneous leishmaniasis has multiple forms and is caused by a variety of different *Leishmania* species. American mucocutaneous leishmaniasis, also known as espundia, is the most chronic and is typically caused by protozoans of the *L. brasiliensis* complex. Infection involves the development of a skin lesion (Figure 8.2) that later metastasizes to the oro-naso-pharyngeal mucosa and leads to progressive disfigurement (Figure 8.3) and pain. However, parasites are scarce and difficult to detect within the lesion.

A second form of chronic cutaneous infection, called 'leishmaniasis recidivans', is associated with persistent cell-mediated immunity and is characterized by the development of recrudescing lupoid nodules which form around healed primary lesions; it is caused by *L. mexicana pifanoi*. Diffuse cutaneous leishmaniasis is associated with infection by *L. tropica* and *L. mexicana* and occurs in the absence of cell-mediated immunity. In this case the primary non-ulcerated nodules eventually metastasize to cover almost the entire skin surface and are rich in parasites. The pathological manifestations of cutaneous leishmaniasis thus seem to depend upon the species or strain of parasite and upon the immunological responses mounted by the host (Turk and Bryceson, 1971; Mauel *et al.*, 1974; Belehu *et al.*, 1980).

The human skin response to infection

There are many published accounts of human cutaneous leishmanial infection that include both clinical details and histological observations of skin biopsies. Such studies report a wide variety of disease manifestations that vary from normal skin to intense cellular infiltration, to granuloma formation and/or necrosis. It is not surprizing, therefore, that many authors have concluded that they are unable to connect such features into a meaningful pattern, or to analyse the histology in relation to parasite load (Ridley *et al.*, 1980). In

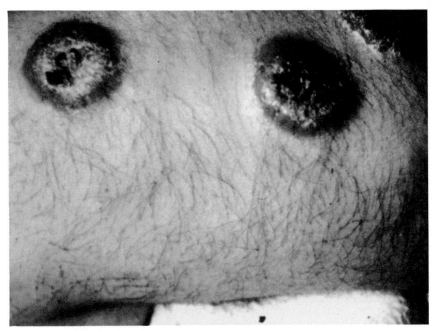

Figure 8.2. Mucocutaneous leishmaniasis: multiple lesions showing central necrosis on the limb of a patient (kindly provided by Dr. R.N. Incani, Department of Parasitology, University of Carabobo, Valencia, Venezuela).

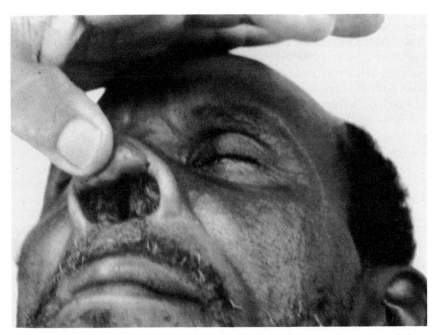

Figure 8.3. Mucocutaneous leishmaniasis: secondary lesion involving destruction of the nasal septum in a human patient (kindly provided by Dr. R.N. Incani, Department of Parasitology, University of Carabobo, Valencia, Venezuela).

general though, the cutaneous infiltrate is described as mononuclear in composition, comprizing macrophages, epithelioid cells, plasma cells and lymphocytes (Hassan *et al.*, 1984).

Immunocytochemical staining has revealed a predominance of helper over-suppressor T cells in both mucosal and cutaneous lesions, while granulomatous reactions have been recorded on occasion near ulcer surfaces, and hyperplasia of the epidermis is a prominent feature (Barral *et al.*, 1987). Ridley and Ridley (1983) identified three basic mechanisms of parasite elimination from histological analysis of human skin biopsies: 1) elimination of parasites within intact macrophages that later evolved as epithelioid cells, 2) elimination of parasites as a result of macrophage lysis and 3) necrosis at the centre of a focal mass of macrophages. They concluded that such responses depended on factors such as parasite load and the geographical origin of the leishmanial isolates but suggested that the three responses might be the outcome of a common immunological mechanism.

Skin responses in laboratory models

Experimental models of leishmaniasis are of course subject to greater control, and by selecting various host/parasite relationships, a considerable spectrum of disease patterns can be reproduced. Inbred BALB/c mice, for example, are relatively susceptible to *L. mexicana amazonensis*, while A/J mice are relatively resistant. Histopathological studies of lesions produced in both strains (Andrade *et al.*, 1982) have shown that the BALB/c mice develop nodular lesions infiltrated with 'foamy' macrophages, progressive depression of a delayed-type, hypersensitivity response and a tendency towards cutaneous metastases, while the resistant mice exhibit localized nodules with a mixed cellular infiltrate containing many lymphocytes and other mononuclear cells, focal fibrinoid necrosis and interstitial and peripheral fibrosis. Interestingly, the authors could find no real evidence of parasite destruction within the macrophages of either mouse strain, although in the resistant A/J mice, parasitized macrophages often appeared necrotic and degenerating leishmanias were frequently seen free in the interstitial tissue.

Monroy *et al.* (1980) saw similar features of necrosis in histological studies of lesions developed in the *L. enriettii*/guinea pig model. In this system, extensive necrosis of parasite-laden macrophages was seen in the lesions of all animals at week 4 to 5 post-infection, while the core of the ulcer contained numerous free parasites but few intact cells. Necrosis was less marked at 8 weeks and absent from resolving 10 week lesions. Morphological evidence of macrophage activation was recorded from week 5. It was suggested that the majority of parasites were released from the cells through necrosis, although intracellular destruction was considered to be important at later stages before resolution of the lesion. McElrath *et al.* (1987) subsequently showed by immuno-cytochemical analysis, that T lymphocytes preferentially entered *L. mexican amazonensis* healing lesions of resistant C57BL/6 mice but were minimal in the non-healing lesions of susceptible BALB/c rodents; they suggested that the T cells either provided lymphokines that activated macrophages to destroy intracellular parasites or played a direct cytotoxic role by killing infected macrophages and allowing local humoral factors to destroy released extracellular parasites. In this context, Sher *et al.* (1983) showed that the parasites of healing strains of *Leishmania* were readily destroyed by lymphokine-activated macrophages *in vitro*, whereas parasites of non-healing strains that produced chronic infections in mice were resistant *in vitro* to the same cells. The evasive capacity of some strains may therefore account for their ability to induce a chronic disease state.

More recently, Ridel *et al.* (1988) have detected a high proportion of cells bearing the K/NK phenotype (killer cells) in skin biopsies from patients infected with *L. brasiliensis guyanensis* and have suggested that these cells play a crucial role in the local control of parasite dissemination. In a different study, Grimaldi *et al.* (1984) examined acute and chronic phases of *L. mexicana mexicana* infections in Swiss-Webster mice. Their data are worthy of mention, since in addition to macrophages, they recorded granulocytes, particularly eosinophils, accumulating in the lesions during the acute phase of the disease and apparently destroying many parasites. The chronic phase was also characterized by an infiltration of granulocytes that paralleled parasite multiplication, but although both eosinophils and neutrophils exhibited low level phagocytosis of leishmanias throughout, the internalized parasites seemed able to avoid immune destruction. Parasitized macrophages were also prominent in chronic lesions.

Adaptations for survival within macrophages

The inability of the macrophage (a professional killer cell par excellence) to destroy *Leishmania* parasites undoubtedly accounts for the hold these parasites have over their hosts and also contributes to the chronicity and poor prognosis of the disease. Studies *in vitro* have indicated that the interaction of leishmanias with host macrophages is a receptor-mediated event (Blackwell *et al.*, 1985) that depends on the microviscosity of the cell membrane (Mukherjee *et al.*, 1988) and that the ecological niche of the parasite is the phagolysosome itself. Indeed, the transformation of promastigotes to amastigotes is thought to be an adaptive process that compensates for the harsh environment within the vacuole (Lewis and Peters, 1977). It has been suggested that evasion centres on resistance by the internalized parasite to lysosomal attack (Alexander and Vickerman, 1975), while other work has indicated that the quantity of lysosomal enzymes released into the phagosome might be important since if this is increased through exposure of the cells to agents, such as chloroquine, that stimulate lysosomal fusion with the vacuole, the killing potential of the cell increases (Alexander, 1981). There is also evidence that killing is a function of a certain stage of the macrophage cell cycle (Alexander, 1981). Intracellular survival thus depends upon a delicate balance between the potency of macrophage cidal mechanisms and the efficacy of the evasive strategies evolved by the parasite itself (Mauel, 1984).

8.3. NEMATODE PARASITES

Nematodes or roundworms are unsegmented, cylindrical, tapered at the extremities and have few appendages; they vary considerably in size, being either microscopic or several centimetres in length, and there seem to be species capable of parasitizing almost every tissue of the body. Some have direct life cycles while others utilize an intermediate host. Infection by medically important nematodes is in many cases via the oral route and representative examples will be described in other chapters of this book; other nematodes either penetrate skin directly or are introduced into the skin by a vector, and it is these, along with a consideration of the inflammatory reactions they elicit, which will be described here.

Onchocerca

Onchocerciasis currently affects some 40 million people in Africa, the Yemen, Mexico, Central and South America; it is caused by the filarial nematode *Onchocerca volvulus* and is transmitted by blackflies of the *Simulium* species. Cattle are also infested but with other species of *Onchocerca*. The pathology of the human disease has four principal manifestations: dermatitis, subcutaneous nodules, sclerosing lymphadenitis and eye disease (reviewed by Connor *et al.*, 1985; MacKenzie *et al.*, 1987), all of which may first occur many months after the original bite by the vector (Figure 8.4). The first two of these conditions concern the skin and will be described further.

Dermatitis

DURING NORMAL INFECTION
Dermatitis is characterized by itching, altered pigmentation and papule formation. In Africans this reaction, which was first described over 100 years ago, is generalized, being diffuse and characteristically maximal on the lower trunk, pelvic girdle and thighs. Onchocercal dermatitis in the Yemen differs in being more localized and it often shows a striking asymmetry that involves a single limb; this condition causes darkening of the skin and is consequently known as 'Sowda' (the Arabic word for black being aswad). Pruritis has been related to the presence, migration and death of microfilariae (early larval stages) in the upper dermis (Figure 8.5); it has been suggested that the shedding of secretory enzymes or the excretion of toxic materials by the larvae may be important factors (Burchard and Bierther, 1978). Light and ultrastructural studies of skin biopsies taken from patients have revealed

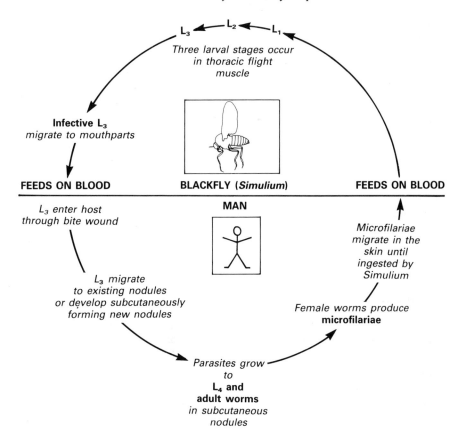

Figure 8.4. Life cycle of *Onchocerca volvulus*.

macrophages and mast cells in the vicinity of dermally located microfilariae, swelling of the endothelial walls of blood vessels and perivascular infiltrates comprizing lymphocytes, macrophages and plasma cells (Burchard and Bierther, 1978).

DURING TREATMENT WITH ANTHELMINTIC DRUGS
When Onchocerciasis patients are treated with the anthelmintic drug diethylcarbamazine (DEC), the dermatitis becomes suddenly worse and a classical Mazzotti reaction is initiated. The severe pruritis may be accompanied by erythema, oedema, urticaria, vesicle and papulae formation as well as scaling after several days (Gibson *et al.*, 1980). Inflammatory cells (lymphocytes, histiocytes, plasma cells and eosinophils) begin to localize around microfilariae within 24 hours, and immunocytochemical studies have demonstrated eosinophil degranulation and the subsequent deposition of major basic protein from the eosinophil granules onto the surfaces of the larvae (Kephart *et al.*, 1984). The larvae in consequence undergo a series of degenerative changes;

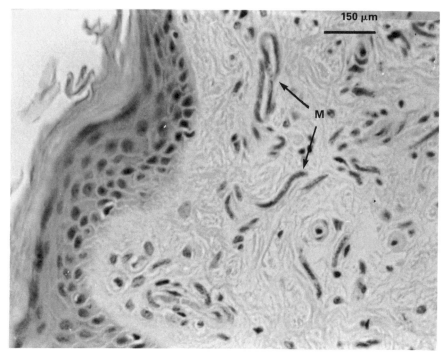

Figure 8.5. Onchocerciasis: section of skin showing microfilariae (M) in the dermis (kindly provided by Dr. M.J. Worms, Department of Parasitology, National Institute for Medical Research, Mill Hill, London, UK).

initially they appear eosinophilic, but later become hyperchromatic, fragmented and finally indistinct. Most of the microfilariae die in the upper region of the dermis, but some migrate into the epidermis where they provoke eosinophil-rich abscesses that persist for several days after drug treatment.

Interestingly, DEC does not kill microfilariae *in vitro* (Hawking *et al.*, 1950), and it has been suggested, therefore, that the drug might either 'unmask' microfilariae in the skin so that they are recognized as foreign bodies and destroyed by the host's immune defense system (Gibson *et al.*, 1976), or modulate host immune responsiveness to microfilarial antigens by activating inflammatory cells so that adherence reactions are facilitated and enhanced (MacKenzie, 1980). Evidence in support of the latter view comes from the demonstration that eosinophils exposed to DEC *in vitro* exhibit dose-related vacuolation, degranulation and granule lysis (MacKenzie, 1980), although it must be noted that others (Kephart *et al.*, 1984) have questioned the relevance of these data on the basis that the concentrations of drug tested were considerably higher than would be expected in tissue or serum after topical or oral drug therapy (Lubran, 1950). Metrifonate, an alternative anthelmintic compound, has also been reported to induce eosinophil accumulation in the

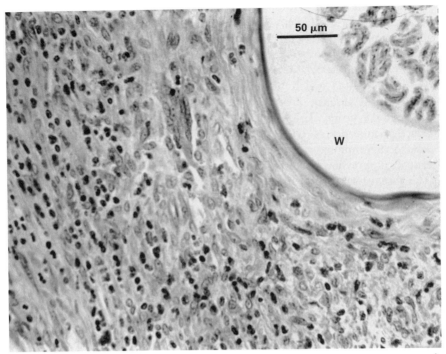

Figure 8.6. Onchocerciasis: section through a skin nodule showing cellular infiltration in the vicinity of an adult filarial worm (W).

cutaneous tissues, and a direct association between these granulocytes and dead microfilariae has been reported (Burchard *et al.*, 1979).

Subcutaneous nodules

Adult *Onchocerca volvulus* become encapsulated in discrete nodules located in the basal region of the dermis and in the subcutaneous layer of the skin. For reasons as yet undetermined, there is a predisposition for the nodules to form over bony prominences. These reactions (Figure 8.6) comprize an outer scar layer, and an inner inflammatory exudate that includes neutrophils, eosinophils, lymphocytes and plasma cells (Gibson *et al.*, 1980); degranulating eosinophils have been recognized in very close proximity to the adult worms (Connor *et al.*, 1976; MacKenzie, 1980; McLaren, 1982), but direct contact is only established following DEC treatment which has been shown from *in vitro* studies to have a profound effect upon eosinophil function (MacKenzie, 1980). Some nodules have a solid matrix of granulation tissue, while others contain a pool of fibrin. Neutrophil granulocytes appear to ingest fibrin from the cuticle surface and at later times, histiocytes and giant cells phagocytose the cuticle and other tissues of the degenerating nematodes. The nodules ultimately become calcified. Microfilariae are released into the nodules by female worms, and some of these degenerate as a consequence of eosinophil degranulation while others are phagocytosed by giant cells.

The complex interrelationship between the adult worm and the inflammatory nodule within which it becomes enclosed is of special interest, not least because in the absence of drug therapy, worms are reported to survive there for up to 18 years. Parkhouse *et al.* (1985) have recently investigated the surface phenotypes of lymphoid cells contained within the nodule, with a view to determining whether or not the worms escape immune rejection through suppression of immune responsiveness. B cells were generally found to be absent from the infiltrate, and although macrophages and dendritic cells were demonstrated consistently, analysis of T-cell subsets did not suggest active T cell-mediated suppression within the nodule. Experimental studies in which Indian ink was injected into the capillaries serving the nodules, have indicated that the worm may subvert the vascular reaction thereby causing controlled haemorrhages that supply its nutritional requirements (Connor *et al.*, 1985).

Loa loa

Another important filarial nematode that provokes notable skin lesions in man is *Loa loa* (comprehensively reviewed by Spencer, 1973). Loiasis has a fairly narrow distribution, occurring only in equatorial rain forest regions of Africa. Infective larval stages are implanted in the subcutaneous tissues by infected tabanid flies, and although little is known about subsequent development, the adult worms mature and migrate actively through the loose connective tissue of the skin. Vigorous and severe local reactions ensue (Calabar swellings) which comprize lymphocytes and eosinophils; these are thought to result from an allergic response to excretory/secretory antigens of the worm, although the parasites themselves are rarely found within such reactions. Interestingly, comparable swellings have been induced in Loiasis patients by the subcutaneous injection of antigens prepared from *L. loa* or species of *Dirofilaria* (another filariid).

Dracunculus medinensis

Dracontiasis or Guinea worm infection results from the presence of adult *Dracunculus medinensis* in the subcutaneous tissues of man (see Muller, 1971); it is a disease of great antiquity being known to Greek and Roman physicians and is contracted by drinking water containing infected *Cyclops* crustacea. Infections occur in Africa, most of the Indian subcontinent and in parts of the Middle East. Cutaneous lesions, associated with the head of the female worm, develop most commonly in the legs or feet of the host. Initially an indurated papule appears, but this soon forms a vesicle containing turbid fluid, eosinophils and neutrophils. When the vesicle ruptures it leaves an ulcer in which the worm head is often visible although the body of the parasite lies in a long, subcutaneous tunnel lined by fibrous tissue and containing inflammatory cells. Worms can be removed mechanically by gently winding the body around a stick, or they may be expelled spontaneously at about 4 weeks. After this time the lesion heals rapidly. Problems may arise however through secondary infection with other pathogenic organisms. Interestingly, male worms are smaller than females and do not elicit comparable lesions.

Skin penetrating, soil transmitted nematodes

Hookworms

Skin penetration by infective larvae of the medically important hookworm nematodes is also of interest because of the condition in man known as 'creeping eruption'. This form of dermatitis is caused by the burrowing of dog and cat hookworms into human skin and is characterized by a progressive papulomatous, vascular, pruritic lesion which develops in the wake of the migrating larvae (Figure 8.7). Fulleborn (1927) considered that the condition resulted from the inability of the worms to penetrate beyond the epidermis in an inappropriate host. In a more recent study, Vetter and Leegwater v.d. Linden (1977a, 1977b) examined the possibility that creeping eruption might be explained on the basis of fundamental differences in the structure of skin between man and domestic animals. In dogs for example, there are more hairs, sebaceous glands and apocrine sweat glands than are present on the limbs and arms of man (areas primarily affected by creeping eruption) where eccrine rather than apocrine sweat glands predominate. In the first of two investigations, Vetter and Leegwater-v.d. Linden (1977a) evaluated the penetration of *Ancylostoma braziliense* into the lateral skin of puppies (appropriate hosts). Most larvae were found to have entered the skin within 30 minutes of exposure, but the hair follicles did not seem to constitute the normal route of entry. Instead the larvae moved into and amongst the squames and then penetrated the epidermis before either entering the dermis directly or passing into the external root sheath of a hair follicle. In the latter case, the worms moved into the sebaceous glands or the apocrine glands before entering the dermis or subcutaneous layers, respectively. Parasites were detected in lymphatic vessels, but not blood vessels, within the sub-dermal layers of the skin as early as 30 minutes post-exposure.

In a second investigation, the same authors (1977b) examined larval penetration into the hairless skin of the metacarpal footpad of puppies, since besides being hairless, this area contains sweat glands of the eccrine type only and is therefore more similar to the skin of man. Interestingly, the eccrine sweat glands did not seem to constitute a route of entry for the larvae; instead the worms were seen exclusively in the epidermis where tunnels observed at 5 hours contained infiltrates of neutrophils. The histopathology induced by

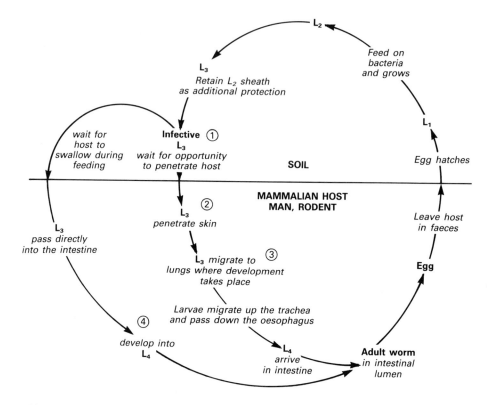

Figure 8.7. Life cycles of soil transmitted nematode species in which infective larvae develop after a free-living stage.

1. Two routes of infection are used depending on species. In *Nippostrongylus brasiliensis* and *Necator americanus* skin penetration is obligatory. In *Heligmosomoides polygyrus*, *Haemonchus contortus*, and *Ostertagia* sp. infection is only by the oral route. *Ancylstoma* sp. can invade by either route.

2. *N. brasiliensis* L_3 pass rapidly through the skin arriving in the lungs by 18 hrs. *N. americanus* stay in the local skin site for 36–48 hrs before migrating to the lungs.

3. *N. brasiliensis* moults to the L_4 stage in the lungs and within 60 hrs of infection most larvae have left the lungs. *N. americanus* arrives in the lungs 2–3 days post infection and develops into advanced L_3 stages, leaving the lungs on day 7. The L_4 moult probably occurs during migration or immediately on arrival in the small intestine. When *Ancylostoma* sp. use the skin penetration route, they pass through the lungs without developing. Further growth and moult to the L_4 stage occurs in the intestine.

4. *H. polygyrus* L_3 penetrate the mucosal tissues and develop in the outside of the muscularis externa, beneath the serosa. Here the $L_3 \rightarrow L_4$ moult occurs and the final moult to the preadult stage. When these have been completed 8–9 days post infection the worms return to the intestinal lumen.

5. Most species live in the intestinal lumen as adults, residing between the villi. Hookworms browse on the mucosal tissues and bite deeply into the intestinal walls, causing severe blood loss. Other species such as *N. brasiliensis* and *H. polygyrus* do not normally feed on blood.

the larvae in this particular skin site was therefore considered similar to that seen in cases of human creeping eruption and was taken to indicate the importance of hairs, sebaceous glands and apocrine sweat glands for successful penetration into the deeper layers of the skin.

Strongyloides spp.

Strongyloides stercoralis is an intestinal nematode of man acquired through skin penetration by free-living, third-stage infective larvae. The larvae move from the cutaneous tissues to the lungs and then on to the small intestine where the adult worms establish and commence egg laying. In the laboratory *S. ratti* constitutes a convenient model system with which to analyse the cutaneous response to parasite invasion. Rats and mice have both served as experimental hosts for this parasite, although it should be noted that the murine model is abnormal in that very few worms develop to adulthood, and even then in only a few selected mouse strains. Moqbel (1980) reported that rats exposed to a primary infection with *S. ratti* showed very few changes in the cutaneous tissues except for slight dilation of the walls of both the capillaries and lymphatics. Penetration was very rapid, larvae being observed in the subcutaneous layers within 30 minutes of exposure. A second population of larvae failed to provoke a significant cutaneous response, although at the sixth challenge, the skin reaction became immediate and intense. Within 30 min the subcutaneous tissues were filled with granulocytes and histiocytes and the larvae themselves became surrounded by granulomata rich in eosinophils that made contact with the cuticle surface. The number of eosinophils was found to have doubled within 2 to 4 h of exposure and to have increased still further after 24 h. The fact that larvae could not be recovered from the lungs of such animals was taken to indicate that they had died in the skin through immunologically-mediated mechanisms.

McHugh *et al.* (1989) recently confirmed the unexpectedly rapid migration of *S. ratti* larvae through both naive and sensitized rat skin and showed, by quantitative histological analysis of sections, that total cell numbers in the dermis were virtually identical in both naive/challenged and sensitized/challenged rodents (Figure 8.8). Immunofluorescence studies revealed that invading larvae changed their surface antigenicity shortly after skin penetration, and it was suggested that this, coupled with their speed of migration, accounted for the absence of a secondary cutaneous reaction in the host.

The *S. ratti*/mouse model differs. Dawkins *et al.* (1981) recognized neutrophils and eosinophils around invading larvae by 12 h in once-infected mice and reported that mononuclear cells became prominent by 48 h. The reaction became maximal on the fifth day, by which time most larvae had left the skin; the small proportion of larvae which remained in the cutaneous tissues were ultimately destroyed. Skin penetration was thought to be direct rather than via the hair follicles. Upon challenge, the cutaneous tissues of sensitized mice became rapidly oedematous and granulocytes were recorded

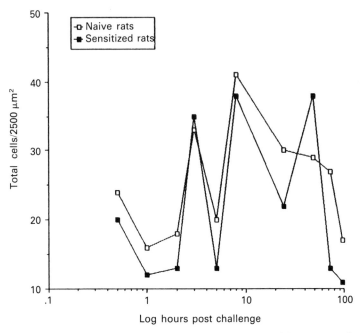

Figure 8.8. Strongyloidiasis: graph illustrating the total number of leucocytes infiltrating the skin site of larval (L₃) invasion in naive and previously sensitized rats.

around the invading larvae within 2 h. By 24 h, the infiltrate was dense, comprizing granulocytes and mononuclear cells and a percentage of the worms was again observed in varying stages of disintegration. The reaction then became predominantly mononuclear in composition and remained intense until day 5 after which time it gradually subsided. The majority of challenge parasites were reported to be killed not in the skin, but in the lungs through the action of mononuclear cells and polymorphs. These results must be viewed with some caution, however, in view of the abnormal nature of the mouse model outlined above.

Despite detailed histopathological studies, features of immune-mediated parasite damage and destruction have largely been neglected. An exception is the ultrastructural study of Lee (1976) on *Nippostrongylus brasiliensis*. Challenge larvae were still observed in the dermis of sensitized mice 2 to 3 h after presentation to the skin, and most exhibited a coiled appearance; this was taken to indicate immobilization through the action of immunity. On closer inspection the larvae were seen to be surrounded by bundles of collagen as well as macrophages and fibrocytes and many worms exhibited manifestations of cuticular damage; trauma was suggested to be mediated by collagenase secreted from host defense cells. This interpretation would perhaps be consistent with the notion that concealed collagens constitute important targets of the immune response to nematodes (Pritchard *et al.*, 1988). Internal damage to *N. brasiliensis* was recognized only after the cuticle had been breached and the larvae were ultimately walled-off by collagen fibres. The value of these purely morphological observations would have been greatly enhanced by a quantitative assessment of the numbers of worms lost in the skin. Moreover, since the mouse/*N. brasiliensis* model again constitutes a somewhat abnormal experimental system, it is vital that these data are confirmed and extended to other nematode species.

Host-protective and parasite-protective mechanisms in the skin

Extensive studies have been carried out on the *in vitro* interaction of nematode larvae with various effector cells. These have given further clues about the classes of antibody that mediate the attachment of different cell types to the surfaces of larger worms and revealed that very small larvae, such as microfilariae, can be taken into and destroyed within phagocytic vacuoles (reviewed by McLaren, 1985). *In vitro* assays have also revealed that one important evasive strategem used by nematode parasites concerns stage-specific antigenicity (MacKenzie *et al.*, 1978). This means that new cuticular antigens are exposed to the host with each successive moult so that established immune responses no longer recognize the worm. Other nematodes, such as *Strongyloides ratti*, use this strategy during early moults but appear not to replace their surface antigens at late moults so that the surface cuticle of the adult worm becomes antigenically inert (McHugh *et al.*, 1989).

8.4. PLATYHELMINTH PARASITES

The Platyhelminthes, or flatworms, are among the most primitive groups of animals to exhibit bilateral symmetry, a feature characterizing the basic body plan of all higher forms of animal life; they range in size from less than one millimetre to several metres in length and may be free-living or parasitic. The parasitic group includes the trematodes (or flukes) and the cestodes (or tapeworms), many of which are responsible for important diseases of man and domestic animals; such worms may be acquired through the ingestion of contaminated, inadequately cooked meat, fish, vegetation or crayfish or through skin penetration, and it is this latter group which concerns us here.

Skin penetration by schistosome larvae

Schistosomes are blood-dwelling flukes that differ from all other trematodes in that the sexes are separate. Thus, there is a sexual phase of reproduction in the definitive host and an asexual phase of reproduction in an intermediate snail host. Infection occurs when larval cercariae bearing forked tails emerge from aquatic molluscs into water and contact the skin of the definitive host. The penetration process involves attachment via spines and suckers, surface exploration, and finally, entry into the epidermis (Haas and Schmidt, 1982) via a combination of muscular and secretory events; the cercaria possesses three sets of gland cells whose secretions facilitate passage through the stratum corneum and epidermis (Stirewalt, 1974). It is thought that prostaglandins and essential free fatty acids are critical to the completeness of skin penetration (Salafsky *et al.*, 1984) and experiments showing that dog skin surface components that lack free fatty acids fail to stimulate penetration indicate that these molecules are probably host-derived (Haas *et al.*, 1987). The free-living cercaria loses its tail during the course of host invasion and is then known as a parasitic schistosomulum. Skin-stage schistosomula traverse the epidermis of the host within 30 min, but further migration seems to be temporarily halted by the basement membrane which apparently constitutes a formidable barrier; it is usually breached by day 2. Most schistosomula then move through the dermis, enter post-capillary venules and are carried to the lungs and then the liver. Some individuals may alternatively exit the skin via the efferent lymphatics (Wilson, 1988).

Primary response to infection with *Schistosoma mansoni*

The fact that several common laboratory rodents are susceptible to infection with schistosome species that infect man, has facilitated detailed and elegant investigations into the host/parasite relationship and immunopathology associated with this disease. Thus, many investigators have examined the cutaneous inflammatory response to cercarial penetration and schistosomular migration in both naive and previously sensitized rodents and primates. Experiments in which the invading parasites of a primary schistosome population were recovered from the skin, lungs and liver by tissue mincing and incubation techniques and then counted (Clegg and Smithers, 1968), prompted the view that those parasites which fail to reach maturity (30–50%) died in the skin (Smithers and Gammage, 1980). This particular recovery technique has since been shown to have low efficiency, however, and has for the most part, been superseded by compressed organ autoradiography (Georgi, 1982), in which the cercariae are labelled isotopically with ^{75}Se-selenomethionine and their migration tracked by autoradiographic analysis of the tissues and organs through which they pass. Thus, it is now known that more than 90% of successful skin penetrants migrate on to the lungs of naive mice (Mangold and Dean, 1983), rats (Knopf *et al.*, 1986) and guinea pigs (Kamiya and McLaren, 1987), and that it is in this organ rather than the skin, that a large proportion of the primary worm burden ultimately expires.

Histological studies have revealed that the invading parasites do invoke an inflammatory response in the cutaneous tissues, that is in general neutrophil-enriched (Lichtenberg *et al.*, 1976; Incani and McLaren, 1984; Ward and McLaren, 1988); this reaction seems to be associated a) with the elimination of cast cercarial tails, which are often seen within neutrophil micro-abscesses in the epidermis (Incani and McLaren, 1984) and b) with the extensive penetration tracks induced by the actively migrating schistosomula (Incani and McLaren, 1984; Pearce and McLaren, 1986).

Acquired immunity to *Schistosoma mansoni*

Studies of acquired immunity to schistosomiasis have centred on both *in vitro* and *in vivo* models. *In vitro* cytotoxicity assays have long implicated the young skin stage schistosomulum as the likely target of resistance (Smithers *et al.*, 1977), principally because 24 h and older parasites are completely refractory to the majority of immune effector mechanisms reproduced in the test tube (McLaren, 1980; McLaren and Terry, 1982). Histological studies have revealed that animals sensitized by exposure to either a live, or a radiation-attenuated schistosome infection, mount an impressive cutaneous response upon challenge that is both accelerated and enhanced with eosinophils replacing neutrophils after day 2 (von Lichtenberg *et al.*, 1976; Hsu *et al.*, 1979; Savage and Colley, 1980; Bentley *et al.*, 1981; Incani and McLaren, 1984; Pearce and McLaren, 1986; Ward and McLaren, 1988). Other experiments with animal models have largely revealed, however, that lung- and liver-stage juvenile schistosomes are the major targets of immune-dependent challenge elimination in sensitized rats and guinea pigs, respectively (Mangold and Knopf, 1981; Ford *et al.*, 1984; McLaren *et al.*, 1985; Knopf *et al.*, 1986; Oshman *et al.*, 1986), and that minimal (rats) or no (guinea pigs) schistosomular attrition occurs in the skin of these rodents.

The situation in sensitized mice is less clear cut. All published reports agree that lung-stage challenge parasites constitute the principal target of resistance. For reasons that are as yet unclear, however, some workers find that skin stage schistosomula become trapped within focal reactions in the subcutaneous layers of the skin, yet transform into lung-stage parasites before being eliminated (McLaren *et al.*, 1985; Kamiya *et al.*, 1987; McLaren and Smithers, 1987; Ward and McLaren, 1988), while other workers have equally conclusive data to show that the larvae reach the lungs and are then killed in the pulmonary vasculature (Dean *et al.*, 1984; Mangold and Dean, 1986; Mangold *et al.*, 1986; Wilson *et al.*, 1986). Since the former situation is pertinent to the subject of this chapter, it will be recounted in more detail here. The latter phenomenon is described further in Chapter 9.

Mice vaccinated with radiation-attenuated cercariae

CBA/Ca mice vaccinated with gamma-irradiated cercariae of *S. mansoni* develop, at 96 h post-challenge, numerous and extensive subcutaneous focal reactions (Figure 8.9), which enclose one or more challenge parasites (Figure 8.10a); such larvae have become longer and more slender than typical schistosomula and have lost their mid-body spines, features that indicate morphological transformation from skin-stage to lung-stage larvae (McLaren, 1980). The reactions themselves comprize roughly equal numbers of eosinophils

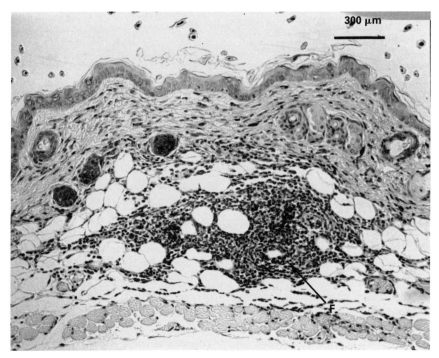

Figure 8.9. Schistosomiasis: subcutaneous focal reaction (F) in the skin of a vaccinated mouse at day 4 post-challenge.

and mononuclear cells (Figure 8.10b). That the worms are indeed trapped within such foci in vaccinated individuals is emphasized by two facts: firstly, naive/challenged mice exhibit only a few small reactions which rarely if ever include larvae; and secondly, as mentioned above, virtually all challenge larvae are known to have arrived in the lungs of naive mice by 96 h (Mangold and Dean, 1983; Kamiya *et al.*, 1987). Challenge larvae trapped within the skin reactions of vaccinated mice show varying manifestations of damage which involve internal rather than surface tissues. Ultrastructural studies have identified the subtegumental muscle cells of the worm as the principal target of immune-mediated damage, with the non-contractile portions of the cells exhibiting vacuolation and the myofibrils becoming progressively disorganized until they ultimately degenerate completely. Interestingly, neither the surface tegument nor the gut caecae of such parasites show obvious abnormalities (Ward and McLaren, unpublished data).

The effector mechanisms responsible for inducing these effects have been investigated by means of passive protection and leucocyte ablation experiments. If vaccinated mice are subjected at the time of challenge to treatment with a monoclonal antibody raised in rats against mouse leucocytes (a protocol which has been shown by quantitative histology to significantly reduce the cutaneous

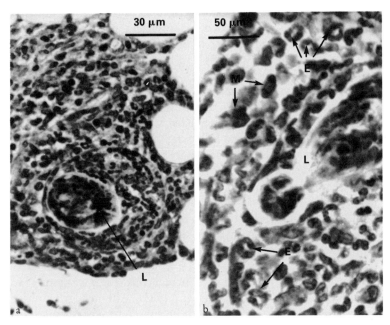

Figure 8.10. Schistosomiasis: challenge larvae (L) trapped within cutaneous focal reactions of the kind shown in Figure 8.9. The inflammatory infiltrates comprize roughly equal numbers of eosinophils (E) and mononuclear cells (M).

inflammatory response to challenge in vaccinated mice) or to 550 rads of whole body irradiation (which depletes radio-sensitive cells), then the expression of acquired resistance is reduced by 65% and 80%, respectively (McLaren *et al.*, 1987; Delgado and McLaren). Such data emphasize the important contribution made by cutaneous effector cells to immune expression in vaccinated mice and in particular the contribution made by radio-sensitive leucocytes (i.e. granulocytes/lymphocytes).

In other experiments, serum harvested from multiply vaccinated donor mice has been found to confer significant protection upon naive recipients when transferred at the time of challenge and to induce the development of subcutaneous focal reactions in the serum recipients which parallel exactly those seen in straight-forward vaccinated animals; the cellular composition of the foci and their role in the entrapment of lung-stage larvae in the skin is also identical (McLaren and Smithers, 1988). Serum thus protects naive mice through the recruitment of cutaneous effector cells.

There is, therefore, a consensus of data accrued from a variety of experimental techniques to show that, in this mouse model of vaccine resistance, up to 60% of the challenge larvae are first trapped by and then killed within eosinophil- and macrophage-rich cutaneous inflammatory foci. The manifestations of damage, which seem to be directed specifically against the subtegumental muscle cells, are reminiscent of those described in schistosomula subjected to activated macrophages *in vitro* (McLaren and James, 1985) or lung worms exposed to granulocyte-derived cationic proteins in the test tube (McLaren *et al.*, 1984). In this context, both eosinophils and macrophages are present in the skin foci, and eosinophils are radio-sensitive cells which are totally depleted from the cutaneous tissues by the two leucocyte ablation protocols mentioned above. Also pertinent to this argument is the fact that P-strain mice, which have a genetic defect in macrophage activation, fail to become immune following vaccination (James *et al.*, 1984). Thus, muscle trauma inflicted by these cells might very well be the reason why challenge larvae are unable to migrate out of the skin, despite having transformed to the next stage of development (i.e. the lung worm). It also seems feasible that those mice which kill challenge parasites in the lungs, mount similar if not identical immune responses, but that these mechanisms simply operate in the next organ along the migration pathway.

It is important to mention here that mice vaccinated by exposure to X-rather than gamma-irradiated cercariae of *S. mansoni* have also been shown by histological studies to develop subcutaneous focal reactions around challenge parasites in the skin (Hsu *et al.*, 1983). Although the precise role of these foci in parasite killing has yet to be substantiated by experiments of the kind described above, it seems likely that both kinds of radiation-attenuated parasites elicit comparable immune mechanisms.

Mice sensitized with normal cercariae

The role of skin-phase resistance in the immunity that can be stimulated in mice by sensitization with live, as opposed to radiation-attenuated, parasites of *S. mansoni*, is still a matter for debate. Some workers consider that eggs deposited in the liver by the primary worm population induce granuloma formation, hypertension and the development of a collateral circulation that enables challenge worms to escape from the liver and lodge in other sites (Wilson *et al.*, 1983). Low worm recoveries from sensitized animals at liver perfusion are thus taken to indicate resistance, whereas in fact the worms are simply located in sites from which they cannot be recovered by this technique. A number of investigators have shown, however, that 12- to 16-week infected mice do develop an anamnestically accelerated, eosinophil-enriched cutaneous inflammatory reaction to the invasion of challenge cercariae (von Lichtenberg *et al.*, 1976; Savage and Colley, 1980; Bentley *et al.*, 1981; Incani and McLaren, 1984) and that a proportion of challenge parasites become trapped and eliminated in granulocytic epidermal micro-abscesses (Incani and McLaren, 1984).

Depletion of cutaneous inflammation through treatment of the mice with anti-leucocyte polyclonal (Mahmoud *et al.*, 1975) or monoclonal antibodies (McLaren *et al.*, 1987), significantly reduces immune expression. Furthermore, P-strain mice, mentioned previously to have a genetic defect in macrophage activation, do not become immune following sensitization with a live schistosome infection (James and Cheever, 1985). Finally, mice given intradermal injections of schistosome antigen plus BCG to stimulate delayed-type hypersensitivity responses have been found to express comparable levels of resistance to those achieved following exposure to radiation-attenuated cercariae (James, 1985; 1986).

The protective schistosome antigen recognized in intradermally vaccinated mice is a 97 kDa protein identified as paramyosin (Lanar *et al.*, 1986) which has recently been localized by immune-labelling studies to inclusion bodies within the worm tegument (Matsumoto *et al.*, 1988); paramyosin may then be yet another example of a cryptic parasite antigen crucial to protective immunity. Taken together, these data support the notion that cutaneous cellular responses do contribute to both infection and vaccine resistance to *Schistosoma mansoni* in mice.

Skin responses of primates to *Schistosoma mansoni*

Recent histological studies of the skin reactions to challenge mounted by naive and chronically infected baboons have revealed important differences from mice. Neither control animals nor animals harbouring a 10 week *S. mansoni* infection mounted a significant inflammatory response against the challenge schistosomula (Seitz *et al.*, 1988), a result which may reflect the fact that baboons take several months to generate high levels of anti-parasite antibodies (Butterworth *et al.*, 1976; Susuki and Damian, 1981). At 8 months post-infection however, the animals mounted an impressive, eosinophil-enriched reaction, and eosinophils were identified attached to degenerating schistosomula in the epidermis (Seitz *et al.*, 1988). As the immune status of these animals was not recorded by portal perfusion, the number of challenge parasites killed in the skin remains unknown.

Acquired resistance to *Schistosoma japonicum*

Although *S. mansoni* has received most attention in laboratory investigations, simply through ease of life cycle maintenance, some workers have examined the skin reactions associated with invasion by *Schistosoma japonicum*. Maeda *et al.* (1982) reported that the skin reactions to challenge exhibited by 8-week infected mice were neutrophil-enriched and that these cells attached to and damaged challenge parasites in the subcutaneous tissues. Although the lung-recovery technique was used in this study to indicate loss of challenge parasites in the skin, the inefficiency of this protocol has been mentioned previously. It is perhaps relevant to mention, therefore, that mice vaccinated with ultraviolet-attenuated cercariae have been shown to make protective responses principally against lung-stage, but not skin-stage parasites (Moloney *et al.*, 1987). The situation in rats and rabbits is different however, since vaccination with ultraviolet-attenuated cercariae does stimulate immunity against the early skin-stage parasites in these animals (Moloney *et al.*, 1987). The skin reactions to challenge developed by Rhesus monkeys vaccinated with X-irradiated cercariae were found to be characterized by an infiltration of eosinophils in the epidermis and dermis and apparent schistosomulicidal activity by these granulocytes (Hsu *et al.*, 1971, 1975, 1979). In addition, mast cells present in whealing reaginic reactions were demonstrated by cytochemical staining methods to be coated with IgE antibodies (Hsu *et al.*, 1979, 1981). IgG was prominent in Arthus-like reactions, while IgG-complexes were found on challenge schistosomula, on blood vessel walls and in granulocytes located near to the parasites (Hsu *et al.*, 1981).

Cutaneous inflammation is not always responsible for immune attrition

One important lesson to be learned from these studies is that it is unwise to speculate on the role of the skin in immune attrition simply from histological analysis. A good example is the fact that guinea pigs vaccinated with gamma-irradiated cercariae of *S. mansoni* develop a massive cutaneous basophil-hypersensitivity reaction around challenge schistosomula, and cutaneous eosinophils have been seen degranulating against the parasite surface (Pearce and McLaren, 1986); yet we now know from quantitative experimental studies of the kind described above, that skin-phase attrition is negligible in these particular rodents. A multifaceted approach to the elucidation of attrition sites is therefore crucial.

8.5.   ARTHROPOD PARASITES

Arthropods have a well-known reputation for thwarting the efforts of man to establish safe and stable environments. During the sixth century B.C., malaria and plague flourished in newly established cities, and the only known solution was to move away until the epidemic had subsided. Although it was long recognized that arthropods lived on the skin surfaces of mammals, the relationship between insects and disease-causing organisms was not established until the nineteenth century. Red water fever, for example, was a recognized disease of cattle in the USA in 1796, but the fact that its causative agent, the protozoan *Babesia bigemina*, was transmitted by ticks, was not recognized until 1893. Indeed, this pioneering work constituted a model for the subsequent discovery that many arthropods, in addition to causing serious injury or sensitization in their own right, serve either as intermediate hosts for parasites or as vectors for pathogenic organisms. We shall consider selected examples here.

Arachnid ectoparasites

Ticks and mites are medically important ectoparasites because they can cause disease directly (e.g. tick paralysis and mite scabies), transmit infectious diseases (e.g. tick-borne typhus and mite-borne rickettsial pox), or act as reservoirs of infectious disease agents that pass via eggs or larvae to succeeding generations of arthropods.

Ticks

Ticks can be classified into two principal taxonomic groups, the Ixodidae or hard ticks and the Argasidae (Figure 8.11) or soft ticks; these differ both in structure and biology. The skin lesions and immune responses induced by ticks have attracted considerable attention. Primary reactions observed at tick bite sites are sometimes slight, as in the case of *Dermacentor variabilis* in mice (den Hollander and Allen, 1985), or may take the form of an acute inflammatory abscess, as with *Rhipicephalus appendiculatus* in rabbits and cattle (Walker and Fletcher, 1986); they are usually dominated by neutrophils and mononuclear cells, although in guinea pigs, eosinophils and basophils may also contribute to primary cutaneous reactions.

CUTANEOUS REACTIONS TO PRIMARY AND SECONDARY INFECTIONS
Over the last 15 years it has been recognized that cattle and experimental animals develop long-lasting resistance following a single exposure to small

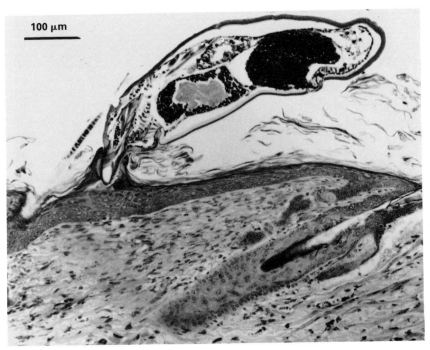

100 μm

Figure 8.11. Tick infestation: section showing the soft tick *Ornithodorus tartakovskyi* attached to and feeding on the skin of a guinea pig (kindly provided by Dr. M.J. Worms, Department of Parasitology, National Institute for Medical Research, Mill Hill, London, UK).

numbers of hard ticks; manifestations of immunity include impaired feeding by the secondary ectoparasites, thwarted development, rejection and subsequent death. The secondary reaction elicited in the skin of the host is usually enhanced and heavily infiltrated with granulocytes. Indeed, in guinea pigs it represents an extreme example of cutaneous basophil hypersensitivity (Allen, 1973) with these granulocytes constituting up to 90% of the infiltrate (Figure 8.12) recruited to the feeding site (Brown *et al.*, 1983; McLaren *et al.*, 1983a, 1983b).

Ultrastructural studies using the guinea pig model have revealed that basophils in lesions elicited by hard ticks exhibit massive anaphylactic degranulation (McLaren et al., 1983a), whereas basophils in lesions induced by soft ticks exhibit a variety of granule-release mechanisms, including vesicle externalization, single or compound exocytosis of whole granules, or complete disintegration of the cell (McLaren *et al.*, 1983b). Eosinophils also contribute significantly to some guinea pig skin reactions. It is of interest that although soft tick lesions induced in guinea pigs essentially resemble those elicited by hard ticks, they do not lead to immune rejection. This has been related to the fact that fast feeding Argasid ticks probably complete their blood meal

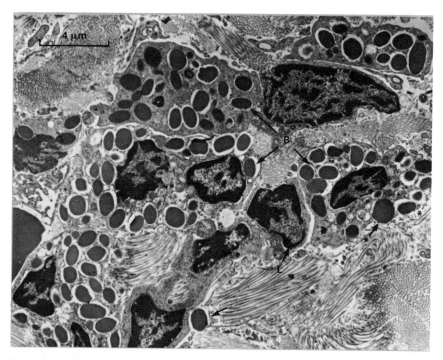

Figure 8.12. Tick infestation: electron micrograph showing a cutaneous basophil hypersensitivity reaction in the skin of a sensitized guinea pig following tick challenge. The basophils (B) contain haloed granules, some of which are seen in the process of release from the cell (*), while others lie free within the dermis (->). Lymphocytes (L) may also be recognized.

prior to basophil arrival, whereas the cutaneous reaction becomes maximal during the prolonged blood meal taken by Ixodids. Challenge tick populations induce more variable basophil responses in rabbits and cattle; in some model systems these cells dominate the bite sites (Brown *et al.*, 1984) while in others neutrophils and mononuclear cells predominate, with eosinophils also showing a significant increase in numbers (Gill and Walker, 1985; Gill, 1986; Walker and Fletcher, 1986).

EVIDENCE THAT PROTECTION IS MEDIATED VIA CUTANEOUS EFFECTOR CELLS

That basophils and eosinophils recruited to the bite site actively contribute to immune rejection is indicated from a number of experimental studies. Firstly, naive guinea pigs protected passively with serum harvested from actively sensitized individuals, have been found to express significant cutaneous basophil responses upon challenge (McLaren *et al.*, 1983a; Brown *et al.*, 1984). Secondly, guinea pigs actively sensitized with *Amblyomma americanum*, but depleted of basophils by administration of specific anti-basophil antiserum, failed to express resistance to tick challenge (Brown *et al.*, 1982). Moreover, guinea pigs depleted of eosinophils in the same way showed partial resistance to challenge, a result which implies cooperation between these two cell types in effecting tick rejection (Brown *et al.*, 1982). Thirdly, basophils can be recruited into guinea pig skin and induced to degranulate by sensitization and challenge with a non-tick protein antigen, keyhole limpet cyanin, and this reaction also leads to tick rejection (Brown and Askenase, 1985).

The immune response of both guinea pigs (Bagnall, 1975; Wikel, 1982) and livestock (Tatchell and Bennett, 1969; Berdyev and Khudainazarona, 1976; Willadsen *et al.*, 1979) can be blocked by the administration of anti-histamine drugs, while the injection of histamine into guinea pig skin at *Boophilus microplus* attachment sites early during feeding, causes tick detachment (Kemp and Bourne, 1980). More recently, however, Brown and Askenase (1985) reported that guinea pigs actively sensitized with a different tick, *A. americanum*, and then treated with H1 and H2 histamine receptor antagonists during the challenge tick infestation period, expressed normal resistance. Further, naive guinea pigs treated with anti-histamines could still be protected with immune serum. Basophil-derived mediators other than histamine would therefore seem to be important in this model system.

Wikel and Allen (1977) demonstrated that sensitized guinea pigs depleted of complement with cobra venom factor, showed a significant reduction in immune rejection of *Dermacentor andersoni*. Immunofluorescence techniques subsequently revealed tick salivary gland antigens, complement and IgG at the dermo-epidermal junction of resistant guinea pigs, both in the vicinity of and at a distance from the site of parasite attachment (Allen *et al.*, 1979). This prompted the suggestion that antigen-antibody reactions and complement activation might together contribute to the development of skin lesions and attract basophils to the bite site. Tick antigens were additionally observed in dendritic suprabasal cells, thought to be Langerhan's cells, located in the epidermis. Depletion of Langerhan's cells by exposure of the animal to ultraviolet irradiation prior to primary infestation significantly reduced the acquisition of resistance, while the same treatment applied to resistant individuals caused a marked reduction in immune rejection (Nithiuthai and Allen, 1984). These results were taken to indicate a functional requirement for Langerhan's cells for both the induction and expression of tick resistance.

CONCEALED ANTIGENS

More recent interest has focused on the possibility that 'concealed antigens' play a role in immunity to ticks (reviewed by Willadsen and Kemp, 1988). Vaccination trials with tick internal organs, usually the gut, of semi-engorged individuals have given promising results, and ticks that have fed on cattle vaccinated in this way exhibit severe damage to their gut cells. It has been suggested that immunity may operate through antibody attack on one or more critical cell functions of the tick digest cells, probably mediated by a cell surface molecule, and that following initial damage, antibody leaks into the haemocoel where a range of other target cells become vulnerable. Evidence in support of this notion comes from the observation that endocytosis by tick digest cells is inhibited in individuals fed on sera from vaccinated cattle and that this event is followed by morphological evidence of damage.

Mites

Man and his domestic animals are infested by several species of mite; some, like *Demodex folliculorum*, are thought to be almost universal in older children and adults, yet are broadly speaking asymptomatic. *Sarcoptes scabies*, on the other hand, burrows into skin and provokes a cutaneous reaction but is non-infectious, while other mite species transmit important diseases, such as scrub typhus and rickettsial pox. The cutaneous responses to mite infestation and their associated immune responses are not as well characterized as those induced by ticks. Although it is generally thought that mites migrate through the stratum corneum, a recent report states that they also migrate through the epidermis and even occasionally to the dermo-epidermal interface (van Neste and Lachapelle, 1981).

In man, the primary scabetic lesion is marked by hyperkeratosis, acanthosis, oedema and vesiculation of the epidermis, and a perivascular and diffuse cellular infiltrate in the dermis; mononuclear cells predominate in such reactions although eosinophils and mast cells also contribute. Secondary lesions show acanthosis and perivascular inflammatory infiltrates; these are also dominated by mononuclear cells (Falk and Eide, 1981) which have been identified from immunofluorescence studies on sectioned tissue to be principally T lymphocytes (Falk and Matre, 1982). The severity of the cutaneous response has been reported to correlate with the number of circulating eosinophils and the concentration of serum IgE. In the latter context, IgE has been detected by immunocytochemical assays on mast cells located both in the dermis and subcutaneous tissues of scabies lesions, in the vessel walls of the upper dermis, in biopsies containing mites and in inflammatory papules without mites, on the surfaces of the mites themselves and on mite faeces (Frentz *et al.*, 1977; Hayashi *et al.*, 1986); such reactions did not occur in biopsies of normal skin from the same patient.

Psoroptes ovis is a non-burrowing, surface-feeding mite which is the causative agent of psoroptic scabies or mange in cattle and sheep. Lesions induced by this parasite in naive and previously exposed cattle have recently been examined by Stromberg and Fisher (1986). The principal pathology associated with infestation was a chronic, exudative, superficial, perivascular dermatitis that was qualitatively similar in both groups of animals; primary lesions appeared late but became rapidly progressive, whereas secondary reactions occurred early but progressed slowly. Dermatitis was characterized by the accumulation of eosinophils and to a lesser extent lymphocytes, macrophages, plasma cells and mast cells, around venules in the superficial dermis. There was also marked endothelial hypertrophy, vascular dilatation and congestion and occasional margination of neutrophils. Eosinophils and mononuclear cells were, nevertheless, the most important cellular constituents of the infiltrate. Mites on naive calves exhibited exponential growth and high fecundity and reached high population densities, but secondary populations showed low growth rates, low fecundity and 100 to 1000 times lower densities. Acquired resistance was thus suggested to operate, at least in part, through decreased fecundity of the female parasites.

Insect ectoparasites

Lice

Lice parasitize birds and many species of mammal; they have mouth parts specially adapted for sucking blood and tissue fluids or for chewing epithelioid structures associated with skin. Two genera are associated with man, *Pediculus* and *Phthirus*. Members of the latter genus include the body louse and the head louse, both of which, besides being intrinsic pests, can also transmit important infectious diseases through mechanical means (cholera and impetigo) or pass on rickettsial (epidemic typhus and trench fever) and bacterial (relapsing fever) infections through biological transmission.

As with other resident blood-sucking arthropods, there is usually an initial period in the host-parasite relationship during which the numbers of lice increase, followed by a period of decline. This may result from the fact that inflammatory responses recognized in the skin following infestation, are characterized by extensive cutaneous arteriolar vasoconstriction that reduces the blood flow to such an extent that the lice are prevented from feeding and die through starvation (Nelson and Bainborough, 1963; Nelson *et al.*, 1972).

In the laboratory, a convenient louse (*Polyplax serrata*)/mouse model has been established to study these aspects further. Interestingly, mice of the CFW strain quickly become resistant to this parasite, whereas C57BL mice do not (Clifford *et al.*, 1967), a feature which has facilitated comparative studies of the respective skin responses. Nelson *et al.* (1979) found that CFW mice exhibited more rapid reactions than non-resistant mice and that these contained five times as many neutrophils and half as many eosinophils at week 2 post-infestation and twice as many mast cells and fibroblasts at later times. It was suggested that a chemotactic factor for neutrophils might be absent in the C57BL strain which prevented the full expression of acquired resistance. It is important to mention that epidermal hyperplasia and grooming behaviour were also less intense in the C57BL mice, indicating that these animals experienced less irritation and were in consequence less inclined to scratch; this perhaps reflects differences in the biosynthesis and secretion of prostoglandin E and of the threshold response to histamine and other mediators.

Fleas

Fleas are blood-sucking ectoparasites of particular medical and historical importance because of their involvement in the transmission of plague and endemic typhus. They can, in addition, serve as mechanical vectors for a number of helminths, such as dog and rat tapeworms, or cause intense irritation and ulceration simply by attaching to or burrowing into the skin (e.g. the chigger flea—Tunga). Tungid species existed before the exploration of the New World and have, through infesting the feet of transcontinental travellers, crossed from South America to Africa. *Tunga monositus* has been studied under laboratory conditions in mice (Lavoirpierre *et al.*, 1979). During the first 24 h of attachment to the skin, the parasite feeds on fluid exudates of the host and induces the infiltration of neutrophils into the cutaneous tissues. The flea becomes fully embedded into the skin by day 3 and then begins to feed on the neutrophils themselves. From days 10 to 14, the lesion is marked by granulation tissue and the flea injests fibroblasts, collagen and macrophages, but after this time, feeding involves whole blood taken initially from within the vessels but later, at about day 42, from the blood pool which develops locally. Mating and oviposition then occur. The parasite may be sloughed from the skin at around day 56, at which time healing commences, although rejection may be delayed until the end of the third month. Paradoxically then, *T. monositus*, appears to depend upon the inflammatory and repair response of the host for survival and reproduction. In a more recent investigation, Johnston and Brown (1985) showed that tick bites in guinea pigs that had been pre-sensitized on the opposite flank with the flea, *Xenopsylla cheopis*, contained a markedly enhanced population of eosinophils as compared to primary tick feeding sites in naive rodents; a result which suggested the possible cross reactivity of flea and tick antigens.

Tsetse flies

Tsetse flies are especially important for the transmission of trypanosomiasis (sleeping sickness) to man and his livestock; they induce a local skin reaction, known as a chancre, at the feeding site. In an experimental study, *Glossina morsitans morsitans*, infected with *Trypanosoma congolense*, induced a chancre as early as day 5 post feeding (Akol and Murray, 1982). The first manifestation was a small nodule 2 to 3 mm in diameter, which increased to 100 mm by day 10 to 13, at which time it appeared as a hot, raised, indurated swelling. Histopathological investigations revealed an intense cellular inflammatory reaction, dominated initially by polymorphonuclear leucocytes but soon replaced by mononuclear cells comprizing mainly small to medium sized lymphocytes. Emery and Moloo (1980) recorded essentially similar histological data for chancres elicited in goats by *Glossina morsitans morsitans* infected with *Trypanosoma brucei*. Interestingly, chancre formation and the appearance of parasites in the cutaneous tissues of cattle, preceded both generalized parasitaemia and other clinical symptoms by several days, and it was concluded that the skin acted not only as a focus for the establishment of infection, but also as a site for localized parasite proliferation prior to dissemination in the bloodstream.

It is noteworthy that the bite of an uninfected tsetse fly does not elicit a chancre reaction in cattle, whereas the intradermal injection of metacyclic, but not bloodstream forms of *T. congolense*, does. In contrast, uninfected *Glossina morsitans* does induce a skin response in guinea pigs. Brown and Cipriano (1985) found the cutaneous reaction in naive individuals to be dominated by mononuclear cells but included a week granulocytic component in which eosinophils were prominent. Feeding sites in thrice sensitized guinea pigs were also dominated by mononuclear cells, but in this case granulocytes made a much greater contribution to the infiltrate and basophils predominated. Tsetse feeding sites were further marked by haemorrhages apparently related to the probing behaviour of the fly.

8.6. SUMMARY AND OVERVIEW

The skin clearly constitutes a formidable barrier to infestation by many different parasitic organisms. Primary responses are multifarious and often associated with the repair of mechanical damage induced by the invader. Secondary reactions are usually, but not always, accelerated and enhanced and often include a significant granulocytic component. The nature and extent of cutaneous inflammation depends to a large extent upon the characteristics of the host/parasite system in question and can fulfil many roles; it can contribute directly to the immune elimination of schistosome larvae and some nematode larvae or effect tick rejection. Other parasites, on the other hand, have adapted to live within the skin lesions they provoke; *Onchocerca*, for example, inhabits cutaneous nodules, and *Leishmania* parasites actually survive within leucocytes which must be considered the most professional of all the phagocytes. Certain fleas seem to depend upon the inflammatory repair response of the host for nutrition and successful reproduction, while some Trypanozoans are thought to proliferate within skin chancres prior to dissemination in the bloodstream. The undoubted success of parasitic organisms as a group can almost certainly be attributed to the fact that many avoid the potentially damaging effects of cutaneous and other inflammatory responses by employing sophisticated evasive strategies; they are thus able to establish and maintain the delicate balance which ensures survival of both partners in a complex relationship. Internal or concealed parasite antigens are currently the subject of much experimental interest, not least because they may be unprotected by evasive mechanisms and therefore constitute more accessible targets for synthetically engineered vaccines.

REFERENCES

Akol, G.W. and Murray, M. (1982), Early events following challenge of cattle with tsetse infected with *Trypanosoma congolense*: development of the local skin reaction, *Veterinary Record*, **27**, 295–302.

Alexander, J. (1981), *L. mexicana*: inhibition and stimulation of phagosome–lysosome fusion in infected macrophages, *Experimental Parasitology*, **52**, 261–70.

Alexander, J. and Vickerman, K. (1975), Fusion of host cell secondary lysosomes with the parasitophorous vacuole of *L. mexicana*-infected macrophages, *Journal of Protozoology*, **22**, 502–8.

Allen, J.R. (1973), Tick resistance: basophils in skin reactions of resistant guinea pigs, *International Journal of Parasitology*, **3**, 195–200.

Allen, J.R., Khalil, H.M. and Graham, J.E. (1979), The location of tick salivary antigens, complement and immunoglobulin in the skin of guinea pigs infested with *Dermacentor andersoni* larvae, *Immunology* **38**, 467–72.

Andrade, Z.A., Reed, S.G., Roters, S.B., and Sadigursky, M. (1982), Immunopathology of experimental cutaneous leishmaniasis, *American Journal of Pathology*, **114**, 137–48.

Bagnall, B.G. (1975), Cutaneous immunity to the tick *Ixodes holocyclus*, PhD Thesis, University of Sydney.

Barral, A., Jesus, A.R., Almeida, R.P., Carvalho, E.M., Barral-Netto, M., Costa, J.M.L., Badaro, R., Rocha, H. and Johnson, W.D. (1987), Evaluation of T-cell subsets in the lesion infiltrates of human cutaneous leishmaniasis, *Parasite Immunology*, **9**, 487–97.

Belehu, A., Louis, J.A., Pugin, P. and Miescher, P.A. (1980), Immunopathological aspects of leishmaniasis, *Seminars in Immunopathology*, **2**, 399–415.

Bentley, A.G., Carlisle, A.S. and Phillips, S.M. (1981), Ultrastructural analysis of the cellular response to *Schistosoma mansoni*. II. Inflammatory responses in rodent skin, *American Journal of Tropical Medicine and Hygiene*, **30**, 515–24.

Berdyev, A. and Khudainazarona, S.N. (1976), A study of acquired resistance to adults of *Hyalomma asiaticum asiaticum* in experiments on lambs, *Parazitologiya*, **10**, 519–25.

Blackwell, J.M., Ezekowits, R.A.B., Roberts, M.B., Channon, J.Y., Sim, R.B. and Gordon, S. (1985), Macrophage complement and lectin-like receptors bind

Leishmania in the absence of serum, *Journal of Experimental Medicine*, **162**, 324–31.

Bos, J.D. and Kapsenberg, M.L. (1986), The skin immune system, *Immunology Today*, **7**, 235–40.

Brown, S.J. and Askenase, P.W. (1985), Rejection of ticks from guinea pigs by anti-hapten-antibody-mediated degranulation of basophils at cutaneous basophil hypersensitivity sites: role of mediators other than histamine, *Journal of Immunology*, **134**, 1160–5.

Brown, S.J. and Cipriano, D.M. (1985), Induction of systemic and local basophil and eosinophil responses in guinea pigs by the feeding of the tsetse fly *Glossina morsitans*, *Veterinary Parasitology*, **17**, 337–48.

Brown, S.J., Worms, M.J. and Askenase, P.W. (1983), *Rhipicephalus appendiculatus*: larval feeding sites in guinea pigs actively sensitised and receiving immune serum, *Experimental Parasitology*, **55**, 111–20.

Brown, S.J., Bagnall, B.G. and Askenase, P.W. (1984), *Ixodes holocyclus*: kinetics of cutaneous basophil responses in naive, actively and passively sensitised guinea pigs, *Experimental Parasitology*, **57**, 40–7.

Brown, S.J., Barker, R.W. and Askenase, P.W. (1984), Bovine resistance to *Amblyomma americanum* ticks: an acquired immune response characterised by cutaneous basophil infiltrates, *Veterinary Parasitology*, **16**, 147–65.

Brown, S.J., Galli, S.J., Gleich, G.J. and Askenase, P.W. (1982), Ablation of immunity to *Amblyomma americanum* by anti-basophil serum: co-operation between basophils and eosinophils in expression of immunity to ectoparasites (ticks) in guinea pigs, *Journal of Immunology*, **129**, 790–6.

Burchard, G.D. and Bierther, M. (1978), Electron microscopical studies on Onchocerciasis. I. Mesenchyme reaction in untreated onchocercal dermatitis and ultrastructure of the microfilariae, *Tropenmedicine und Parasitologie*, **29**, 451–61.

Burchard, G.D., Albiez, E.J. and Bierther, M. (1979), Electron microscopical studies on Onchocerciasis. II. Skin and microfilariae after treatment with Metrifonate, *Tropenmedicine und Parasitologie*, **30**, 97–102.

Butterworth, A.E., Sturrock, R.F., Houba, V. and Taylor, R. (1976), *Schistosoma mansoni* in baboons. Antibody-dependent cell-mediated damage to ^{51}Cr-labelled schistosomula, *Clinical and Experimental Immunology*, **25**, 95–103.

Clegg, J.A. and Smithers, S.R. (1968), Death of schistosome cercariae during penetration of the skin. II. Penetration of mammalian skin by *Schistosoma mansoni*, *Parasitology*, **58**, 111–28.

Clifford, C.M., Bell, J.F., Moore, G.J. and Raymond, G. (1967), Effects of limb disability on lousiness in mice. IV. Evidence of genetic factors in susceptibility to *Polyplax serrata*, *Experimental Parasitology*, **20**, 56–67.

Connor, D.M., Neafie, R.C. and Meyers, W.M. (1976), Loiasis, in Binford, C.H. and Connor, D.M. (Eds.), *Pathology of Tropical and Extraordinary Diseases—An Atlas*, pp. 356–9, Washington DC: Armed Forces Institute of Pathology.

Connor, D.H., George, G.H. and Gibson, D.W. (1985), Pathologic changes of human Onchocerciasis: implications for future research, *Reviews of Infectious Diseases*, **7**, 809–19.

Dawkins, H.J.S., Muir, G.M. and Grove, D.I. (1981), Histopathological appearances in primary and secondary infections with *Strongyloides ratti* in mice, *International Journal for Parasitology*, **11**, 97–103.

Dean, D.A., Mangold, B.L., Georgi, J.R. and Jacobson, R.H. (1984), Comparisons of *Schistosoma mansoni* migration patterns in normal and irradiated cercaria-immunised mice by means of autoradiographic analysis. Evidence that worm elimination occurs after the skin phase in immunised mice, *American Journal of Tropical Medicine and Hygiene*, **33**, 89–96.

Delgado, V.S. and McLaren, D.J. (1990), Evidence that radio-sensitive cells are central to skin-phase immunity in CBA/Ca mice vaccinated with radiation-attenuated cercariae of *Schistosoma mansoni* as well as in naive mice protected with vaccine serum, *Parasitology*, **100**, 45–56.

den Hollander, N. and Allen, J.R. (1985), *Dermacentor variabilis*: acquired resistance to ticks in BALB/c mice, *Experimental Parasitology*, **59**, 118–29.

el Hassan, A.M., Veress, B. and Kutty, M.K. (1984), The ultrastructural morphology of human cutaneous leishmaniasis of low parasite load, *Acta Dermato-Venereologica*, **64**, 501–5.

Emery, D.L. and Moloo, S.K. (1980), The sequential cellular changes in the local skin reaction produced in goats by *Glossina morsitans morsitans* infected with *Trypanosoma (Trypanozoon) brucei*, *Acta Tropica (Basel)*, **37**, 137–49.

Falk, E.S. and Eide, T.J. (1981), Histologic and clinical findings in human scabies, *International Journal of Dermatology*, **20**, 600–5.

Falk, E.S. and Matre, R. (1982), *In situ* characterisation of cell infiltrates in the dermis of human scabies, *American Journal of Dermatopathology*, **4**, 9–15.

Ford, M.J., Bickle, Q.D., Taylor, M.G. and Andrews, B.J. (1984), Passive transfer of resistance and the site of immune-dependent elimination of the challenge infection in rats vaccinated with highly irradiated cercariae of *Schistosoma mansoni*, *Parasitology*, **89**, 461–82.

Frentz, G., Veien, N.K. and Erikson, K. (1977), Immunofluorescence studies in scabies, *Journal of Cutaneous Pathology*, **4**, 191–3.

Fulleborn, F. (1927), Durch Hakenwurmlarven des Hundes (*Uncinaria stenocephala*) beim Menschen erzeugte "Creeping Eruption". *Abhandlungen aus dem Gebiet der Auslandkunde*, Hamburgische Universität, **26**, Reihe D. Medizin, **2**, 121–33.

Georgi, J.R. (1982), *Schistosoma mansoni*: quantitation of skin penetration and early migration by differential external radio assay and autoradiography, *Parasitology*, **84**, 263–81.

Gibson, D.W., Heggie, C. and Connor, D.W. (1980), Clinical and pathological aspects of Onchocerciasis. *Pathology Annual*, **15**, 195–240.

Gibson, D.W., Connor, D.H., Brown, H.L., Fuglsang, H., Anderson, J., Duke, B.O.L. and Buck, A.A. (1976), Onchocercal dermatitis: ultrastructural studies of microfilariae and host tissues, before and after treatment with Diethylcarbamazine (Hetrazan), *The American Journal of Tropical Medicine and Hygiene*, **25**, 74–87.

Gill, H.S. (1986), Kinetics of mast cell, basophil and eosinophil populations at *Hyalomma anatolicum anatolicum* feeding sites on cattle and the acquisition of resistance, *Parasitology*, **93**, 305–15.

Gill, H.S. and Walker, A.R. (1985), Differential cellular responses at *Hyalomma anatolicum anatolicum* feeding sites on susceptible and tick resistant rabbits, *Parasitology*, **91**, 591–607.

Grimaldi, G., Soares, M.J. and Moriarty, P.L. (1984), Tissue eosinophilia and *Leishmania mexicana mexicana* eosinophil interactions in murine cutaneous leishmaniasis, *Parasite Immunology*, **6**, 397–408.

Hass, W. and Schmidt, R. (1982), Characterisation of chemical stimuli for the penetration of *Schistosoma mansoni* cercariae. I. Effective substances, host specificity, *Zeitschrift für Parasitenkunde*, **66**, 293–307.

Hass, W., Granzer, M. and Garcia, G.G. (1987), Host identification by *Schistosoma japonicum* cercariae, *Journal of Parasitology*, **73**, 568–77.

Hawking, F.J., Sewell, P. and Thurston, J.P. (1950), The mode of action of Hetrazan on filarial worms, *British Journal of Pharmacology and Chemotherapy*, **5**, 217–38.

Hayashi, M., Uchiyama, M., Nakajima, H. and Nagai, R. (1986), The immunohistopathologic study of Scabies by the PAP method-identification of IgE positive mast cells, *The Journal of Dermatology*, **13**, 70–3.

Hsu, S.Y.Li., Lust, G.L. and Hsu, H.F. (1971), The fate of challenge schistosome cercariae in a monkey immunised by cercariae exposed to high doses of X-irradiation, *Proceedings of The Society for Experimental Biology and Medicine*, **136**, 727–31.

Hsu, S.Y.Li., Hsu, H.F. and Hanson, H.O. (1981), Immunoglobulins and complement in the skin of Rhesus monkeys immunised with X-irradiated cercariae of *Schistosoma japonicum*, *Zeitschrift für Parasitenkunde*, **66**, 133–43.

Hsu, S.Y.Li., Hsu, H.F., Johnson, S.C., Xu, S.T. and Johnson, S.M. (1983), Histopathological study of the attrition of challenge cercariae of *Schistosoma mansoni* in the skin of mice immunised by chronic infection and by use of highly X-irradiated cercariae, *Zeitschrift für Parasitenkunde*, **69**, 627–42.

Hsu, S.Y.Li., Hsu, H.F., Penick, G.D., Lust, G.L., Osborne, J.W. and Cheng, H.F. (1975), Mechanisms of immunity to schistosomiasis: histopathologic study of lesions elicited in Rhesus monkeys during immunisations and challenge with cercariae of *Schistosoma japonicum*, *Journal of The Reticuloendothelial Society*, **18**, 167–85.

Hsu, S.Y.Li., Hsu, H.F., Penick, G.D., Hanson, H.O., Schiller, H.J. and Cheng, H.F. (1979), Immunoglobulin E, mast cells and eosinophils in the skin of Rhesus monkeys immunized with X-irradiated cercariae of *Schistosoma japonicum*, *International Archives of Allergy and Applied Immunology*, **59**, 383–93.

Incani, R.N. and McLaren, D.J. (1984), Histopathological and ultrastructural studies of cutaneous reactions elicited in naive and chronically infected mice by invading schistosomula of *Schistosoma mansoni*, *International Journal for Parasitology*, **14**, 259–76.

James, S.L. (1985), Induction of protective immunity against *Schistosoma mansoni* by a non-living vaccine is dependent on the method of antigen presentation, *Journal of Immunology*, **134**, 259–76.

James, S.L. (1986), Induction of protective immunity against *Schistosoma mansoni* by a non-living vaccine. III. Correlation of resistance with induction of activated macrophages, *Journal of Immunology*, **136**, 3872–7.

James, S.L. and Cheever, A.W. (1985), Comparison of immune responses between high and low responder mice in the concomitant immunity and vaccine models of resistance to *Schistosoma mansoni*, *Parasitology*, **91**, 301–15.

James, S.L., Correa-Oliveira, R. and Leonard, E.J. (1984), Defective vaccine induced immunity to *Schistosoma mansoni* in P strain mice. II. Analysis of cellular responses, *Journal of Immunology*, **113**, 1587–93.

Johnston, C.M. and Brown, S.J. (1985), Cutaneous and systemic cellular responses induced by the feeding of the argasid tick *Ornithodorus parkeri*, *International Journal for Parasitology*, **15**, 621–8.

Kamiya, H. and McLaren, D.J. (1987), *Schistosoma mansoni*: migration potential of normal and radiation-attenuated parasites in naive guinea pigs, *Experimental Parasitology*, **63**, 98–107.

Kamiya, H., Smithers, S.R. and McLaren, D.J. (1987), *Schistosoma mansoni*: autoradiographic tracking studies of isotopically-labelled challenge parasites in naive and vaccinated CBA/Ca mice, *Parasite Immunology*, **9**, 515–29.

Kemp, D.H. and Bourne, A. (1980), *Boophilus microplus*: the effect of histamine on the attachment of cattle tick larvae—studies *in vivo* and *in vitro*, *Parasitology*, **80**, 487–96.

Kephart, G.M., Gleich, G.J., Connor, D.H., Gibson, D.W. and Ackerman, S.J. (1984), Deposition of eosinophil granule major basic protein onto microfilariae of *Onchocerca volvulus* in the skin of patients treated with Diethylcarbamazine, *Laboratory Investigation*, **50**, 51–61.

Knopf, P.M., Cioli, D., Mangold, B.L. and Dean, D.A. (1986), Migration of *Schistosoma mansoni* in normal and passively immunised laboratory rats, *American Journal of Tropical Medicine and Hygiene*, **35**, 1173–84.

Lanar, D.E., Pearce, E.J., James, S.L. and Sher, A. (1986), Identification of paramyosin as schistosome antigen recognised by intradermally vaccinated mice, *Journal of Immunology*, **136**, 2644–8.

Lavoirpierre, M.M.J., Radovsky, F.J. and Budweiser, P.D. (1979), The feeding process of a Tungid flea, *Tunga monositus* (Siphonaptera; Tungidae), and its relationship to the host inflammatory and repair response, *Journal of Medical Entomology*, **15**, 187–217.

Lee, D.L. (1976), Ultrastructural changes in the infective larvae of *Nippostrongylus brasiliensis* in the skin of immune mice, *Rice University Studies*, **62**, 175–82.

Lewis, D.H. and Peters, W. (1977), The resistance of intracellular Leishmania parasites to digestion by lysosomal enzymes, *Annals of Tropical Medicine and Parasitology*, **67**, 457–62.

Lubran, M. (1950), Determination of Hetrazan in biological fluids, *British Journal of Pharmacology*, **5**, 210–6.

MacKenzie, C.D. (1980), Eosinophil leucocytes in filarial infections, *Transactions of the Royal Society for Tropical Medicine and Hygiene*, **74**, (supplement), 51–8.

MacKenzie, C.D., Preston, P.M. and Ogilvie, B.M. (1978), Immunological properties of the surface of parasitic nematodes, *Nature (London)*, **276**, 826–8.

MacKenzie, C.D., Williams, J.F., Guderian, R.H. and O'Day, J. (1987), Clinical responses in human Onchocerciasis: parasitological and immunological implications, in *Filariasis, Ciba Foundation Symposium*, **127**, pp. 46–72, Chichester: J. Wiley.

Maeda, S., Irie, Y. and Yasuraoka, K. (1982), Resistance of mice to secondary infection with *Schistosoma japonicum*, with special reference to neutrophil enriched response to schistosomula in the skin of immune mice, *Japanese Journal of Experimental Medicine*, **52**, 111–8.

Mahmoud, A.A.F., Warren, K.S. and Peters, P.A. (1975), A role for the eosinophil in acquired resistance to *Schistosoma mansoni* infection as determined by anti-eosinophil serum, *Journal of Experimental Medicine*, **142**, 805–13.

Mangold, B.L. and Knopf, P.M. (1981), Host protective humoral immune responses to *Schistosoma mansoni* in the rat. Kinetics of hyperimmune serum-dependent sensitivity and elimination of schistosomes in a passive transfer system, *Parasitology*, **83**, 559–74.

Mangold, B.L. and Dean, D.A. (1983), Autoradiographic analysis of *Schistosoma mansoni* migration from skin to lungs in naive mice. Evidence that most attrition occurs after the skin phase, *American Journal of Tropical Medicine and Hygiene*, **32**, 785–9.

Mangold, B.L. and Dean, D.A. (1986), Passive transfer with serum and IgG antibodies of irradiated cercaria-induced resistance against *Schistosoma mansoni* in mice, *Journal of Immunology*, **136**, 2644–8.

Mangold, B.L., Dean, D.A., Coulson, P.S. and Wilson, R.A. (1986), Site requirements and kinetics of immune-dependent elimination of intravascularly administered lung stage schistosomula in mice immunised with highly irradiated cercariae of *Schistosoma mansoni*, *American Journal of Tropical Medicine and Hygiene*, **35**, 332–44.

Matsumoto, Y., Perry, G., Levine, R.J.C., Blanton, R., Mahmoud, A.A.F. and Aikawa, M. (1988), Paramyosin and actin in schistosomal teguments, *Nature*, **333**, 76–8.

Mauel, J. (1984), Mechanisms of survival of protozoan parasites in mononuclear phagocytes, *Parasitology*, **88**, 579–92.

Mauel, J., Behin, R., Noerjasin, B. and Doyle, J.J. (1974), Survival and death of leishmania in macrophages, in *Parasites in the Immmunised Host: Mechanisms of Survival, Ciba Foundation Symposium*, **25**, pp. 225–42, Amsterdam: Associated Scientific Publishers.

McElrath, M.J., Kaplan, G., Nusrat, A. and Cohn, Z.A. (1987), Cutaneous leishmaniasis. The defect in T cell influx in BALB/c mice, *Journal of Experimental Medicine*, **165**, 546–59.

McHugh, T.D., Jenkins, T. and McLaren, D.J. (1989), *Strongyloides ratti*: studies of cutaneous reactions elicited in naive and sensitised rats and of changes in surface antigenicity of skin penetrating larvae, *Parasitology*, **98**, 95–103.

McLaren, D.J. (1980), *Schistosoma mansoni*: the parasite surface in relation to host immunity, in Brown, K.N. (Ed.), *Tropical Medicine Research Studies*, **1**, pp. 1–229, Chichester: J. Wiley.

McLaren, D.J. (1982), Role of granulocytes in immune defense against parasites, *Zentralblatt für Bacteriologie und Parasitenkunde*, **12**, 57–73.

McLaren, D.J. (1985), Parasite defense mechanisms, in Venge, P. and Lindbom, A. (Eds.), *Inflammation: Basic Mechanisms, Tissue Injuring Principles and Clinical Models*, pp. 219–54, Sweden: Almquist and Wiksell.

McLaren, D.J. and Terry, R.J. (1982), The protective role of acquired host antigens during schistosome maturation, *Parasite Immunology*, **4**, 129–48.

McLaren, D.J. and James, S.L. (1985), Ultrastructural studies of the killing of schistosomula of *Schistosoma mansoni* by activated macrophages *in vitro*, *Parasite Immunology*, **7**, 315–31.

McLaren, D.J. and Smithers, S.R. (1987), The immune response to schistosomes in experimental hosts, in Rollinson, D. and Simpson, A.J.G. (Eds.), *The Biology of Schistosomes from Genes to Latrines*, pp. 231–63, London: Academic Press.

McLaren, D.J. and Smithers, S.R. (1988), Serum from CBA/Ca mice vaccinated with irradiated cercariae of *Schistosoma mansoni* protects naive recipients through the recruitment of cutaneous effector cells, *Parasitology*, **97**, 287–302.

McLaren, D.J., Worms, M.J. and Askenase, P.W. (1983a), Cutaneous basophil-associated resistance to ectoparasites (ticks). II. Electron microscopy of *Rhipicephalus appendiculatus* feeding sites in actively sensitised guinea pigs and recipients of immune serum, *Journal of Pathology*, **139**, 291–308.

McLaren, D.J., Peterson, C.G.B. and Venge, P. (1984), *Schistosoma mansoni*: further studies of the interaction between schistosomula and granulocyte-derived cationic proteins *in vitro*, *Parasitology*, **88**, 491–553.

McLaren, D.J., Pearce, E.J. and Smithers, S.R. (1985), Site potential for challenge attrition in mice, rats and guinea pigs vaccinated with irradiated cercariae of *Schistosoma mansoni*, *Parasite Immunology*, **7**, 29–44.

McLaren, D.J., Strath, M. and Smithers, S.R. (1987), *Schistosoma mansoni*: evidence that immunity in vaccinated and chronically infected CBA/Ca mice is sensitive to treatment with a monoclonal antibody that depletes cutaneous effector cells, *Parasite Immunology*, **9**, 667–82.

McLaren, D.J., Worms, M.J., Brown, S.J. and Askenase, P.W. (1983b). Quantitative and ultrastructural studies of cutaneous basophil responses elicited in guinea pigs by the Argasid tick *Ornithodorus tartakovskyi*, *Experimental Parasitology*, **56**, 153–68.

Moloney, N.A., Hinchcliffe, P. and Webbe, G. (1987), Passive transfer of resistance to mice with sera from rabbits, rats or mice vaccinated with ultraviolet-attenuated cercariae of *Schistosoma japonicum*, *Parasitology*, **94**, 497–508.

Monroy, A., Ridley, D.S., Heather, C.J. and Ridley, M.J. (1980), Histological studies of the elimination of *Leishmania enriettii* from skin lesions in the guinea pig, *British Journal of Experimental Pathology*, **61**, 601–10.

Moqbel, R. (1980), Histopathological changes following primary, secondary and repeated infections of rats with *Strongyloides ratti*, with special reference to tissue eosinophils, *Parasite Immunology*, **2**, 11–27.

Mukherjee, S., Ghosh, C. and Basu, M.K. (1988), *Leishmania donovani*: role of microviscosity of macrophage membrane in the process of parasite attachment and internalisation, *Experimental Parasitology*, **66**, 18–26.

Muller, R. (1971), Dracunculus and Dracunculiasis, in *Advances in Parasitology*, **9**, 73–151.

Nelson, W.A. and Bainborough, A.R. (1963), Development in sheep of resistance to the ked *Melophagus ovinus (L.)*. III. Histopathology of sheep skin as a clue to the nature of resistance, *Experimental Parasitology*, **13**, 118–27.

Nelson, W.A., Bell, J.F. and Stewart, S.J. (1979), *Polyplax serrata*: cutaneous cytologic reactions in mice that do (CFW strain) and do not (C57Bl strain) develop resistance, *Experimental Parasitology*, **48**, 259–64.

Nelson, W.A., Clifford C.M., Bell, J.F. and Hestekin, B. (1972), *Polyplax serrata*: histopathology of the skin of louse-infested mice. *Experimental Parasitology*, **31**, 194–202.

Nithiuthai, S. and Allen, J.A. (1984), Effects of ultraviolet irradiation on the acquisition and expression of tick resistance in guinea pigs, *Immunology*, **51**, 153–60.

Oshman, R., Knopf, P.M., von Lichtenberg, F. and Byram, J.E. (1986). Effects of protective immune serum on the yields of parasites and pulmonary cell reactions in schistosome infected rats, *American Journal of Tropical Medicine and Hygiene*, **35**, 523–30.

Parkhouse, R.M.E., Bofill, M., Gomez-Priego, A. and Janossy, G. (1985), Human macrophages and T-lymphocyte subsets infiltrating nodules of *Onchocerca volvulus*, **62**, 13–8.

Pearce, E.J. and McLaren, D.J. (1986). *Schistosoma mansoni*: the cutaneous response to cercarial challenge in naive guinea pigs and guinea pigs vaccinated with highly irradiated cercariae, *International Journal for Parasitology*, **5**, 465–79.

Pritchard, D.I., McKean, P.G. and Rogan, M.T. (1988), Cuticular collagens—a concealed target for immune attack in hookworms, *Parasitology Today*, **4**, 239–41.

Ridel, P.R., Esterre, P., Dedet, J.P., Pradinau, R., Santoro, F. and Capron, A. (1988), Killer cells in human cutaneous leishmaniasis, *Transactions of the Royal Society of Tropical Medicine and Hygiene*, **82**, 223–5.

Ridley, D.S. and Ridley, M.J. (1983), The evolution of the lesion in cutaneous leishmaniasis, *Journal of Pathology*, **141**, 83–96.

Ridley, D.S., Marsden, P.D., Cuba, C.C. and Barreto, A.C. (1980), A histological classification of mucocutaneous leishmaniasis in Brazil and its clinical evaluation, *Transactions of the Royal Society of Tropical Medicine and Hygiene*, **74**, 508–14.

Roitt, I.M., Brostoff, J. and Male, D.K. (1985), *Immunology*, Edinburgh: Churchill Livingstone.

Salafsky, B., Wang, Y.S., Fusco, A.C. and Antonacci, J. (1984), The role of essential fatty acids and prostoglandins in cercarial penetration (*Schistosoma mansoni*), *Journal of Parasitology*, **70**, 656–700.

Savage, A.M. and Colley, D.G. (1980), The eosinophil in the inflammatory response to cercarial challenge of sensitised and chronically infected CBA/J mice, *American Journal of Tropical Medicine and Hygiene*, **29**, 1268–78.

Seitz, H.M., Cottrell, B.J. and Sturrock, R.F. (1988), A histological study of skin reactions of baboons to *Schistosoma mansoni* schistosomula, *Transactions of the Royal Society of Tropical Medicine and Hygiene*, **81**, 385–90.

Sher, A., Sacks, D.L. and Scott, P.A. (1983), Host and parasite factors influencing the expression of cutaneous leishmaniasis, in *Cytopathology of Parasitic Disease*, *Ciba Foundation Symposium*, **99**, pp. 174–89, London: Pitman Books.

Smithers, S.R. and Gammage, K. (1980), Recovery of *Schistosoma mansoni* from the skin, lungs and hepatic portal system of naive mice and mice previously exposed to *S. mansoni*: evidence for two phases of immune attrition in immune mice, *Parasitology*, **80**, 289–300.

Smithers, S.R., McLaren, D.J. and Ramalho Pinto. (1977), Immunity to schistosomes: the target. *American Journal of Tropical Medicine and Hygiene*, **26**, 11–9.

Spencer, H. (1973), Nematode diseases II, in *Tropical Pathology*, pp. 512–59, New York: Springer Verlag.

Stirewalt, M.A. (1974), *Schistosoma mansoni*: cercaria to schistosomula, *Advances in Parasitology*, **12**, 115–82.

Stromberg, P.C. and Fisher, W.F. (1986), Dermatopathology and immunity in experimental *Psoroptes ovis* (Acari: Psoroptidae) infestation of naive and previously exposed Hereford cattle, *American Journal of Veterinary Research*, **47**, 1551–60.

Susuki, T. and Damian, R.T. (1981), Schistosomiasis mansoni in baboons. IV. The development of antibodies to *Schistosoma mansoni* adult worm, egg and cercarial antigens in acute and chronic infections, *American Journal of Tropical Medicine and Hygiene*, **30**, 825–35.

Tatchell, R.J. and Bennett, G.F. (1969), *Boophilus microplus*: antihistaminic and tranquilising drugs and cattle resistance, *Experimental Parasitology*, **26**, 369–77.

Turk, J.L. and Bryceson, A.D.M. (1971), Immunological phenomena in leprosy and related diseases, *Advances in Immunology*, **13**, 209–66.

van Neste, D. and Lachapelle, J.M. (1981), Host parasite relationships in hyperkeratotic (Norwegian) scabies: pathological and immunological findings, *British Journal of Dermatology*, **105**, 667–78.

Vetter, J.C.M. and Leegwater v.d. Linden, M.E. (1977a), Skin penetration of infective hookworm larvae. I. The path of migration of infective larvae of *Ancylostoma braziliense* in canine skin, *Zeitschrift für Parasitenkunde*, **53**, 255–62.

Vetter, J.C.M. and Leegwater v.d. Linden, M.E. (1977b), Skin penetration of infective hookworm larvae. II. The path of migration of infective larvae of *Ancylostoma braziliense* in the metacarpal foot pads of dogs, *Zeitschrift für Parasitenkunde*, **53**, 263–6.

von Lichtenberg, F., Sher, A., Gibbons, N. and Doughty, B.L. (1976), Eosinophil-enriched inflammatory response to schistosomula in the skin of mice immune to *Schistosoma mansoni*, *American Journal of Pathology*, **84**, 479–500.

Walker, A.R. and Fletcher, J.D. (1986), Histological study of the attachment sites of adult *Rhipicephalus appendiculatus* on rabbits and cattle, *International Journal for Parasitology*, **16**, 399–413.

Ward, R.E.M. and McLaren, D.J. (1988), *Schistosoma mansoni*: evidence that eosinophils and/or macrophages contribute to skin phase challenge attrition in the vaccinated CBA/Ca mouse, *Parasitology*, **96**, 63–84.

Wikel, S.K. (1982), Histamine content of tick attachment sites and the effects of H1 and H2 histamine antagonists on the expression of resistance, *Annals of Tropical Medicine and Parasitology*, **76**, 179–85.

Wikel, S.K. and Allen, J.R. (1976), Acquired resistance to ticks. I. Passive transfer of resistance, *Immunology*, **30**, 311–6.

Wikel, S.K. and Allen, J.R. (1977), Acquired resistance to ticks. III. Cobra venom factor and the resistance response, *Immunology*, **32**, 457–65.

Willadsen, P. and Kemp, D.H. (1988), Vaccination with 'concealed' antigens for tick control, *Parasitology Today*, **4**, 196–8.

Willadsen, P., Wood, G.M. and Riding, G.A. (1979), The relationship between histamine concentration, histamine sensitivity and resistance of cattle to the tick *Boophilus microplus*, *Zeitschrift für Parasitenkunde*, **59**, 87–94.

Wilson, R.A. (1988), Development and migration in the mammalian host, in Rollinson, D. and Simpson, A.J.G. (Eds.), *The Biology of Schistosomes from Genes to Latrines*, pp. 115–46, London: Academic Press.

Wilson, R.A., Coulson, P.S. and McHugh, S.M. (1983), A significant part of the concomitant immunity of mice to *Schistosoma mansoni* is a consequence of a leaky hepatic portal system not immune killing, *Parasite Immunology*, **5**, 595–601.

Wilson, R.A., Coulson, P.S. and Dixon, B. (1986), Migration of the schistosomula of *Schistosoma mansoni* in mice vaccinated with radiation-attenuated cercariae, and normal mice: an attempt to identify the site and timing of parasite death, *Parasitology*, **92**, 101–16.

9. Pulmonary immune responses to parasites

R.A. Wilson

9.1. INTRODUCTION

The lungs play a significant role in the life cycle of several groups of parasites, either as an organ on the migration route to some other location or as the site of parasitization (Figure 9.1.). In many respects their structure is organized to deal with inhaled viral and bacterial pathogens and noxious substances. Most parasites, on the other hand, enter via the vasculature or pleural cavity. The majority are helminths, protozoans being conspicuously less successful colonizers. Before the different groups of parasites are considered, the relevant aspects of lung structure will be reviewed. For more detailed information the reader should consult publications by Wheater *et al.* (1979)., Jeffery and Corrin (1984) and Spencer (1985).

9.2. STRUCTURAL FEATURES

The flattened epithelial layers covering the inner surface of the thoracic cavity and outer surface of the lungs are termed the parietal and visceral pleura, respectively. The pleural cavity, the potential space in between, normally contains so little fluid that the two pleural surfaces are closely apposed. They meet at the hilum of the lungs, the region where blood vessels, nerves and airways enter. The lobes of the lungs are imperfectly divided into subunits or lobules by fibrous septa continuous with the visceral pleura.

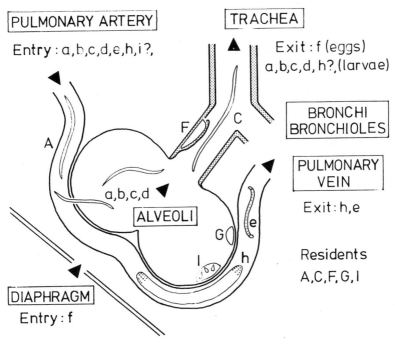

Figure 9.1. Diagrammatic summary of parasites associated with the lungs (not to scale): Aa, *Angiostrongylus*; b, *Ascaris*; Cc, *Dictyocaulus*; d, *"hookworms"*; e, microfilariae; Ff, *Paragonimus*; G, *Pneumocystis*; h, schistosomes; Ii, *Toxoplasma*.

Airways

Bronchi and bronchioles

The tracheal airway divides after entering the thorax to form the primary bronchi. In man about 20 more divisions separate primary bronchi from terminal bronchioles. Each of the latter gives rise to transitional structures, the respiratory bronchioles, alveolar ducts, and alveolar sacs which finally open into alveoli. Collectively this assemblage, centred on a terminal bronchiole, forms the basic respiratory unit of the lung, the acinus. (There are significant differences in structural organization between species at this level; McLaughlin *et al.*, 1961.)

Microscopically, the airways comprise a lining mucosa of epithelial cells supported by elastic connective tissue, the lamina propria. Beneath this is the submucosa in which lie glands, muscle and cartilage. The whole is bounded by a thin adventitial coat which merges with surrounding tissues. The cells of the epithelium are ciliated and columnar in the proximal airways but become cuboidal in more distal regions. Up to eight different types of cell have been identified, amongst which mucus-secreting goblet cells are prominent. In normal airway epithelium, lymphocytes (often IgA positive), globular leucocytes of uncertain function, and mast cells are found, the last increasing in distal regions to densities 150 times those in the trachea. The submucosal glands are sparse in smaller bronchi and are rarely found in the bronchioles, which also lack cartilage. However, a layer of smooth muscle is prominent beneath the lamina propria of the bronchioles. The tonus of these muscle fibres, which spiral round the bronchiole wall, controls the resistance to air flow in the lungs—a feature of great significance in allergic asthma.

Alveoli

The alveoli are separated from each other by thin septa of connective tissue, the interstitium. Each is lined by Type 1 flattened, alveolar epithelial cells and Type 2 cells secreting lung surfactant. The pulmonary capillaries form a network within the interstitium presenting first at one alveolar surface and then in an adjacent alveolus. This means that the blood-air barrier may be as little as 0.15 μm or as much as 12 μm wide. The interstitium contains collagen and elastin fibres, cells, and nerve endings. Fibroblasts, mast cells, and monocytes in transit from blood to alveoli are present in normal non-inflamed tissue. There is a continuous flow of fluid from the capillaries through the interstitium to the distal lymphatic channels. In contrast, the normal alveolar epithelium is relatively impermeable to fluids, antigens or particles. In pathological states the fused basement membranes of the alveolar epithelium and capillary endothelium may be forced apart by infiltrating inflammatory cells, fluid exudates or collagen deposition (Turner-Warwick, 1983).

Alveolar macrophages

Alveolar macrophages comprize >95% of the cells found in the film of liquid lining normal alveoli (lymphocytes make up the residue). They are avidly phagocytic and clear the alveolus of inhaled dust and bacteria; they also remove spent surfactant. There is a continuous recruitment of precursor monocytes from the blood; the life of a resident macrophage, may be as long as three to four weeks (Blusse van Oud Alblas and van Furth, 1979). The fate of the vast majority of macrophages is to be carried up the airways by ciliary transport and swallowed or expectorated. A very small proportion (bearing antigen) may re-enter the interstitium and migrate via the lymphatics to lymph nodes (Lipscomb *et al.*, 1982; Corry *et al.*, 1984). However, alveolar macrophages appear to function only poorly as accessory cells of T-lymphocyte activation, compared to blood or peritoneal macrophages (Holt, 1986). Indeed they appear to possess a powerful suppressive capacity *in vitro* which may be mediated by prostaglandins (Kaltreider *et al.*, 1986). Thus, in common with other mucosal surfaces, the lungs are protected from over-reacting to environmental, non-harmful substances through mechanisms which down regulate undesirable responses. A potential *in vivo* role in antigen presentation for those alveolar macrophages which migrate to regional lymphoid tissue awaits clarification.

Vasculature

The lungs receive their major blood supply via the pulmonary arteries which divide and course alongside the branching airways (Figure 9.2). Most small pulmonary arteries enter an acinus alongside its terminal bronchiole and give

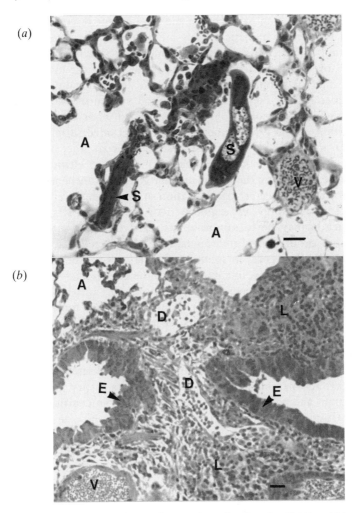

Figure 9.2. (a) Light micrograph of normal mouse lung showing alveoli (A) and blood vessels (V). A schistosomulum larva of *Schistosoma mansoni* (S), caught in the act of entering the air spaces from a blood vessel, spans several alveoli and is much larger than host leucocytes. Bar = 20 μm. (b) Bronchiole lined with epithelium (E), and peribronchiolar lymphoreticular tissue (L) plus lymphatic duct (D) containing lymphocytes (Crabtree and Wilson, 1986). Bar = 20 μm.

rise to the meshwork of capillaries situated in the alveolar walls. These regroup at the periphery of the acinus to form the pulmonary veins carrying blood back to the hilum alongside the airways. In two respects the pulmonary arteries are unusual; they have relatively thin walls and they are capable of elastic expansion and recoil.

Lymphatics and lymph nodes

The pulmonary lymphatics are important in clearing away excess fluid (oedema) from the lung parenchyma and have a large reserve capacity for this purpose. Terminal lymphatic vessels are absent at the level of the alveoli, commencing in the walls of the bronchioles. The lymphatic vessels drain outwards towards the visceral pleura and inwards to the major airways and hilum. In man the lymph passes through hilar, tracheal and mediastinal lymph nodes before joining the systemic circulation via the thoracic duct. In small animals, such as the mouse, only the mediastinal node is prominent.

In addition to the lymph nodes already mentioned, discrete aggregations of follicular lymphoid tissue are associated with the bronchi (BALT; Bienenstock, 1984). They are thought to represent elements in common mucosal systems uniting the gut, mammary glands, bronchi and cervix. BALT follicles have their greatest concentration at airway bifurcations. They have an arterial blood supply and capillary network but lack afferent lymphatics. Their function may be to sample interstitial lymph for antigens since they contain both dendritic and interdigitating macrophages. The lymphocytes of BALT are mostly B cells; in some species the majority bear IgA (cf. Peyer's Patches of the gut). Thus, in common with the gut-associated lymphoid tissue (GALT), BALT can generate a population of IgA-bearing cells. The recruitment of these lymphocytes to the bronchial mucosa endows the pulmonary airways with the capacity to mount a local immune response to inhaled antigens, mediated by secretory IgA.

Less well organized lympho-reticular aggregates are widely scattered through the lung tissue in the pleura, interlobular septa, and interposed between the terminal bronchiole and its artery. In this situation they are well placed to monitor fluid and cells draining from the alveolar interstitium.

9.3.   PARASITES INHABITING THE LUNGS

Paragonimus sp., the lung flukes

Trematodes of the genus *Paragonimus* are parasites of the lungs of man and other mammals in many parts of the world. Infection results from the ingestion of metacercarial cysts in the raw flesh of freshwater crustaceans. The larva excysts in the ileum and penetrates the intestinal wall to reach the peritoneal cavity. Migration to the lungs can be a prolonged process and *P.*

ohirai in rats spends some time in the liver parenchyma (Kawaguchi *et al.*, 1983). The pleural cavity is reached by penetration of the diaphragm, and between three and eight weeks after infection worms enter the lung tissues. Egg laying begins after about 45 days. The mature worms inhabit cysts of modified lung tissue, up to 4 cm in diameter in *P. westermani* infections of man (these cysts can persist for up to 20 years; Barrett-Connor, 1982).

Pathology

It is not surprizing that a parasite with the capacity for tissue migration should generate strong inflammatory reactions during its progress to the bronchioles. Fibrin deposition and inflammation of the pleura occur during passage through the pleural cavity (Choi *et al.*, 1979). Puncture marks appear on the lung surface where worms have penetrated and linear tracks with haemorrhagic edges pass through the lung parenchyma. Neutrophils and eosinophils are found free in the alveoli and in diffuse pleural exudates.

When worms reach the terminal bronchioles they cause modification of host tissues to form the characteristic cyst. Its luminal surface is lined with transformed alveolar or bronchiolar epithelium (Choi *et al.*, 1979; Lee, 1979). The cyst may actually block the bronchiole in which it lies, damaging the more distal regions of an acinus. Mast cells and a few plasma cells mingle with the epithelial lining cells of the cyst, and larger numbers of plasma cells, surrounded by dense collagen deposits, lie towards its outer surface. By six months the cyst has stabilized. The symptoms of infection, a persistent cough and the production of a rusty sputum containing parasite eggs, can persist for years. A proportion of the parasite eggs released into the bronchioles are transported into the alveoli, possibly as a result of bouts of coughing. They provoke a granulomatous reaction composed of macrophages, neutrophils and fibroblasts (Lee, 1979). Sclerosis of blood vessels in the granulomata can occur, and areas of the lung become filled with fluid.

Immunity

ANTIBODY RESPONSES
Circulating IgM, IgG and IgE antibodies to *Paragonimus* have been detected in rats (Ikeda and Fujita, 1980). Whereas IgM and IgG levels rose from three weeks post-infection, IgE was detected at two weeks, peaked at three, and then declined. The IgE peak coincided with migration through the peritoneal cavity. When worms were injected into the pleural cavity, the resulting IgE titre was much lower. The decline in IgE was abrogated by whole body irradiation of rats at two weeks, suggesting the existence of a radiation-sensitive balance between helper and suppressor functions of T cells (Ikeda and Fujita, 1982). Cells producing IgE are present in greatest numbers in

lymph node draining sites on the migration route, particularly the mediastinal node (Ikeda *et al.*, 1982; cf. Ascaris).

CELLULAR RESPONSES

During peritoneal migration in rats, there is a pronounced infiltration of leucocytes peaking at day 30, up to 35% of which are eosinophils. A transient peripheral eosinophilia is also observed, peaking at 25 days. This is followed by a comparable eosinophilia in the pleural cavity, although the total influx of cells is much less (Kawaguchi *et al.*, 1983). An increase in the numbers of mast cells in the lungs of rats infected with *P. iloktsuenesis* is stimulated by arrival of worms in the pleural cavity and lung parenchyma at three to four weeks (Choi and Rah, 1981). The extent of peribronchial lymphatic tissue also increases. The fragmentary information on host responses to *Paragonimus* points to the development of immediate hypersensitivity to migrating worms (cf. *Nippostrongylus*). Unlike infections with *Nippostrongylus*, the reactions rapidly subside once mature parasites are established in the bronchiolar cysts; this could indicate reduced inflammatory potential, or a more active suppression of host responses by the mature parasite. In the chronically infected host, immune mechanisms operating against challenge parasites (whether at the level of the gut wall, peritoneal cavity, or lungs) are unclear.

The metastrongyle nematodes

Nematodes of the family *Metastrongylidae* are parasites of the respiratory system in many species of mammal. Pulmonary responses to *Angiostrongylus cantonensis* and *Dictyocaulus viviparus* are discussed below as representative of the group. The pulmonary pathology associated with metastrongyle infections was reviewed by Stockdale (1976).

Angiostrongylus cantonensis

Angiostrongylus cantonensis is a natural parasite of the rat which becomes infected by ingesting third-stage larvae. The life cycle is shown in Figure 9.3, but in the present context it is important to note that having completed development in the brain, the worms leave via the cerebral veins and pass through the heart to lodge in the pulmonary arteries (Ko, 1981). The thread-like adults, about 2 cm long, are restricted to the larger vessels with no evidence of penetration into the lung parenchyma (Sodeman *et al.*, 1969).

PATHOLOGY

Oviposition occurs in the bloodstream and eggs embolize widely throughout the lung parenchyma. The embolization and impaction of eggs causes thrombosis and infarction (localized death) of lung tissue. Eggs in the lung parenchyma become surrounded by inflammatory infiltrates and the walls of blood vessels are disrupted with consequent haemorrhage. Larvae hatch from

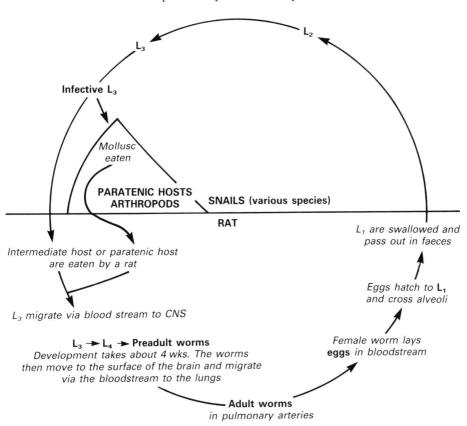

In abnormal hosts such as mice and man, severe complications may arise from the natural tendency of larval stages to seek out the CNS for development. The parasites die on route generating intense inflammation, particularly eosinophilia in the cerebrospinal fluid.

Figure 9.3. Life cycle of *Angiostrongylus cantonensis*.

eggs in the lungs and pass to the lumina of bronchioles via the interstitium and connective tissues, or the alveoli; they leave the host via the trachea and alimentary tract. The alveolar interstitium is infiltrated with inflammatory leucocytes, and granulomatous lesions develop around eggs and larvae in the parenchyma. Immune deposits of IgG and C3 occur within the granulomata (Takai, 1983). The extent and nature of pulmonary responses is strain-dependent (Yoshimura *et al.*, 1979a). It is unclear whether the granulomatous responses are generated by antibody-antigen interactions or delayed-type hypersensitivity.

Pathological reactions to *Angiostrongylus* infection also occur in the walls of the pulmonary arteries (Takai, 1983). Deposits of immune complexes containing IgG ad C3 and fibrin are present, and infiltration of plasma cells occurs. The vascular changes result in increased heart weight, the pulmonary side being swollen. There is also leucocytic infiltration between the heart muscle fibres (cf. *Schistosoma* eggs in the lungs).

IMMUNITY

A small dose of 5–50 normal *Angiostrongylus cantonensis* larvae will stimulate protective immunity in rats (Au and Ko, 1979). The immunity is incomplete, some primary worms surviving or secondary worms becoming established upon challenge (Yong and Dobson, 1983). The immunity can be passively transferred to naive rats with serum and sensitized cells (Yong and Dobson, 1982b; Yong *et al.*, 1983) although the procedure is not as protective as naturally acquired immunity. Antibody-mediated immunity appears to involve circulating IgM and IgG classes. In marked contrast to *Ascaris* or *Nippostrongylus* infections, little IgE antibody seems to be produced against *Angiostrongylus* (Yoshimura and Soulsby, 1976; Yong and Dobson, 1982b). Paradoxically, peripheral eosinophilia has been observed in rats immediately after infection, the response being enhanced after each successive reinfection (Yong and Dobson, 1984).

There is little evidence that antigen presentation in the lungs plays a significant part in the induction of immunity to *Angiostrongylus*. Termination of an infection by chemotherapy before parasites reach the lungs is more effective at inducing protection than transplantation of juveniles to the pulmonary arteries (Yoshimura *et al.*, 1979b). Two exposures to 200 radiation-attenuated larvae were very effective, inducing 97% protection to challenge; few, if any, immunizing larvae reached the lungs (Lee, 1969). The conclusion that the earlier migratory stages stimulate protection is reinforced by the observation of lymphocyte proliferative activity in the cervical lymph nodes draining the head region (Yoshimura and Soulsby, 1976).

It is also likely that the effector arm of the protective immune response operates earlier in migration than the lungs. Once established, adult worms can survive in the pulmonary artery for over a year in an otherwise protected host (Yong and Dobson, 1982a). The only direct benefit to the host of the pulmonary inflammatory events detailed above must lie in the minimization of tissue damage. There would of course be indirect effects on the population dynamics of the parasite via a constraint on the flow of infective larvae into the external environment.

Dictyocaulus viviparus

For nearly 30 years, *Dictyocaulus viviparus*, the lungworm of cattle, has had the distinction of being the only parasite against which a commercial vaccine (Dictol, Glaxo) is available (see Section 14.5). Infection results from the ingestion of third-stage larvae which exsheath in the intestine and penetrate the gut wall. They are said to enter the lymphatics and travel via the mesenteric lymph nodes to enter the venous circulation via the thoracic duct. The fourth-stage larvae arrive in the lungs about 7 days post-infection (Jarrett and Sharp, 1963) and reach the airways by breaking out into the alveoli. Mature worms are found from 22 days in the lumen of bronchi and bronchioles. Eggs released by the female are carried up the trachea and leave via the alimentary tract, hatching whilst in transit.

PATHOLOGY

Parasitic bronchitis, the disease caused by *D. viviparus* is commonest in young calves turned out to grass in the summer and autumn. Its severity is related to the number of larvae ingested (Jarrett et al., 1957a). With heavy infections (50 000 larvae), there is a marked rise in respiratory rate by 14 days with gasping for breath and short, shallow respiration—death follows in the third week of infection. Lower doses of larvae (5 000) result in respiratory rates as high as 100^{-1} min during the fourth week. The frequent coughing gives the disease its common name of 'husk'. By the fifth to sixth week the post-patent phase of disease begins with an improvement in condition. The number of adult worms in the bronchi diminishes to very low levels, but clinical signs may persist for several months. Some animals may never fully recover, showing the condition of bronchiectasis—a widening of the bronchi and continued coughing of purulent sputum.

The arrival of *Dictyocaulus* larvae in the lungs with consequent pathological changes obviously has dramatic effects on pulmonary function. Intra-alveolar haemorrhage occurs from day seven onwards as larvae break out into the alveoli (Jarrett et al., 1957b). Small clusters of alveoli collapse; their walls are infiltrated by mononuclear cells and the lumina with neutrophils, eosinophils and giant cells. As fourth-stage larvae enter the bronchioles, they provoke a peribronchial eosinophilia with damage to the epithelial lining. Eosinophils dominate the exudates in the bronchi, and there is a production of excess mucus by the glands in the bronchial walls. In heavy infections emphysema (the entry of air into the tissues) develops, followed by pulmonary oedema and the appearance of hyaline material in collapsed alveoli, all features intimately connected with the respiratory difficulties of the host. The broncho-mediastinal lymph nodes are greatly enlarged at this time suggesting active lymphocyte proliferation in response to antigen release by the larvae.

The maturation of adult worms in the bronchi gives rise to further damage to the respiratory epithelium. A second phase of injury to the alveoli ensues as granulomata develop around aspirated eggs and larvae (cf. *Paragonimus*). The granulomata contain many giant cells, presumably derived from fused macrophages. From about 35 days onwards in affected alveoli, there is increasing replacement of Type 1 alveolar cells by cuboidal epithelium (epithelialization); the condition is associated with masses of dead lungworm debris (Jarrett et al., 1957b). Interstitial emphysema is commonly observed, particularly in the septa between lung lobules and in the lymphatic channels.

IMMUNITY

Calves exposed to small numbers of normal larvae show a strong resistance to secondary challenge (Rubin and Lucker, 1956) which declines only slowly over many months. Complement-fixing antibodies (IgM, IgG) have been detected about 40 days after infection, the titre peaking around 100 days, long after infection has waned (Jarrett *et al.*, 1957a). A secondary or tertiary infection stimulates a rapid, more elevated antibody response (Cornwell and Michel, 1960). The relevance of antibodies in immunity was demonstrated by the passive transfer of a high level of protection (95%) to naive calves by the globulin fraction of immune serum (Jarrett *et al.*, 1957a).

The success of the Dictol vaccine against *Dictyocaulus* derives partly from the strong immunity induced by a normal infection and partly from the long shelf life of the irradiated infective larvae. The objective of irradiation was to permit larvae to migrate far enough to induce protection without producing extensive pathology or clinical disease (Jarrett *et al.*, 1960). Two doses of 1000 attenuated larvae, separated by a 30-day interval, were sufficient (Jarrett *et al.*, 1958); the larvae reached the lungs in significant numbers but failed to develop (Jarrett and Sharp, 1963). Lymphocyte proliferation occurred in both mesenteric and broncho-mediastinal lymph nodes, and disintegrating larvae were frequently observed in the latter. Jarrett and Sharp (1963) also described lympho-reticular, broncho-occlusive lesions containing damaged or dead larvae, after a second exposure to irradiated (or normal) parasites. Initially these consisted of inflammatory reactions, rich in eosinophils, in the lumen of a bronchiole. The lesions developed to contain macrophages, plasma cells and actively proliferating lymphoid tissue. Eventually, they reached 5 mm in diameter, occupying an entire lobule, and took on the appearance of a small lymph node with germinal centres.

Pirie *et al.*, (1971) concluded that there was a direct relationship between the number of nodules and the degree of immunity. It would thus appear that the immune effector mechanisms active in the lungs against challenge larvae operate by positive feedback to amplify that protective response and develop into the lymphoid nodules. The nature of the effector mechanism is unclear (Armour, 1987); circumstantial evidence points to cooperation between circulating antibodies and leucocytes, probably eosinophils. It would be interesting to know the class(es) of antibody produced by the abundant plasma cells in the lungs of immune calves.

As a footnote to the use of Dictol, it is worth recording that some workers have advocated the use of chemotherapy in calves showing clinical symptoms of parasitic bronchitis as an alternative to vaccination (Downey, 1980; Oakley, 1980). Urquhart *et al.*, (1981) reviewed this proposal reporting that clinical symptoms after challenge of such treated animals were not significantly ameliorated. (The lungs of these animals had already been damaged by the primary exposure.) They concluded that vaccination offers the only effective method of prophylaxis against *Dictyocaulus* infection in cattle.

9.4. PARASITES WHICH ENTER VIA BLOOD VESSELS AND LEAVE VIA THE AIRWAYS

Nippostrongylus brasiliensis

Nippostrongylus brasiliensis, a natural parasite of the rat, has been intensively studied and will serve as a model (see Chapter 8, Figure 8.7). Transit through the skin is rapid, and by 15–24 h the majority of larvae have migrated via the bloodstream or lymphatics to the lungs (Love *et al.*, 1974). The larvae break out into the alveoli, begin to feed, and moult to the fourth-larval stage. By 59 h over 90% of larvae have migrated up to the trachea and reached the intestine. The lifespan in the intestine is normally short, adult worms being expelled from day 11 onwards. Thereafter, rats are strongly immune to reinfection.

Cellular responses during primary and secondary infection

Taliaferro and Sarles (1939) carried out a thorough histopathological study of host responses to migrating *Nippostrongylus* in the lungs of normal, immune and hyperimmune rats. After a primary exposure, migration through the lungs occurred before immunity had developed, and inflammatory responses were correspondingly mild. There were small areas of haemorrhage and fibrin deposition in the alveoli (except where secondary pneumonia complicated events). Over the second and third days small numbers of eosinophils and neutrophils migrated from blood vessels into alveoli; leucocytes also infiltrated the perivascular connective tissue. By day 14 the appearance of the lungs had returned to normal. After a primary infection many larvae in the lungs have been observed unconnected with host reactions and the majority reach the intestine (Love *et al.*, 1974).

The leucocyte populations of the airways of rats after a primary exposure, sampled by broncho-alveolar lavage, showed a biphasic response (Figure 9.4). The early component reflected migration of larvae through the lungs whereas the second peak at day 16 coincided with the expulsion of adult worms from the intestine and occurred in the absence of lung parasites (Egwang *et al.*, 1984a).

In rats exposed for a second time, clear evidence was found for retarded larval migration through the lungs (Taliaferro and Sarles, 1939); after a third or fourth infection, a prolonged sojourn and retention occurred (Love *et al.*, 1974). The arrival of challenge larvae in the lungs provoked a marked exudative inflammation in the alveoli and in peribronchial and perivascular tissues. Eosinophils were very prominent in the infiltrates, and some larvae became surrounded by granulomatous reactions (termed nodules by Taliaferro and Sarles, 1939). Macrophages within the granulomata fused to form giant cells around trapped worms. Lavage of the lungs after challenge has revealed a

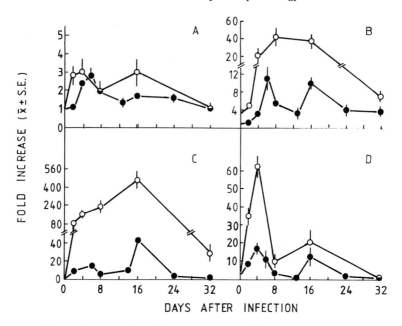

Figure 9.4. Changes in the numbers of leucocytes recovered from the lungs of rats by broncho-alveolar lavage following primary (●) and secondary (○) infection with *Nippostrongylus brasiliensis*. A, macrophages; B, lymphocytes; C, eosinophils; D, neutrophils. Counts expressed as fold increase relative to normal uninfected rats (Egwang *et al.*, 1984a).

marked anamnestic increase in leucocytes of all classes (Figure 9.4). The magnitude for lymphocytes, neutrophils and eosinophils was greater than after a primary infection, and the macrophage infiltration occurred two days earlier (Egwang *et al.*, 1984a).

A secondary exposure to *Nippostrongylus* leaves its imprint on the lungs. Enhanced activity in and increased amounts of peribronchial lymphoid tissue were observed, and large numbers of eosinophils and plasma cells remained in peribronchial and perivascular regions (Taliaferro and Sarles, 1939). The plasma cells were predominantly IgA-secreting and numbers accumulated around the granulomata (Salman and Brown, 1980).

Alveolar macrophages

Alveolar macrophages activated as a result of *Nippostrongylus* infection express Fc receptors for IgA—this represents a potential arming of the cells (Gauldie *et al.*, 1983). There is also an increase in the number of macrophages bearing C3 receptors (Egwang *et al.*, 1984b). Increased receptor expression coincides with the development of the ability to kill L3 larvae *in vitro*, a property not possessed by normal alveolar macrophages, suggesting that the lungs become an increasingly hostile environment for migrating *Nippostrongylus*.

Other immunological processes may be at work in the lungs of animals exposed to *Nippostrongylus*. An increase in the serum level of two acute-phase proteins has been detected in mice (Lamontagne *et al.*, 1984). Large amounts of a third acute-phase product (α1 protease inhibitor) were present in alveolar macrophages; this observation provoked a detailed examination of the properties of these cells after infection. By two days (coincident with lung migration), alveolar macrophages spontaneously release interleukin 1 and hepatocyte stimulating factor and hence may stimulate the liver to produce acute-phase proteins (Gauldie *et al.*, 1983). Indeed, an early tissue migration may be a prerequisite for the production of these proteins by the liver (Gauldie *et al.*, 1985).

There are also some indicators that pulmonary responses may be down regulated by infection. Antibody responses to heterologous antigens by cells in the bronchial lymph nodes are depressed following infection (McElroy *et al.*, 1983). An increase in the number of giant cells recoverable by lavage also occurs after *Nippostrongylus* infection (Egwang *et al.*, 1985). Their formation by the fusion of alveolar macrophages is undoubtedly regulated by lymphokines released by alveolar lymphocytes. The giant cells express fewer C3 and Fc receptors than their alveolar macrophage precursors, reducing their potential capacity for larval killing.

Anaphylactic responses and associated pathology

After rats had been exposed several times to *Nippostrongylus*, their reactions to worms became intense. They showed a laboured breathing, sneezing, bloody discharge from the nose, and respiratory distress upon inhalation anaesthesia. An increase in tissue mast cells has been reported, and five times as much histamine was available for release after a fourth compared to a primary infection (Keller and Jones, 1971). The level of IgE antibodies in the lungs was also higher, but leukotrienes (SRS-A) were probably not of major importance in the pulmonary responses. Macrophages, giant cells and eosinophils were prominent in the intense reactions (Taliaferro and Sarles, 1939). However, beyond a certain point, reactions diminished and granulomata became more compact, perhaps indicating the induction of regulatory mechanisms.

Immune-dependent expulsion of a secondary challenge of *Nippostrongylus* larvae from the small intestine has been intensively investigated (reviewed by Lloyd and Soulsby, 1987) and may have some relevance to pulmonary immunity after multiple exposures (Chapter 10). Compared to the well-defined sequence of immunological events in the intestine, the situation in the lungs is poorly understood. It is unclear whether the increased numbers of mast cells are of the connective tissue or mucosal type. The concomitant increase in circulating IgE provides the requisites for a pulmonary hypersensitivity response upon challenge. Befus *et al.* (1982) demonstrated the increased sensitivity of a tracheal preparation to worm allergens 9–15 days after a primary infection, and local IgE synthesis in bronchial lymph nodes was detected. These observations, plus the work of Keller and Jones (1971) cited earlier, all point to the increasing intensity of immediate hypersensitivity responses in the lungs with successive exposures to *Nippostrongylus*. Whether worms are eliminated by the direct action of anaphylactic mediators, or some more subtle process is unclear. Certainly, the expulsion of worms up the respiratory tract seems intrinsically more improbable than down the intestine.

Host-protective immunity

Compared to *Ascaris*, migration of larval *Nippostrongylus* through the lungs appears to be unimportant in the induction of immunity. Termination of infection by chemotherapy during migration does not induce protection in rats (Ogilvie and Jones, 1971), whereas transplantation of adult worms to the intestine of naive rats does (Ogilvie, 1965). Similarly, infective larvae, attenuated by exposure to a 180 K rad dose of X-rays will accumulate in the lungs up to ten days post-infection. Such larvae have little immunogenic power in spite of their much longer contact with lung tissues than normal larvae (Prochazka and Mulligan, 1965). A possible explanation of these observations is that the recruitment of the T lymphocytes and mast cells required for immune elimination occurs in the gut after a single exposure but in the lungs only after multiple infections.

Hookworms

Hookworms in man

Although human hookworm larvae migrate via the lungs (Chapter 8, Figure 8.7), little is known about pulmonary responses. Case reports suggest that asthma, respiratory distress, and a peripheral eosinophilia, with or without pulmonary infiltrates may accompany larval migration (Muhleisen, 1953; Kalmon, 1954). The condition, associated with creeping eruption of the skin, could be due to infection with the dog hookworm *Ancylostoma braziliensis*. A similar condition, called Wakana disease in Japan, results from consumption of green vegetables carrying infective hookworm larvae (Harada, 1962). It begins with itching of the throat and neck. After three days a characteristic cough occurs, usually accompanied by expectoration of sputum for three weeks or more. Patients with the condition have a higher eosinophilia than individuals with conventional hookworm disease. These anecdotal reports suggest a pulmonary allergic response of uncertain, probably low, frequency to hookworm infection in man.

Pathology and vaccination

Host responses to hookworms of the genus *Ancylostoma* illustrate graphically the dangers of extrapolation from one parasite-host interaction to another. *Ancyclostoma caninum* causes haemorrhagic anaemia in young dogs, and death can occur within 14 days of infection. One of the earliest signs is observed when the larvae of a primary infection reach the lungs. The severity of tissue disruption and pulmonary haemorrhage depends on the number of larvae migrating (Miller, 1971). Following the success of the Dictol vaccine against *Dictyocaulus viviparus*, a similar irradiated vaccine was developed against *A. caninum* (Miller, 1978). The best protection of young dogs was achieved by the subcutaneous injection of radiation-attenuated larvae; a similar number of normal larvae were less satisfactory (Miller, 1966). It was suggested that arrest, extended sojourn and entrapment of the attenuated larvae stimulated maximum immunity, and the lungs were the most likely site. Moulting of these attenuated larvae from the third to fourth stage occurred prematurely in the lungs with concomitant release of antigens, perhaps over an extended period. A vaccine was eventually marketed for a brief period in the USA and then withdrawn for commercial rather than scientific reasons (Miller, 1987).

Experiences with the vaccine in the field suggest that the lungs are also the site of immune effector responses (Miller 1978). In the resistant animal, larvae appear not to escape from the alveoli in any numbers, inducing a hypersensitivity reaction rather than haemorrhage. A few larvae, particularly if irradiated, can cause disproportionately severe reactions in the lungs of immune dogs. Respiratory distress with severe coughing occurred after vaccination of many adult dogs exposed naturally to infection in endemic areas. The condition resolved spontaneously within three to seven days, but its occurrence implies the development of an immune responsiveness in the lungs reminiscent of ascaris pneumonia.

Necator americanus in mice

A laboratory model of *Necator* in BALB/c mice has recently been described (Wells and Behnke, 1988a). Migration is slower than with *Nippostrongylus*, larvae arriving in the lungs after 36–48 h and reaching the intestine by seven days. The leucocyte responses in the lungs have been sampled by lavage (Wells and Behnke, 1988b). The total numbers of leucocytes recovered were 2.5 times and 5 times normal 14 days after a primary and a secondary infection, respectively. In the naive mouse, macrophages comprized > 98% of cells. Eosinophils increased dramatically, 400-fold after a primary and 900-fold after a secondary infection, becoming the most numerous leucocytes in some samples. Neutrophils and lymphocytes also increased, the latter being the most persistent components of the pulmonary inflammation. It was concluded that reinfected mice have the capacity to trap challenge parasites during their passage through the skin and development in the lungs.

Strongyloides sp.

Although belonging to a different order of nematodes, *Strongyloides* sp. are included here because of similarities in migration. There is controversy about the precise route after filariform third-stage larvae penetrate the skin, but some at least enter the lungs in arterial blood and leave via the trachea. Migration of *S. ratti* is rapid (cf. *Nippostrongylus*), larvae being detected in the lungs from 24 h (Dawkins *et al.*, 1981) and reaching the intestine from 72–103 h post-infection (Moqbel, 1980).

Cellular responses during primary and secondary infection

There appears to be little pulmonary response to a primary infection in either mice or rats. A slight increase in the numbers of lymphocytes and neutrophils was observed but eosinophils were not evident (Moqbel, 1980). In contrast, after a secondary challenge, pulmonary reactions in rats were very severe. Discrete areas of lymphoid infiltration were seen with the appearance of germinal centres. Large numbers of eosinophils accumulated, especially in peribronchial and perivascular tissues (Moqbel, 1980). Although infiltrating cells were present in the vicinity of larvae, no eosinophils were seen in direct contact. The responses of mice to a secondary challenge of *S. ratti* were initially dominated by neutrophils, with mononuclear cells prevalent by 72 h (Dawkins *et al.*, 1981). The non-appearance of larvae in the lungs of multiply-infected rats led Moqbel (1980) to suggest that the site of immune elimination shifts from the intestine to the lungs and finally to the skin in hyperimmune animals.

Host-protective immunity

A single exposure of mice to normal larvae of *S. ratti* produces a strong immunity to challenge (Dawkins and Grove, 1982). Immunization of rats by transplantation of adult worms to the intestine also protects against challenge (Korenaga *et al.*, 1983a), whereas termination of an infection by chemotherapy during migration has less effect (Korenaga *et al.*, 1983b). The results of vaccination with radiation-attenuated larvae are inconclusive (Katz, 1987). The balance of evidence points away from migrating larvae, and hence the lungs, being involved in the induction of protective immunity, at least after a primary exposure.

The immune effector mechanisms which bring about elimination of challenge parasites in the lungs are unclear although there are analogies with *Nippostrongylus* in rats. Elevated levels of circulating IgE have been detected in rats with *S. ratti* and man with *S. stercoralis* infections (Katz, 1987). T-cell-dependent recruitment of mast cells to the intestine has been reported (Genta and Ward, 1980), but goblet cell hyperplasia has not (Mimori *et al.*, 1982). Olsen and Schiller (1978) suggested that worms were expelled from the rat intestine as a result of immediate hypersensitivity responses in the mucosa. Dawkins *et al.* (1982) detected 15 min, 5 h and 24 h hypersensitivity responses in the footpads of mice exposed to *S. ratti*; the 15 min (immediate) reaction was the most prominent after challenge. There is also evidence from passive transfer experiments that circulating antibodies have a role in protection (Carlson and Goulson, 1981; Dawkins and Grove, 1981). From the foregoing we might anticipate the operation of IgE-mediated hypersensitivity in the lungs; there are reports of respiratory distress in human strongyloidiasis although not of the intensity of e.g. *Ascaris* infections. Perhaps lung-phase elimination of *Strongyloides* is a transient phenomenon in the development of dermal protection.

The ascarid nematodes

This grouping includes *Ascaris suum* and *A. lumbricoides*, the intestinal roundworms of pig and man; second-stage larvae hatch from resistant eggs ingested by the host. They rapidly penetrate the wall of the small intestine and pass predominantly via hepatic portal vessels to the liver where numbers peak one to two days post-infection (Soulsby 1965). Migration in the bloodstream to the lungs via the pulmonary artery occurs between days three and five. Moulting to the third-larval stage (day four to five) initiates growth; the larvae break out of blood vessels to enter the alveoli and progressively move up the bronchioles, bronchi and trachea. Moulting to the fourth-larval stage occurs between eight and ten days as larvae begin to pass from the trachea to the intestine.

Pathology

Ascaris larvae migrating through the lungs of an unsensitized host generate pathological changes proportional to their numbers. Macroscopically, haemorrhages are visible on the lung surface; in severe cases these can progress to a general haemorrhagic pneumonia and rapid death (Soulsby, 1965). Lighter infections cause a partial collapse of alveoli, with entry of erythrocytes and oedema fluid and a generalized infiltration of leucocytes, all features typical of mechanical damage. In pigs, there is a progressive infiltration of eosinophils as migration of *A. suum* larvae proceeds, accompanied by a peripheral eosinophilia (Eriksen, 1980); the number of mast cells in the lungs increases *after* the larvae have passed through. In humans exposed maliciously to a single large dose of *A. suum*, Phills *et al.*, (1972) observed massive pulmonary infiltration on chest X-rays 10–14 days later, followed by a peripheral eosinophilia in the recovery stage.

A secondary challenge of previously exposed animals with *Ascaris* appears to cause less tissue damage and much less intra-alveolar haemorrhage. This may be due to the reduced number of larvae which reach the lungs (20% of those after a primary infection; Eriksen, 1980). At points where larvae enter the alveoli, there is an earlier, more intense leucocytic infiltration implying the operation of an anamnestic immune response. Eriksen (1980) recorded a lower peripheral eosinophilia after secondary (13%) than after primary (21%) exposure of pigs.

Ascaris pneumonia

The migration of *Ascaris* larvae through the human lung can cause a clinical condition known as ascaris pneumonia or seasonal pneumonitis. It is associated with hypereosinophilia (Goel *et al.*, 1982) and may be one cause of Loeffler's syndrome (Loeffler, 1956). Mild respiratory distress is the commonest feature, but the condition may require hospitalization (Phills *et al.*, 1972) and can cause death. It has been detected in human volunteers exposed to very small numbers of *Ascaris* eggs (Vogel and Minning, 1942). The symptoms can range from malaise and a persistent cough through chest pain, breathlessness and blood-streaked sputum (Gelpi and Mustafa, 1967) to marked respiratory failure with severe depression of arterial oxygen tension (Phills *et al.*, 1972). Wheezing and asthma-like symptoms, suggestive of airway constriction or obstruction, are frequent. The peripheral eosinophilia may be transient or persistent, a fact probably related to the previous history of exposure; Goel *et al* (1982) noted that very young children comprized a large proportion of patients with *Ascaris*-related hypereosinophilia. Eriksen (1980) reported that pigs showed respiratory distress six to ten days after a primary exposure but that after a secondary challenge the distress was less marked.

The incidence of ascaris pneumonia in endemic areas seems very variable. Gelpi and Mustafa (1967) described 95 cases in Saudi Arabia where transmission is seasonal, whereas Spillman (1975) found only four individuals with Loeffler's syndrome in more than 10 000 people examined in Colombia where transmission is continuous. A high proportion of the Colombians had peripheral eosinophilia, and Spillman (1975) concluded that continuous exposure to *Ascaris* led to a desensitization to *Ascaris* allergens.

Antibody responses

The allergic responses and eosinophilia indicate that immediate hyper-sensitivity reactions may play a role in immunity to *Ascaris*. Elevated IgE and IgM were reported in human patients with *Ascaris* whose IgA and IgG levels were normal (Phills *et al.*, 1972). Similarly, O'Donell and Mitchell (1980) found that only one patient out of five with anti-*Ascaris* IgE had a detectable IgG response. They also demonstrated that *Ascaris* body fluids contain a range of allergens. Guinea pigs are also known to produce an IgE response to *A. suum* (Dobson *et al.*, 1971). The reaction of this host to infection is local in character for the first 11–12 days (Khoury and Soulsby, 1977a). Lymphocyte proliferative responses peak sequentially in mesenteric, hepatic and mediastinal nodes as larvae migrate from the intestine via the liver to the lungs. The B lymphocytes and plasma cells produce predominantly IgM and IgE antibody, the latter isotype being particularly prominent in the mediastinal nodes draining the lungs. A similar situation recurs after secondary challenge with marked IgE responses in the mediastinal node. Significant IgA responses were also detected in mesenteric and mediastinal nodes (Khoury and Soulsby, 1977b).

Host-protective immunity

It is uncertain whether antigen presentation in the lungs plays a crucial role in the induction of protective immunity. Exposure of pigs to *Ascaris* eggs induces protection against a challenge three weeks later (Lunney *et al.*, 1986). Feeding of eggs attenuated by ultraviolet (UV) light will also induce protection (Tromba, 1978a, 1978b) as will termination of an infection by chemotherapy 2–4 or 6–8 days after ingestion of viable eggs (Stewart *et al.*, 1984/5). It seems likely that larvae must undertake a partial migration to stimulate immunity. The studies of Tromba (1978a, 1978b) suggest that *attenuated* larvae migrate only as far as the liver. Contrasting with this, Stromberg and Soulsby (1977) used culture products from L3–L4 (lung-stage) larvae to induce protective immunity in guinea pigs. One fact does seem clear; in immune animals much of the elimination of a secondary challenge takes place in the liver, not the lungs, as evidenced by the appearance of milk spot lesions in pigs and guinea pigs (Soulsby 1987).

A tentative conclusion can be drawn regarding the role of the lungs in immunity to *Ascaris*. Partly because *Ascaris* products are allergenic and partly due to the lung-phase of migration, a potent parasite-specific IgE response is generated. In naive or partially immune hosts (man may come into this category in regions where immunity waxes and wanes due to seasonal transmission), the *Ascaris* larvae which reach the lungs trigger immediate hypersensitivity responses. Respiratory distress with pulmonary, and later peripheral eosinophilia, is a consequence. With continued exposure immunity builds up, operating at progressively earlier sites on the migration route (cf. *Strongyloides*) and pulmonary symptoms diminish.

9.5. PARASITES MAKING AN INTRAVASCULAR MIGRATION THROUGH THE LUNGS

Whilst many blood-dwelling Protozoa (*Plasmodium*, *Trypanosoma*) circulate continuously through the lungs, they have little or no direct interaction with pulmonary tissues. The migration of two groups of helminths is, however, influenced by passage through the pulmonary vasculature. The microfilaria larvae of many filarial nematodes circulate in the bloodstream awaiting ingestion by arthropod vectors, and the larvae of schistosome flukes pass through, in-transit from the skin to the hepatic portal system.

Filarial infections and tropical pulmonary eosinophilia

Adult filarial nematodes may inhabit the bloodstream, lymphatics, connective tissues, body cavities and other locations, depending on species. Their microfilaria larvae frequently show periodicity in the peripheral blood, numbers fluctuating on a regular 24 h basis (Hawking, 1975). When not circulating, microfilariae accumulate in the terminal pulmonary arterioles upstream of the alveolar capillary beds. This behavioural feature may account for the prominent role played by the lungs in the clearance of microfilariae from the blood of an immune host.

Antimicrofilarial responses

Individuals carrying chronic infections of the human filariae *Wuchereria bancrofti* or *Brugia malayi* show a marked immunologic hypo-responsiveness to filarial antigens (Ottesen, 1984). In endemic areas, the reactivity of young children with no clinically detectable infection is maximal; as infection becomes established after repeated exposure, these responses subside. Some infected individuals with circulating microfilariae progressively lose them, acquiring the capacity to clear microfilariae from the blood with no obvious clinical manifestations, others sustain a life-long chronic infection. Circulating antibodies directed against the microfilarial surface are produced in the former group to mediate this 'silent' attrition. The presence of adult worms but absence of microfilariae in the blood is referred to as occult filariasis.

Studies with the rodent filaria Acanthocheilonema (*Dipetalonema*) *viteae* in hamsters indicate that the mechanism of clearance involves cooperation between opsonizing antibody and adherent leucocytes (Weiss and Tanner, 1979). Microfilariae implanted subcutaneously at various stages of infection were rapidly eliminated only if the chambers holding them had a pore size (3–5 μm) large enough to admit leucocytes; diffusion of serum components alone was ineffective. Antibodies of IgM class were involved in opsoninization and eosinophils were the predominant adherent cells; neutrophils, lymphocytes and macrophages also participated. That sera from human patients with *Wuchereria* or *Brugia* infections will promote leucocyte adherence to microfilariae argues for a similar mechanism *in vivo* in man (Piessens and MacKenzie, 1982).

Pulmonary pathology

The pathological process accompanying microfilarial clearance has been intensively studied in dogs infected with *Dirofilaria immitis* (Wong, 1987). In naturally or experimentally infected animals clearance occurred predominantly in the lungs. The reactions generated by trapped larvae could be divided into two types (Castleman and Wong, 1982). In early, acute reactions the microfilariae were normally intravascular, occasionally in the interstitium, and had many associated eosinophils, lymphocytes, macrophages and a few neutrophils. In later lesions, degenerating microfilariae were surrounded by macrophages and epithelioid cells. Dense aggregates of plasma cells were frequent in the thickened interalveolar septa. In chronic occult infections the septa were thickened by bundles of collagen fibres and smooth muscle cells, reminiscent of interstitial lung disease.

Wong (1987) has suggested that the clearance mechanism first requires the adherence of antibodies and cells. (The preferential accumulation of microfilariae in pulmonary arterioles may facilitate this adherence.) As a result of immune-mediated damage the larvae become incapable of traversing the pulmonary capillaries. Capillary endothelial cells may also be damaged by soluble inflammatory mediators, reducing the lumen diameter and promoting embolization of the larvae and associated leucocytes. Finally the degradation of killed larvae stimulates a granulomatous response to persistent antigens (cf. the retention of schistosome larvae in pulmonary capillaries).

In less than 1% of individuals with filariasis, a state of hyper-responsiveness develops to filarial antigens, particularly those from microfilariae (Ottesen, 1984). The clinical condition of tropical pulmonary eosinophilia (TPE) is characterized by bouts of coughing, usually at night, producing little sputum and accompanied by breathlessness and chest pain (Neva and Ottesen, 1978). The symptoms may alleviate, then recur spontaneously weeks or months later. If untreated the condition evolves into chronic interstitial lung disease (cf. *Dirofilaria*). Immunologically, anti-filarial antibodies of all classes are elevated; circulating IgE and eosinophil levels are extremely high (Ottesen *et al.*, 1979). Microfilariae are absent from the blood, much of their clearance having taken place in the lungs (Ottesen, 1984).

The pulmonary inflammatory reactions produce restrictive changes in lung respiratory function (Neva and Ottesen, 1978). In the early stages macrophage-rich infiltrates are found in the alveoli and pulmonary interstitium. Later, eosinophils become prominent and the infiltrates may eventually organize into granulomas with increasing fibrosis (Udwadia, 1975). Ottesen *et al.*, (1981) pointed out that although basophils and mast cells were sensitized with parasite-specific IgE antibody, most patients with filarial infection did not manifest allergic reactions. Their sera contained blocking antibodies of IgG isotype which inhibited histamine release from mast cells in the presence of specific antigen. These antibodies have a potential *in vivo* for modulating the allergic responsiveness to parasite antigen. Hussain and Ottesen (1985) compared the sera from patients with TPE with those from patients with circulating microfilariae (MF) or chronic pathology with elephantiasis (CP). The TPE patients had equal amounts of the IgE and IgG to filarial antigens or more IgE; the MF and CP patients had a higher relative IgG level. It thus appears that TPE could result from a failure of the blocking mechanism in a few individuals.

Schistosomes

Flukes (trematodes) of the genus *Schistosoma* inhabit the hepatic portal vasculature or vesical plexus of the bladder wall. It appears that migration of *S. mansoni* (the most intensively studied species) is entirely intravascular in the direction of blood flow (Wilson, 1987). Schistosomula thus arrive in the lungs from the skin-infection site via the pulmonary artery and leave via the pulmonary vein. Potentially, a larva may make several circuits of the vasculature before entering, by chance, a systemic artery leading to the hepatic portal vein. Its first arrival in the lungs is followed by a period of development lasting about four days which adapts it for onward migration (Crabtree and Wilson, 1980). This is a prolonged process, the larva taking about 30–35 h to negotiate a pulmonary capillary (Wilson and Coulson, 1986). Because of the multiple circuits around the vasculature, the lung phase of migration extends from 3 to perhaps 20 days post-infection in the murine host.

Loss of parasites during passage through the lungs

Even in naive hosts only a proportion of schistosomula successfully complete the migration to the hepatic portal system (10–20% in rats; 20–50% in mice). It is now clear from several studies on the kinetics of migration that the majority of unsuccessful parasites are eliminated in the lungs (Knopf *et al.*, 1986; Wilson *et al.*, 1986). Ultrastructural examination of the lungs has shown that many of the larvae burst out of blood vessels into the alveoli. This may be an accidental consequence of their attempts to progress along the lumen of the very distensible pulmonary capillaries. The mixed neutrophil-mononuclear cell infiltrates observed in the vicinity of alveolar parasites in mice are indicative of mechanical damage rather than immune-mediated inflammation (Crabtree and Wilson, 1986). Schistosomula in the alveoli have a very limited ability to re-enter the tissues and continue their migration (Coulson and Wilson, 1988). Hence, for most the accident terminates migration, although they may linger in the lungs for days or weeks; their ultimate fate may be death from starvation or expectoration (Crabtree and Wilson, 1986). Whilst a peripheral eosinophilia has not been observed in mice over the time course of migration, bronchoalveolar lavage has revealed a marked but transient pulmonary eosinophilia at 21 days post-infection (Menson *et al.*, 1989).

Mechanisms of acquired immunity

The complex question of specific, acquired immunity to schistosomes in various hosts is beyond the scope of this chapter. However, mice (and rats) can be vaccinated by exposure to radiation-attenuated cercariae and develop partial immunity to challenge without overt pathology (Minard *et al.*, 1978). The attenuated larvae must undergo some migration in order to stimulate immunity. The persistence of attenuated schistosomula in the lungs for two to three weeks after vaccination led to the suggestion that antigen presentation in that organ was crucial to the induction of protection (Mastin *et al.*, 1985). However, Mountford *et al.* (1988) have demonstrated the persistence of larvae in lymph nodes draining the skin-infection site. The relative contribution of events in the skin and lungs to the induction of immunity remains to be established. There is strong evidence from studies in mice that the mechanism of immunity stimulated by irradiated parasites involves a T-cell-mediated activation of macrophages (James, 1986). The site of challenge elimination is contentious. However, in some mouse strains it occurs predominantly in the lungs (Wilson *et al.*, 1986). Sampling of the airways by lavage after vaccination has revealed a marked increase in the number of lymphocytes and macrophages present, a situation persisting for at least ten weeks (Figure 9.5; Aitken *et al.*, 1988). The mechanism of elimination is radiation-resistant suggesting that it does not involve the recruitment of leucocytes to the lungs, but rather relies on conditions established in the lungs by vaccination (Aitken *et al.*, 1987).

Aitken *et al.* (1988) have postulated that after vaccination, a population of schistosome-specific T cells is recruited to the lung tissues and remains 'on station' pending the arrival of challenge schistosomula. When this occurs, there is a secondary infiltration of leucocytes into the airways (Figure 9.6) and discrete foci of cells accumulate around the larvae (Figure 9.7). Evidence from lavage (Menson *et al.*, 1989) and electron microscopy (Crabtree and Wilson, 1986) show the cells to be predominantly mononuclear, eosinophils being sparse. The leucocytes infiltrate between the alveolar epithelium and capillary endothelium around a migrating schistosomulum (Figure 9.8a) and the intense focal inflammation leads to the almost complete destruction of the capillary (Figure 9.8b; Crabtree and Wilson, 1986). The parasites

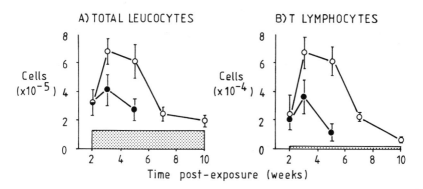

Figure 9.5. Recoveries of (A) leucocytes and (B) T lymphocytes from the lungs of mice by broncho-alveolar lavage at times after exposure to normal cercariae of *Schistosoma mansoni* (●) or vaccination with an equivalent number of radiation-attenuated parasites (○). Shaded areas show mean recoveries from uninfected mice (Aitken *et al.*, 1988).

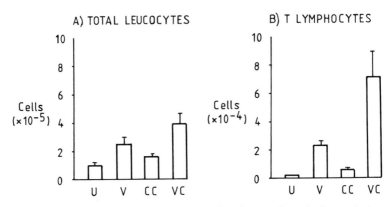

Figure 9.6. Recoveries of (A) leucocytes and (B) T lymphocytes from the lungs of mice 2 weeks after challenge infection with 200 cercariae of *Schistosoma mansoni*. U, uninfected; V, vaccinated only; CC, challenge controls; VC, vaccinated and challenged mice. There is an increase in total leucocytes, and particularly in T lymphocytes, recovered from VC mice relative to other groups (Aitken *et al.*, 1988).

surprizingly seem to be unharmed by the focus of leucocytes. If extracted and transferred to the hepatic portal system of a naive animal, they will mature, whereas if left *in situ* they will not (Coulson and Wilson, 1988).

This mechanism of acquired immunity operating to retard the migration of schistosomula along capillaries has obvious parallels with the antibody-mediated clearance of microfilariae from the circulation in the lungs. However, the surface of the lung schistosomulum is notoriously unreactive with antibodies. It is therefore likely that the pulmonary inflammation is a response to secreted antigens, rather than opsonization. Indeed, the operation of a delayed hypersensitivity response would require antigen recognition by T cells in the context of Class II MHC molecules to initiate lymphokine production for macrophage activation.

Nevertheless, it has been reported that in multiply vaccinated mice, antibodies may mediate elimination of challenge parasites in the lungs (Mangold and Dean, 1986). There is also the well-documented dependence of immunity in rats on circulating antibody (Ford *et al.*, 1984). These observations raise the possibility that antibody/complement-mediated mechanisms (probably of the Arthus type) can also operate against schistosomula in pulmonary capillaries.

9.6. MISCELLANY: ACCIDENTAL PARASITES IN THE LUNGS

Grouped under this heading are a miscellaneous collection of parasites having no obligatory association with the lungs. Their main site of infection may be elsewhere in the body or they may be parasites in the wrong host. Nevertheless

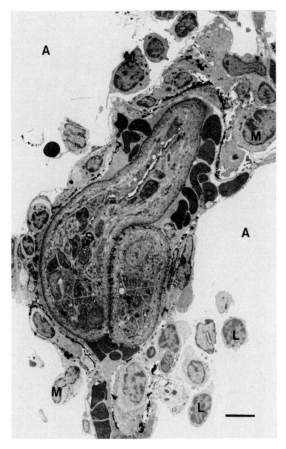

Figure 9.7. Schistosomulum (S) of *S. mansoni* in the lungs of a vaccinated mouse, 7 days after challenge. Lymphocytes (L) and macrophages (M) are present in the alveoli (A) around the intravascular larva and are also infiltrating the pulmonary interstitium (Crabtree and Wilson, 1986). Bar = 5 μm.

they do occur in the lungs with pathological and immunological consequences, in many instances, life-threatening.

In cases of chronic human schistosomiasis, schistosome eggs may gain access to the vena cava. They are carried via the heart and embolize in pulmonary blood vessels. Over a long period distal branches of the arterioles are blocked causing a rise in arterial pressure (von Lichtenberg, 1987). The syndrome of cor pulmonale, an enlargement of the right side of the heart, may develop (cf. *Angiostrongylus*). Host responses to eggs embolized in the lungs after intra-venous administration have been intensively studied as a model for the more extensive hepatic reactions (von Lichtenberg, 1962).

The mature schistosome egg contains a miracidium larva and releases a variety of antigenic macromolecules. The host responds by forming an intense

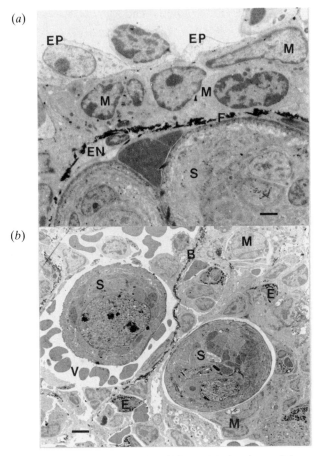

Figure 9.8. a) As Figure 9.7. The alveolar epithelium (EP) is separated from the capillary endothelium (EN) by an infiltrate of mononuclear (M) cells. Fibrous protein (F) is deposited in the basement membrane. Bar = 1μm. b) Schistosomulum (S) in vaccinated mouse lung 17 days post-challenge. The larva is partly intra-vascular (V) and partly intra-alveolar. Only the basement membrane (B) of the capillary is clearly visible. The cellular infiltrate around the alveolar part of the parasite contains macrophages (M) and eosinophils (E) (Crabtree and Wilson, 1986). Bar = 2μm.

focal granulomatous inflammation around each egg. There is abundant evidence, reviewed by Warren (1987) that this is a delayed hypersensitivity reaction mediated by specific T helper lymphocytes, which in turn release lymphokines to activate macrophages. In mice, granuloma size peaks about 12 weeks after infection and then gradually diminishes, reflecting modulation of the immune response. In *S. mansoni* infections modulation is mediated by another T-cell subset; in *S. japonicum* infections, antibodies (possibly anti-idiotypic) have a major role (Stavitsky, 1987). The final stage of granuloma formation involves collagen synthesis and deposition by fibroblasts. In late chronic infections collagen degradation dominates, favouring ultimate resolution of the granuloma (Warren, 1987).

Toxocara canis is the ubiquitous roundworm of dogs parasitizing the small intestine. In the canine host it behaves in a similar manner to *Ascaris* in man or pig, the egg hatching in the intestine and the larvae undergoing a tissue phase of migration before returning to the gut. Some larvae are also believed to distribute themselves through other internal organs.

T. canis eggs can hatch in the intestines of many non-canid hosts where they undergo no development but distribute through most internal tissues causing a condition called visceral larva migrans. This phenomenon has been studied using mice as experimental hosts (Kayes and Oaks, 1976). Within 24 h of egg ingestion, larvae are found in the lungs; they migrate rapidly from the pulmonary parenchyma, few if any being left by 14 days. During their passage through the lungs they elicit an inflammatory exudate which persists for at least seven weeks, long after the larvae have left. The pulmonary infiltrates have been characterized by broncho-alveolar lavage and shown to be very eosinophil rich (80% of cells recovered, versus 10 to 25% in the blood). The response peaks at about day 11 and has diminished considerably by day 17 (Kayes *et al.*, 1987). *Toxocara*-specific T lymphocytes are numerous in the lungs compared to the spleen. The pulmonary response to *T. canis* larvae thus exhibits the features of a granulomatous hypersensitivity. The reaction has been adoptively transferred to naive mice by intratracheal injections of cells recovered from syngeneic animals by lung lavage ten days after infection. It could not be transferred by spleen cells or serum administered intravenously (Kayes *et al.*, 1988). Murine visceral larva migrans may thus provide a model of the transient pneumonitis observed in children who have ingested *T. canis* eggs.

The filarial nematode *Dirofilaria immitis* is normally parasitic in the right side of the heart of dogs. However, there are now more than 20 papers describing accidental infection in man (Jagusch *et al.*, 1984). The cases were often detected by chest X-ray followed by surgery for suspected malignancy or on post-mortem. On close examination embolic occlusion of the pulmonary artery was seen with areas of thrombosis, infarction, and haemorrhage spreading out into the lungs (Brenes *et al.*, 1985). One or more *Dirofilaria* worms were present in the centre of the lesion. Untreated, the lesion may gradually resolve after worm death, leaving only a mass of scar tissue.

Infections with the protozoan *Entamoeba histolytica*, if pathogenic, normally involve the tissues of the colon and liver. Pulmonary amoebiasis can result as a complication of a liver abscess. There may be either rupture of the abscess followed by metastatic lesions downstream in the lungs, or a liver abscess near the diaphragm can erode through into the thoracic cavity. A broncho-pleural fistula may develop to the extent that the patient expectorates the contents of the liver abscess (Barrett-Connor, 1982). Pulmonary abscesses are a relatively rare phenomenon in amoebic infection with an incidence of perhaps 0.1%. When invasion of the thorax does occur it gives rise to the condition of empyema—pus in the pleural cavity. The lung abscess which forms by a process of liquifying necrosis has sticky haemorrhagic contents and a wall of fibrosed lung tissue (Spencer, 1985).

Hydatid cysts of the tapeworm *Echinococcus* occur in many organs of the body. They are much more prevalent in the lungs of individuals infected with the sylvatic than the pastoral form of the disease (Wilson *et al,*. 1968). The sylvatic cycle normally involves wolves and moose; man becomes infected via dogs fed moose offal. Not surprizingly the disease is confined to native inhabitants of Alaska and Canada. The dimensions of sylvatic cysts are smaller (4.1 cm diameter, 36 ml volume) than those of the pastoral form (7.2 cm, 196 ml). Bursting of the latter cysts in the lungs can literally drown the infected individual.

Hydatid cysts in the lungs are most commonly found by chest X-ray. Very little is known about host responses but the boundaries of intact cysts are surrounded by inflammation (Cuthbert, 1975). As they develop, lung tissue is compressed. The main danger from hydatid infection is the rupture of the cyst during exploratory biopsy or surgery. This can lead to daughter cyst formation or systemic anaphylaxis if the host has been sensitized to produce specific IgE antibodies. Inadequate anthelmintic treatment, causing release of antigens without parasite death, can exacerbate the situation by further raising the IgE level (Werczberger *et al.*, 1979).

9.7. PULMONARY PARASITE INFECTIONS IN THE IMMUNOCOMPROMIZED HOST

Over the past decade there has been an increasing use of immunosuppressive therapy following organ transplantation and in the treatment of inflammatory conditions. Infection with the AIDS virus has also become a significant public health problem. In individuals whose immune system is thus compromised, parasites, which under normal circumstances have little impact, become highly pathogenic, and the lungs seem particularly vulnerable to attack.

Toxoplasma gondii

The lungs are a major site of *Toxoplasma* replication in cats; the sequelae are a mild cough producing no sputum and mild interstitial inflammation (Ford, 1984). In cats treated with corticosteroids, an intense proliferation of *Toxoplasma* occurred with parasites located in the alveoli, interstitium, bronchiolar glands, bronchiolar epithelium, and even small blood vessels (Dubey and Frenkel, 1974). Bacterial infections of the bronchi were concurrent in about one third of cats examined. *Toxoplasma* infection was not confined to the lungs. Similar findings have been reported in human cases of toxoplasmosis following chemotherapy, corticosteroid treatment, splenectomy or irradiation; the lungs, myocardium and brain were all affected (Gleason and Hamlin 1974).

The course of a *Toxoplasma* infection in mouse lung has been characterized by broncho-alveolar lavage (Ryning and Remington, 1977). Normal alveolar macrophages supported the intracellular multiplication of *Toxoplasma*; activated macrophages did not. From 40–70 days after intra-peritoneal inoculation of *Toxoplasma* into mice, there was an approximate five-fold increase in leucocytes, largely macrophages and lymphocytes, recovered by lavage. The proportion of lymphocytes rose from 6–16% in the normal animal to 30–45% in the infected mouse. Ryning and Remington (1977) concluded that in the immunocompetent animal, macrophages (activated by lymphokines) are the effector cells in the resistance of the lung to *Toxoplasma* infection.

Pneumocystis

Infections with *Pneumocystis carinii* are an important cause of interstitial pneumonia in immunocompromized individuals. The taxonomic status of *Pneumocystis* is controversial, but it is considered a protozoan by many workers (Spencer, 1985) and so included in this chapter. It is widespread in the human population without causing disease but behaves as an opportunistic pathogen when conditions are favourable. It causes pneumonia in children born prematurely or in a malnourished state. It affects adults with natural or acquired immunodeficiency resulting from malignancy (particularly Hodgkin's disease), use of cytotoxic or immunosupressive drugs, and infection with the AIDS virus.

Infection is believed to result from the inhalation of *Pneumocystis* cysts or trophozoites (Masur and Jones, 1978). The normal habitat is the alveolus and the trophozoites are closely applied to the surface of Type 1 alveolar lining cells (Walzer, 1986). The value of this attachment to the parasite is unclear. The trophozoites will also attach to alveolar macrophages, and if opsonized by antiserum, will be ingested and destroyed (Masur and Jones, 1978). Infection of cortisone-treated rats on a low protein diet, with *Pneumocystis* has been used as a model for the course of the disease in man (Lanken *et al.*, 1980). After one month, focal necrosis of Type 1 pneumocytes occurred and Type 2 cells increased in size and number—a typical host response to alveolar damage. In the treated rats, inflammatory reactions were conspicuously absent.

In human infections, the alveolar interstitium is infiltrated by lymphocytes and large numbers of plasma cells and macrophages (Spencer, 1985). The diseased alveoli have a characteristic foamy filling and contain infiltrating neutrophils. Degenerative changes in Type 1 cells lead to a denudation of the basement membrane (Walzer, 1986). Increased capillary permeability results in pulmonary oedema and the leakage of serum proteins to produce the foamy alveolar contents. After cessation of corticosteroid treatment, rats mount a vigorous immune response with serum antibody production, infiltration of lymphocytes, phagocytosis of *Pneumocystis* trophozoites by alveolar macrophages, and development of interstitial fibrosis (Walzer, 1986). Nevertheless some parasites can persist for long periods and relapses in human patients do occur (Kovacs *et al.*, 1984).

Strongyloides stercoralis

A feature of *Strongyloides stercoralis* infections in man (but not of *S. ratti* in rats) is the development of the transmission stages into infective filariform larvae whilst still in the gut. Auto-infection via the intestinal mucosa or perianal skin is thus possible (Genta, 1984). *Strongyloides stercoralis* infections in man are normally tightly controlled by immune responses and hence clinically silent; however, a small parasite burden may be maintained for many years. If an infected individual is immunosuppressed for any of the reasons listed above, auto-infection increases to massive proportions and severe pulmonary disease caused by disseminated strongyloidiasis ensues (Barrett-Connor, 1982). Animal models of the disseminated disease in immunosuppressed patas monkeys (Harper *et al.*, 1984) and dogs (Genta, 1986) have been described. The major feature of both models was widespread pulmonary haemorrhage. Whether this was due to simple mechanical damage caused by migrating larvae is unclear (Harper *et al.*, 1984). However, the haemorrhage was confined to the lungs and was a common cause of death in the monkeys.

Filaroides hirthi

A second example of disseminated infection by nematodes in immunosuppressed animals has recently been described. *Filaroides hirthi* is a common metastrongyle parasite in the lungs of dogs (particularly beagles; Georgi and Anderson, 1975). A massive fatal infection has been reported in a corticosteroid-treated dog, suggesting the occurrence of auto-infection (August *et al.*, 1980). Huge numbers of adults and larvae were found in the lungs of two immunosuppressed dogs by Genta and Schad (1984). The adults provoked little inflammation, but larvae in the alveoli (and other tissues) caused a granulomatous reaction rich in mononuclear cells and giant cells, suggesting the operation of stage-specific immune mechanisms. In the normal dog these presumably regulate the parasite burden by severely limiting auto-infection.

9.8. CONCLUSIONS

Most of the parasites detailed in this chapter reach the lungs via other tissues, such as the intestine/liver, skin or peritoneal cavity/diaphragm. Consequently, there is a potential for the induction of immune responses at such sites and their later deployment in the pulmonary tissues against lung parasites. The overall contribution of pulmonary immune responses in the protection of a host is therefore difficult to evaluate. That such mechanisms are effective can be judged by the hyper-infections (of e.g. *Toxoplasma* or *Pneumocystis*) observed in the lungs of immunocompromized hosts.

Parasites in intravascular locations are the most accessible to immune effector responses, but it does not follow that they are more vulnerable; the adults of *Angiostrongylus cantonensis* and *Dirofilaria immitis* can survive long periods in the pulmonary arteries. Larvae in capillary beds appear to be more readily attacked, possibly because of their relative immobility. Microfilariae may be opsonized by antibodies, facilitating binding of complement and/or effector leucocytes. Conversely, a cell-mediated delayed hypersensitivity response may operate against schistosome larvae. In this situation, the ensuing inflammatory reactions appear to act by blocking migration, rather than via cytotoxic killing.

The majority of helminth parasites enter the pulmonary airways either to take up residence or in transit to the alimentary tract. Once outside the tissues it is obviously more difficult for immune responses to affect their fate. Alveolar macrophages are clearly no match for helminths although the possibility exists of activating them to increase their efficiency. This provides an adequate defence against protozoans, such as *Toxoplasma* and *Pneumocystis*, in normal circumstances. The luminal location of parasites may explain, by analogy with the intestine, the frequent occurrence of IgE-mediated immediate hypersensitivity responses in the lungs (*Ascaris*, *Nippostrongylus*, *Strongyloides* infections). The increased permeability of epithelial surfaces which follows the degranulation of mast cells would facilitate entry of circulating antibodies and leucocytes to the airways.

Deployment of IgE-mediated responses brings with it the problem of regulation; in some infections an excessive reaction leads to respiratory difficulties due to broncho-constriction (*Ascaris*, filarial infections). The marked pulmonary eosinophilia which normally accompanies such reactions may contribute to the observed pathogenesis. It is clear that not all parasites in the airways provoke immediate hypersensitivity responses (e.g. *Paragonimus*), although whether this benefits the host or parasite is uncertain.

The lungs are richly endowed with lymphoid tissue, arguing for a role in the induction of immune responses. Studies on *Dictyocaulus*, *Ancylostoma caninum*, *Schistosoma* and *Ascaris* have provided evidence for antigen presentation in pulmonary tissues or draining lymph nodes. There may, however, be a further component in the elicitation of protective pulmonary immunity. The parasites which release antigens in the lungs may also serve as an inflammatory stimulus to recruit T and B lymphocytes and other leucocytes to the lung tissues. Here, the cells may play a purely local role in protection by arming the whole organ against the subsequent arrival of challenge parasites. The documented build up of leucocytes in the lung tissues following vaccination with attenuated larvae of *Dictyocaulus* or *Schistosoma* supports this hypothesis. A great deal of current research is directed to the identification of protective antigens. It may prove more difficult to devise vaccination strategies for presenting those antigens in a way which emulates the attenuated larvae by recruiting the appropriate sensitized lymphocytes to the lung tissues.

REFERENCES

Aitken, R., Coulson, P.S. and Wilson, R.A. (1988), Pulmonary leukocytic responses are linked to the acquired immunity of mice vaccinated with irradiated cercariae of *Schistosoma mansoni*, *Journal of Immunology*, **140**, 3573–9.

Aitken, R., Coulson, P.S., Dixon B. and Wilson, R.A. (1987), Radiation-resistant acquired immunity of vaccinated mice to *Schistosoma mansoni*, *American Journal of Tropical Medicine and Hygiene*, **37**, 570–7.

Armour, J. (1987), Lungworms of cattle, sheep and pigs with special reference to their immunology, immunopathology, and immunoprophylaxis, in Soulsby, E.J.L., (Ed.), *Immune Responses in Parasitic Infections*, **Vol 1 Nematodes**, pp. 155–80, Boca Raton: CRC Press.

Au, A.C.S. and Ko, R.C. (1979), Changes in worm burden, haematological and serological response in rats after single and multiple *Angiostrongylus cantonensis* infections, *Zeitschrift für Parasitenkunde*, **58**, 233–42.

August, J.R., Powers, R.D., Bailey, W.S. and Diamond, D.L. (1980), *Filaroides hirthi* in a dog: fatal hyperinfection suggestive of autoinfection, *Journal of the American Veterinary Medical Association*, **176**, 331–4.

Barrett-Connor, E. (1982), Parasitic pulmonary disease, *American Review of Respiratory Disease*, **126**, 558–63.

Befus, A.D., Johnston, N., Berman, L. and Bienenstock, J. (1982), Relationship between tissue and IgE antibody production in rats infected with the nematode *Nippostrongylus brasiliensis*, *International Archives of Allergy and Applied Immunology*, **67**, 213–8.

Bienenstock, J. (1984), Bronchus-associated lymphoid tissue, in Bienenstock, J. (Ed.). *Immunology of the Lung and Upper Respiratory Tract*, pp.96–118, New York: McGraw-Hill.

Blusse van Oud Alblas, A. and van Furth, R. (1979), Origin, kinetics, and characteristics of pulmonary macrophages in the normal steady state, *Journal of Experimental Medicine*, **149**, 1504–18.

Brenes, R., Beaver, P.C., Monge, E. and Zamosa, L. (1985), Pulmonary dirofilariasis in a Costa Rican man, *American Journal of Tropical Medicine and Hygiene*, **34**, 1142–3.

Carlson, J.R. and Goulson, H.T. (1981), *Strongyloides ratti*: transfer of immunity in rats by lymph node cells or serum, *Current Microbiology*, **5**, 307–10.

Castleman, W.L. and Wong, M.M. (1982), Light and electron microscopic pulmonary lesions associated with retained microfilariae in canine occult dirofilariasis, *Veterinary Pathology*, **19**, 355–64.

Choi, C-U. and Rah, B-J. (1981), Effect of *Paragonimus iloktsuenensis* infection on tissue mast cell of lung in albino rats, *Chung-Ang Journal of Medicine*, **6**, 417–20.

Choi, W-Y., Lee, O-R. and Jin, Y-K. (1979), Lung findings in experimental paragonimiasis, *Korean Journal of Parasitology*, **17**, 132–46.

Cornwell, R.L. and Michel, J.F. (1960), The complement fixing antibody response of calves to *Dictyocaulus viviparus*. 1. Exposure to natural and experimental infection, *Journal of Comparative Pathology*, **70**, 482–93.

Corry, D., Kulkarni, P. and Lipscomb, M.F. (1984), The migration of bronchoalveolar macrophages into hilar lymph nodes, *American Journal of Pathology*, **115**, 321–8.

Coulson, P.S. and Wilson, R.A. (1988), Examination of the mechanisms of pulmonary phase resistance to *Schistosoma mansoni* in vaccinated mice, *American Journal of Tropical Medicine and Hygiene*, **38**, 529–39.

Crabtree, J.E. and Wilson, R.A. (1980), *Schistosoma mansoni*: a scanning electron microscope study of the developing schistosomulum, *Parasitology*, **81**, 553–64.

Crabtree, J.E. and Wilson, R.A. (1986), The role of pulmonary cellular reactions in the resistance of vaccinated mice to *Schistosoma mansoni*, *Parasite Immunology*, **8**, 265–85.

Cuthbert, R. (1975), Sylvatic pulmonary hydatid disease: a radiological survey, *Journal of the Association of Canadian Radiologists*, **26**, 132–8.

Dawkins, H.J.S. and Grove, D.I., (1981), Transfer by serum and cells of resistance to reinfection with *Strongyloides ratti* in mice, *Immunology*, **43**, 317–22.

Dawkins, H.J.S. and Grove, D.I. (1982), Immunisation of mice against *Strongyloides ratti*, *Zeitschrift für Parasitenkunde*, **66**, 327–33.

Dawkins, H.J.S., Muir, G.M. and Grove, D.I. (1981), Histopathological appearances in primary and secondary infections with *Strongyloides ratti* in mice, *International Journal for Parasitology*, **11**, 97–103.

Dawkins, H.J.S., Carroll, S.M. and Grove, D.I. (1982), Humoral and cell-mediated immune responses in murine strongyloidiasis, *Australian Journal of Experimental Biology and Medical Science*, **60**, 717–29.

Dobson, C., Morseth, D.J. and Soulsby, E.J.L. (1971), Immunoglobulin E-type antibodies induced by *Ascaris suum* infections in guinea pigs, *Journal of Immunology*, **106**, 128–33.

Downey, N.E., (1980), Effect of treatment with levamisole or fenbendazole on primary experimental *Dictyocaulus viviparus* infection, and on resistance, *The Veterinary Record*, **107**, 271–5.

Dubey, J.P. and Frenkel, J.K. (1974), Immunity to feline toxoplasmosis: modification by administration of corticosteroids, *Veterinary Pathology*, **11**, 505–12.

Egwang, T.G., Gauldie, J. and Befus, D. (1984a), Broncho-alveolar leucocyte responses during primary and secondary *Nippostrongylus brasiliensis* infection in the rat, *Parasite Immunology*, **6**, 191–202.

Egwang, T.G., Gauldie, J. and Befus, D. (1984b), Complement-dependent killing of *Nippostrongylus brasiliensis* infective larvae by rat alveolar macrophages, *Clinical and Experimental Immunology*, **55**, 149–56.

Egwang, T.G., Richards, C.D., Stadnyk, A.W., Gauldie, J. and Befus, A.D. (1985), Multinucleate giant cells in murine and rat lungs during *Nippostrongylus brasiliensis* infections. A study of the kinetics of the response *in vivo*, cytochemistry, IgG- and C3-mediated functions, *Parasite Immunology*, **7**, 11–18.

Eriksen, L. (1980), Host-parasite relations in *Ascaris suum* infection in pigs. in Nielsen, N.C. (Ed.), *Proceedings of the Congress of the International Pig Veterinary Society*.

Ford, R.B. (1984), Infectious respiratory disease, *Veterinary Clinics of North America. Small Animal Practice*, **14**, 985–1006.

Ford, M.J., Bickle, Q.D., Taylor, M.G. and Andrews, B.J. (1984), Passive transfer of resistance and the site of immune-dependent elimination of the challenge infection in rats vaccinated with highly irradiated cercariae of *Schistosoma mansoni*, *Parasitology*, **89**, 327–44.

Gauldie, J., Richards, C. and Lamontagne, L. (1983), Fc receptors for IgA and other immunoglobulins on resident and activated alveolar macrophages, *Molecular Immunology*, **20**, 1029–37.

Gauldie, J., Lamontagne, L. and Stadnyk, A. (1985), Acute phase response in infectious disease, *Survey and Synthesis of Pathology Research*, **4**, 126–51.

Gelpi, A.P. and Mustafa, A. (1967), Seasonal pneumonitis with eosinophilia: a study of larval ascariasis in Saudi Arabs. *American Journal of Tropical Medicine and Hygiene*, **16**, 646–57.

Genta, R.M. (1984), Immunobiology of strongyloidiasis, *Tropical and Geographical Medicine*, **36**, 223–9.

Genta, R.M. (1986), *Strongyloides stercoralis*: parasitological, immunological and pathological observations in immunosuppressed dogs, *Transactions of the Royal Society of Tropical Medicine and Hygiene*, **80**, 34–41.

Genta, R.M. and Ward, P.A. (1980), The histopathology of experimental strongyloidiasis, *American Journal of Pathology*, **99**, 207–19.

Genta, R.M. and Schad, G.A. (1984), *Filaroides hirthi*: hyperinfective lungworm infection in immunosuppressed dogs, *Veterinary Pathology*, **21**, 349–54.

Georgi, J.R. and Anderson, R.C. (1975), *Filaroides hirthi*: sp. n. (Nematoda : Metastrongyloidea) from the lung of the dog, *Journal of Parasitology*, **61**, 337–9.

Gleason, T.H. and Hamlin, W.B. (1974), Disseminated toxoplasmosis in the compromised host, *Archives of Internal Medicine*, **134**, 1059–62.

Goel, R.G., Bhan, M.K., Fazal, M.I. and Srivastava, R.N. (1982), Hypereosinophilia pulmonary symptoms and ascariasis, *Indian Paediatrics*, **19**, 323–6.

Harada, Y. (1962), Wakana disease and hookworm allergy, *Yonago Acta Medica*, **6**, 109–18.

Harper, J.S., Genta, R.M., Gam, A., London, W.T. and Neva, F.W. (1984), Experimental disseminated strongyloidiasis in *Erythrocebus patas*. 1. Pathology, *American Journal of Tropical Medicine and Hygiene*, **33**, 431–43.

Hawking, F. (1975), Circadian and other rhythms of parasites, *Advances in Parasitology*, **13**, 123–82.

Holt, P.G. (1986), Down-regulation of immune responses in the lower respiratory tract: the role of alveolar macrophages, *Clinical and Experimental Immunology*, **63**, 261–70.

Hussain, R. and Ottesen, E.A. (1985), IgE responses in human filariasis. III. Specificities of IgE and IgG antibodies compared by immunoblot analysis, *Journal of Immunology*, **135**, 1415–20.

Ikeda, T. and Fujita, K. (1980), IgE in *Paragonimus ohirai*-infected rats : relationship between titer, migration route and parasite age, *Journal of Parasitology*, **66**, 197–204.

Ikeda, T. and Fujita, K. (1982), IgE in *Paragonimus ohirai* infected rats : Effect of X-irradiation, *Journal of Parasitology*, **68**, 955–7.

Ikeda, T., Oikawa, Y. and Fujita, K. (1982), Kinetics and localisation of parasite-specific IgE in *Paragonimus ohirai*-infected rats. *International Journal of Parasitology*, **12**, 395–8.

Jagusch, M.F., Mere Roberts, R., Rea, H.H. and Priestley, D.R.A. (1984), Human pulmonary dirofilariasis, *New Zealand Medical Journal*, **97**, 556–8.

James, S.L. (1986), Activated macrophages as effector cells of protective immunity to schistosomiasis, *Immunologic Research*, **5**, 139–48.

Jarrett, E.E.E. and Miller, H.R.P. (1982), Production and activities of IgE in helminth infection, *Progress in Allergy*, **31**, 178–233.

Jarrett, W.F.H. and Sharp, N.C.C. (1963), Vaccination against parasitic disease : reactions in vaccinated and immune hosts in *Dictyocaulus viviparus* infection, *Journal of Parasitology*, **49**, 177–89.

Jarrett, W.F.H., McIntyre, W.I.M. and Urquhart, G.M., (1957b), The pathology of experimental bovine parasitic bronchitis, *Journal of Pathology and Bacteriology* **73**, 183–93.

Jarrett, W.F.H., McIntyre, W.I.M., Jennings, F.W. and Mulligan, W. (1957a), The natural history of parasitic bronchitis with notes on prophylaxis and treatment, *The Veterinary Record*, **69**, 1329–38.

Jarrett, W.F.H., Jennings, F.W., McIntyre, W.I.M., Mulligan, W. and Urquhart, G.M., (1960), Immunological studies on *Dictyocaulus viviparus* infection. Immunity produced by the administration of irradiated larvae, *Immunology*, **3**, 145–51.

Jarrett, W.F.H., Jennings, F.W., Martin, B., McIntyre, W.I.M., Mulligan, W., Sharp, N.C.C. and Urquhart, G.M. (1958), A field trial of a parasitic bronchitis vaccine, *The Veterinary Record*, **70**, 451–4.

Jeffery, P.K. and Corrin, B. (1984), Structural analysis of the Respiratory Tract, in Bienenstock, J. (Ed.), *Immunology of the Lung and Upper Respiratory Tract*, pp.1–27, New York: McGraw-Hill.

Kalmon, E.H. (1954), Creeping eruption associated with transient pulmonary infiltrations, *Radiology*, **62**, 222–6.

Kaltreider, H.B., Caldwell, J.L. and Byrd, P.K. (1986), The capacity of normal murine alveolar macrophages to function as antigen presenting cells for the initiation of primary antibody-forming cell responses to sheep erythrocytes *in vitro*, *American Review of Respiratory Diseases*, **133**, 1097–1104.

Katz, F.F. (1987), Immunity to *Strongyloides*, in Soulsby, E.J.L. (Ed.) *Immune Responses in Parasitic Infections*, **Vol 1 Nematodes**, pp.89–110, Boca Raton: CRC Press.

Kawaguchi, H., Takayanagi, T., Suzuki, H., Sato, S. and Kato, S. (1983), Immune responses of rat to the migrating larval worm of *Paragonimus miyazaki*, *Japanese Journal of Parasitology*, **32**, 143–9.

Kayes, S.G. and Oaks, J.A. (1976), Effect of inoculum size and length of infection on the distribution of *Toxocara canis* larvae in the mouse, *American Journal of Tropical Medicine and Hygiene*, **25**, 573–81.

Kayes, S.G., Jones, R.E. and Omholt, P.E. (1987), Use of bronchoalveolar lavage to compare local pulmonary immunity with the systemic immune response to *Toxocara canis*-infected mice, *Infection and Immunity*, **55**, 2132–6.

Kayes, S.G., Jones, R.E. and Omholt, P.E. (1988), Pulmonary granuloma formation in murine toxocariasis: transfer of granulomatous hypersensitivity using bronchoalveolar lavage cells, *Journal of Parasitology*, **74**, 950–6.

Keller, R. and Jones, V.E. (1971), Immunological and pharmacological analysis of the primary and secondary reagin response to *Nippostrongylus brasiliensis* in the rat, *Immunology*, **21**, 565–74.

King, S.J. and Miller, H.R.P. (1984), Anaphylactic release of mucosal mast cell protease and its relationship to gut permeability in *Nippostrongylus*-primed rats, *Immunology*, **51**, 653–60.

Khoury, P.B. and Soulsby, E.J.L. (1977a), *Ascaris suum* : immune response in the guinea pig. 1. Lymphoid cell responses during primary infections, *Experimental Parasitology*, **41**, 141–59.

Khoury, P.B. and Soulsby, E.J.L. (1977b), *Ascaris suum* : lymphoid cell responses during secondary infections in the guinea pig, *Experimental Parasitology*, **41**, 432–45.

Knopf, P.M., Cioli, D., Mangold, B.L. and Dean, D.A. (1986), Migration of *Schistosoma mansoni* in normal and passively immunised laboratory rats, *American Journal of Tropical Medicine and Hygiene*, **35**, 1173–84.

Ko, R.C. (1981), Host-parasite relationship of *Angiostrongylus cantonensis*. 2. Angiotropic behaviour and abnormal site development, *Zeitschrift für Parasitenkunde*, **64**, 195–202.

Korenaga, M., Nawa, Y., Mimori, T. and Tada, I. (1983a), *Strongyloides ratti*: the role of enteral antigen stimuli by adult worms in the generation of protective immunity in rats, *Experimental Parasitology*, **55**, 358–63.

Korenaga, M., Nawa, Y., Mimori T. and Tada, I. (1983b), Effects of preintestinal larval antigenic stimuli on the generation of intestinal immunity in *Strongyloides ratti* infections in rats, *Journal of Parasitology*, **69**, 78–82.

Kovacs, J.A. *et al.*, (1984), *Pneumocystis carinii* pneumonia: a comparison between patients with acquired immunodeficiency syndrome and patients with other immunodeficiencies, *Annals of Internal Medicine*, **100**, 663–71.

Lamontagne, L.R., Gauldie, J., Befus, A.D., McAdam, K.P.W.J., Baltz, M.L. and Pepys, M.B. (1984), The acute phase response in parasite infection. *Nippostrongylus brasiliensis* in the mouse, *Immunology*, **52**, 733–41.

Lanken, P.N., Minda, M., Pietra, G.G. and Fishman, A.P. (1980), Alveolar response to experimental *Pneumocystis carinii* pneumonia in the rat, *American Journal of Pathology*, **99**, 561–88.

Lee, O-R. (1979), A histopathologic study of the lungs infected with *Paragonimus westermani* in the dog, *Korean Journal of Parasitology*, **17**, 19–44.

Lee, S.H. (1969), The use of irradiated third-stage larvae of *Angiostrongylus cantonensis* as antigen to immunise albino rats against homologous infection, *Proceedings of the Helminthological Society of Washington*, **36**, 95–7.

Lipscomb, M.F., Lyons, C.R., O'Hara, R.M. and Stein-Streilein, J. (1982), The antigen-induced selective recruitment of specific T lymphocytes to the lung, *Journal of Immunology*, **128**, 111–5.

Lloyd, S. and Soulsby, E.J.L. (1987), Immunology of gastrointestinal nematodes of ruminants, in Soulsby, E.J.L. (Ed.), *Immune Responses in Parasitic Infections*, **Vol. 1 Nematodes**, pp.299–324, Boca Raton: CRC Press.

Loeffler, W. (1956), Transient lung infiltrations with blood eosinophilia, *International Archives of Allergy and Applied Immunology*, **8**, 54–9.

Love, R.J., Kelly, J.D. and Dineen, J.K. (1974), *Nippostrongylus brasiliensis*: effects of immunity on the pre-intestinal and intestinal larval stages of the parasite, *International Journal for Parasitology*, **4**, 183–91.

Lunney, J.K., Urban, J.F. and Johnson, L.A. (1986), Protective immunity to *Ascaris suum*: analysis of swine peripheral blood cell subsets using monoclonal antibodies and flow cytometry, *Veterinary Parasitology*, **20**, 117–31.

Mangold, B.L. and Dean, D.A. (1986), Passive transfer with serum and IgG antibodies of irradiated cercaria-induced resistance against *Schistosoma mansoni* in mice, *Journal of Immunology*, **136**, 2644–8.

Mastin, A.J., Bickle, Q.D. and Wilson, R.A. (1985), An ultrastructural examination of irradiated, immunising schistosomula of *Schistosoma mansoni* during their extended stay in the lungs, *Parasitology*, **91**, 101–10.

Masur, H. and Jones, T.C. (1978), The interaction *in vitro* of *Pneumocystis carinii* with macrophages and L-cells, *Journal of Experimental Medicine*, **147**, 157–70.

McElroy, P.J., Szewczuk, M.R. and Befus, A.D., 1983, Regulation of heterologous IgM, IgG and IgA antibody responses in mucosal-associated lymphoid tissue of *Nippostrongylus brasiliensis* infected mice, *Journal of Immunology*, **130**, 435–41.

McLaughlin, R.F., Tyler, W.S. and Canada, R.O., (1961), Subgross pulmonary anatomy in various mammals and man, *Journal of the American Medical Association*, **175**, 694–7.

Menson, E.N., Coulson, P.S. and Wilson, R.A. (1989), *Schistosoma mansoni*: circulating and pulmonary leucocyte response related to the induction of protective immunity in mice by irradiated parasites, *Parasitology*, **98**, 43–56.

Miller, T.A., (1966), Comparison of the immunogenic efficiencies of normal and X-irradiated *Ancylostoma caninum* larvae in dogs, *Journal of Parasitology* **52**, 512–9.

Miller, T.A. (1971), Vaccination against the canine hookworm diseases, *Advances in*

Parasitology, **9**, 153–83.

Miller, T.A. (1978), Industrial development and field use of the canine hookworm vaccine, *Advances in Parasitology*, **16**, 333–42.

Miller, T.A. (1987), Immunity to hookworms, in Soulsby, E.J.L. (Ed.), *Immune Responses in Parasitic Infections*, **Vol 1 Nematodes**, pp.111–26, Boca Raton: CRC Press.

Mimori, T., Nawa, Y., Korenaga, M. and Tada, I. (1982), *Strongyloides ratti*: Mast cell and goblet cell responses in the small intestine of infected rats, *Experimental Parasitology*, **54**, 366–70.

Minard, P., Dean, D.A., Jacobson, R.H., Vannier, W.E. and Murrel, K.D. (1978), Immunisation of mice with cobalt-60 irradiated *Schistosoma mansoni* cercariae, *American Journal of Tropical Medicine and Hygiene*, **27**, 76–86.

Moqbel, R. (1980), Histopathological changes following primary, secondary and repeated infections of rats with *Strongyloides ratti*, with special reference to tissue eosinophils, *Parasite Immunology*, **2**, 11–27.

Mountford, A.P., Coulson, P.S. and Wilson, R.A. (1988), Antigen localisation and the induction of resistance in mice vaccinated with irradiated cercariae of *Schistosoma mansoni*, *Parasitology*, **97**, 11–25.

Muhleisen, J.P. (1953), Demonstration of pulmonary migration of the causative organism of creeping eruption, *Annals of Internal Medicine*, **38**, 595–600.

Neva, F.A. and Ottesen, E.A. (1978), Tropical (filarial) eosinophilia, *The New England Journal of Medicine*, **298**, 1129–31.

Oakley, G.A. (1980), The comparative efficacy of levamisole and diethylcarbamazine citrate against *Dictyocaulus viviparus* infection in cattle, *The Veterinary Record*, **107**, 166–70.

O'Donnell, I.J. and Mitchell, G.F. (1980), An investigation of the antigens of *Ascaris lumbricoides*, using a radioimmunoassay and sera of naturally infected humans, *International Archives of Allergy and Applied Immunology*, **61**, 213–9.

Ogilvie, B.M. (1965), Role of adult worms in immunity of rats to *Nippostrongylus brasiliensis*, *Parasitology*, **55**, 325–35.

Ogilvie, B.M. and Jones, V.E. (1971), *Nippostrongylus brasiliensis*: a review of immunity and the host/parasite relationship in the rat, *Experimental Parasitology*, **29**, 138–77.

Olsen, C.E. and Schiller, E.L. (1978), *Strongyloides ratti* infections in rats. 1. Immunopathology, *American Journal of Tropical Medicine and Hygiene*, **27**, 521–26.

Ottesen, E.A. (1984), Immunological aspects of lymphatic filariasis and onchocerciasis in man, *Transactions of the Royal Society of Tropical Medicine and Hygiene*, **78**, (Supplement), 9–18.

Ottesen, E.A., Kumaraswami, V., Paranjape, R., Poindexter, R.W. and Tripathy, S.P. (1981), Naturally occurring blocking antibodies modulate immediate hypersensitivity responses in human filariasis, *Journal of Immunology*, **127**, 2014–20.

Ottesen, E.A., Neva, F.A., Paranjape, R.S., Tripathy, S.P., Thiruvengadam, K.V. and Beaven, M.A. (1979), Specific allergic sensitisation to filarial antigens in tropical eosinophilia syndrome, *The Lancet*, **June 2**, 1158–60.

Phills, J.A., Harrold, A.J., Whiteman, G.V. and Perelmutter, L. (1972), Pulmonary infiltrates, asthma and eosinophilia due to *Ascaris suum* infestation in man, *New England Journal of Medicine*, **268**, 965–70.

Piessens, W.F. and Mackenzie, C.D. (1982), Immunology of lymphatic filariasis and onchocerciasis, in Cohen, S. and Warren, K.S. (Eds.), *Immunology of parasitic infections* 2nd ed. pp.622–53, Oxford: Blackwell Scientific Publications.

Pirie, H.M., Doyle, J., McIntyre, W.I.M. and Armour, J. (1971), The relationship between pulmonary lymphoid nodules and vaccination against *Dictyocaulus viviparus*, in Gaafar, S.M. (Ed.), *Pathology of Parasitic Diseases*, Indiana: Purdue University Press.

Prochazka, Z. and Mulligan, W. (1965), Immunological studies on *Nippostrongylus brasiliensis* infection in the rat: experiments with irradiated larvae, *Experimental Parasitology*, **17**, 51–6.

Rubin, R. and Lucker, J.T. (1956), Acquired resistance to *Dictyocaulus viviparus*, the lungworm of cattle, *Cornell Veterinarian*, **46**, 88–96.

Ryning, F.W. and Remington, J.S. (1977), Effect of alveolar macrophages on *Toxoplasma gondii*, *Infection and Immunity*, **18**, 746–53.

Salman, S.K. and Brown, P.J. (1980), A study of the pathology of the lungs of rats after subcutaneous or intravenous injection of active or inactive larvae of *Nippostrongylus brasiliensis*, *Journal of Comparative Pathology*, **90**, 447–55.

Sodeman, T.M., Sodeman Jr. W.A. and Richards, C.S. (1969), The intrapulmonary localisation of *Angiostrongylus cantonensis* in the rat, *Proceedings of the Helminthological Society of Washington*, **36**, 143–6.

Soulsby, E.J.L. (1965), *Textbook of Veterinary Clinical Parasitology*, **Vol 1 Helminths**, Oxford: Blackwell Scientific Publications.

Soulsby, E.J.L. (1987), Immunology, immunopathology, and immunoprophylaxis of *Ascaris* spp. infection, in Soulsby, E.J.L. (Ed.), *Immune Responses in Parasitic Infections*, **Vol 1 Nematodes**, pp.127–140. Boca Raton: CRC Press.

Spencer, H. (1985), *Pathology of the Lung*, Oxford: Pergamon, 4th ed.

Spillman R.K. (1975), Pulmonary ascariasis in tropical communities, *American Journal of Tropical Medicine and Hygiene*, **24**, 791–800.

Stavitsky, A.B. (1987), Immune regulation in *Schistosomiasis japonica*. *Immunology Today*, **8**, 228–33.

Stewart, T.B., Southern, L.L., Gibson, R.B. and Simmons, L.A. (1984/5), Immunisation of pigs against *Ascaris suum* by sequential experimental infections terminated with fenbendazole during larval migration, *Veterinary Parasitology*, **17**, 319–26.

Stockdale, P.H.G. (1976), Pulmonary pathology associated with metastrongyloid infections, *British Veterinary Journal*, **132**, 595–608.

Stromberg, B.E. and Soulsby, E.J.L. (1977), *Ascaris suum*: immunisation with soluble antigens in the guinea pig, *International Journal for Parasitology*, **7**, 287–91.

Takai, A. (1983), Immunopathological studies of rats infected with *Angiostrongylus cantonensis*. 1. Circulating immune complexes and deposition of the immune complexes on the tissues, *Acta Medica et Biologica*, **30**, 105–23.

Taliaferro, W.H. and Sarles, M.P. (1939), The cellular reactions in the skin, lungs and intestine of normal and immune rats after infection with *Nippostrongylus muris*, *Journal of Infectious Diseases*, **64**, 157–92.

Tromba, F.G. (1978a), Effect of ultraviolet radiation on the infective stages of *Ascaris suum* and *Stephanurus dentatus* with a comparison of the relative susceptibilities of some parasitic nematodes to ultraviolet, *Journal of Parasitology*, **64**, 245–52.

Tromba, F.G. (1978b), Immunisation of pigs against experimental *Ascaris suum* infection by feeding ultraviolet-attenuated eggs, *Journal of Parasitology*, **64**, 651–6.

Turner-Warwick, M. (1983), The pulmonary interstitium in pathological states, *Current Problems in Clinical Biochemistry*, **13**, 157–67.

Udwadia, F.E (1975), Pulmonary eosinophilia, in Herzog, H. (Eds.), *Progress in Respiration Research*, **Vol 7**, pp.286, Basel: Karger.

Urquhart, G.M., Jarrett, W.F.H., Bairden, K. and Bonazzi, E.F. (1981), Control of parasitic bronchitis in calves: vaccination or treatment? *The Veterinary Record*, **108**, 180–2.

Vogel, H. and Minning, W. (1942), Beitrage zur Klinik der Lungen—Ascariasis und zur Frage der fluchtigen eosinophiliea Lungeninfiltrate, *Beitrage zur Klinik der Tuberkulose und zur specifischen Tuberkuloseforschung*, **98**, 620–54.

von Lichtenberg, F. (1962), Host response to eggs of *Schistosoma mansoni*. Granuloma formation in the unsensitised laboratory mouse, *American Journal of Pathology* **41**, 711–31.

von Lichtenberg, F. (1987), Consequences of infections with schistosomes, in Rollinson, D. and Simpson, A.J.G. (Eds)., *The Biology of Schistosomes*, pp.185–232, London: Academic Press.

Walzer, P.D., (1986), Attachment of microbes to host cells: relevance of *Pneumocystis carinii*, *Laboratory Investigation*, **54**, 589–92.

Warren, K.S. (1987), Determinants of disease in human schistosomiasis, *Clinical Tropical Medicine and Communicable Diseases*, **Vol 2, Number 2, Schistosomiasis**, 301–13.

Weiss, N. and Tanner, M. (1979), Studies on *Dipetalonema viteae* (Filarioidea). 3. Antibody-dependent cell-mediated destruction of microfilariae *in vivo*. *Tropenmedizin und Parasitologie*, **30**, 73–80.

Wells, C. and Behnke, J.M. (1988a), The course of primary infection with *Necator americanus* in syngeneic mice, *International Journal for Parasitology*, **18**, 47–52.

Wells, C. and Behnke, J.M. (1988b), Acquired resistance to the human hookworm *Nector americanus* in mice. *Parasite Immunology*, **10**, 493–505.

Werczberger, A., Gohlman, J., Wertheim, G., Gunders, A.E. and Chowers, I. (1979), Disseminated echinococcosis with repeated anaphylactic shock-treated with mebendazole, *Chest*, **76**, 482–4.

Wheater, P.R., Burkitt, H.G. and Daniels, V.G. (1979), *Functional Histology*, Edinburgh: Churchill Livingstone.

Wilson, J.F., Diddams, A.C. and Rausch, R.L. (1968), Cystic hydatid disease in Alaska, *American Review of Respiratory Diseases*, **98**, 1–15.

Wilson, R.A. (1987), Cercariae to liver worms: development and migration in the mammalian host, in Rollinson, D., and Simpson, A.J.G. (Ed.), *The Biology of Schistosomes*, pp.115–46. London: Academic Press.

Wilson, R.A. and Coulson, P.S. (1986), *Schistosoma mansoni*: dynamics of migration through the vascular system of the mouse, *Parasitology*, **92**, 83–100.

Wilson, R.A., Coulson, P.S. and Dixon, B. (1986), Migration of the schistosomula of *Schistosoma mansoni* in mice vaccinated with radiation attenuated cercariae, and normal mice: an attempt to identify the timing and site of parasite death, *Parasitology*, **92**, 101–16.

Wong, M.M. (1987), Immunology, immunopathology, and immunoprophylaxis of *Dirofilaria immitis*, in Soulsby, E.J.L. (Ed.), *Immune Responses in Parasitic Infections*, **Vol 1 Nematodes**, pp.251–280. Boca Raton: CRC Press.

Yong, W.K. and Dobson C. (1982a), Population dynamics of *Angiostrongylus cantonensis* during primary infections in rats, *Parasitology*, **85**, 399–409.

Yong, W.K. and Dobson, C. (1982b), Antibody responses in rats infected with *Angiostrongylus cantonensis* and the passive transfer of protective immunity with immune serum, *Zeitschrift für Parasitenkunde*, **67**, 329–36.

Yong, W.K. and Dobson, C. (1983), Immunological regulation of *Angiostrongylus cantonensis* infections in rats: modulation of population density and enhanced parasite growth following one or two superimposed infections, *Journal of Helminthology*, **57**, 155–65.

Yong, W.K. and Dobson, C. (1984), Peripheral blood white cell responses during *Angiostrongylus cantonensis* infection in rats, *International Journal for Parasitology*, **14**, 207–11.

Yong, W.K., Glanville, R.J. and Dobson, C. (1983), The role of the spleen in protective immunity against *Angiostrongylus cantonensis* in rats: splenectomy and passive spleen cell transfers, *International Journal for Parasitology*, **13**, 165–70.

Yoshimura, K. and Soulsby, E.J.L. (1976), *Angiostrongylus cantonensis*: lymphoid cell responsiveness and antibody production in rats, *American Journal of Tropical Medicine and Hygiene*, **25**, 99–107.

Yoshimura, K., Aiba, H. and Oya, H. (1979b), Transplantation of young adult *Angiostrongylus cantonensis* into the rat pulmonary vessels and its application to the assessment of acquired resistance, *International Journal for Parasitology*, **9**, 97–103.

Yoshimura, K., Aiba, H., Hirayama, N. and Yosida, T.H. (1979a), Acquired resistance and immune responses of eight strains of inbred rats to infection with *Angiostrongylus cantonensis*, *Japanese Journal of Veterinary Science*, **41**, 245–59.

10. Immunological and inflammatory responses in the small intestine associated with helminthic infections

Redwan Moqbel and Angus J. MacDonald

The primary role of the small intestine is the absorption of simple and complex nutrients. As such, the gut is exposed to a wide spectrum of foreign substances ranging from the noxious to those which are absolutely essential to the maintenance of good health, natural growth and development. The capacity of the normal small intestine to distinguish between innocuous substances that traverse its length from pathogenic, irritant or allergenic material is the consequence of precise homeostatic regulation, ensuring for the most part efficient digestive function and freedom from disease. Contact with antigenic material in the gastro-intestinal tract, results in immunological responses which range from tolerance, at the one extreme, to complete protection at the other. In disease conditions, immunological effector mechanisms may be initiated following pathogen-induced injury to the integrity of the intestinal tissue. Such effector mechanisms appear to feature also in immune responses against a number of intestinal helminths. However, in man, it is often the case that ubiquitous gastro-intestinal helminthiases, while eliciting a range of inflammatory responses, appear to give rise to persistence of infection and/or immunological tolerance of the presence of parasites (Chapter 13).

The importance of a fully functional intestinal immune system is readily apparent. Damage to intestinal defense mechanisms or infection by parasites can have profound effects on health. This may be reflected in the millions of people afflicted annually by various forms of diarrhoea (often fatal) and gut-associated diseases which decimate populations in times of war or natural disaster. Immunodeficiency diseases which involve the gastrointestinal system also have remarkable debilitating effects on affected subjects. This review will concentrate on defense mechanisms which are initiated following exposure to infection and which induce in the host a state of protection against a secondary infection. The precise protective events within the mucosal layer of parasitized mammalian gut are complex and some of these will be reviewed here. A major part of our current knowledge of the immunological and inflammatory mechanisms controlling and modulating gastro-intestinal immunity during helminthic infection is based upon work on experimental animal host/parasite models. It is necessary, however, to be aware that there are marked differences in intestinal immune responses between mammalian species, and extreme caution is warranted when extrapolating from rodent models to human or even ungulate (ruminant) infections.

10.1. MUCOSAL DEFENSE MECHANISMS

It has been suggested that, at least in man, mucosal defense mechanisms in the bowel may be divided into three components (Figure 10.1). These are immune exclusion, immune regulation and immune elimination (Brandtzaeg *et al.*, 1985). Immune exclusion represents the first line of defense against antigens and refers mainly to the protection of mucosal surfaces by secretory IgA and IgM acting in conjunction with other innate, non-specific factors. This line of defense contributes significantly to the overall maintenance of homeostasis in the gut.

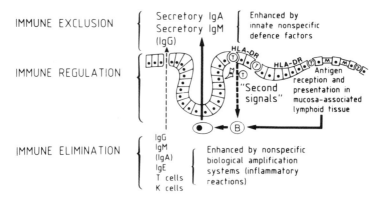

Figure 10.1. Schematic diagram of immune exclusion, immune regulation and immune elimination. M = microfold cells lining the dome of the Peyer's patches. T = T lymphocytes. B = B lymphocytes. HLA-DR = Class II major histocompatibility complex (reproduced from Brandtzaeg *et al.*, 1985).

Immune regulation is a complex second-line defense mechanism which involves specific cellular responses, both in the Peyer's patches and the relevant draining lymphoid tissues. These responses include the interplay between specifically generated sub-populations of T cells following antigen-specific activation and differentation of these cells. Generation of antigen-specific IgA, IgM, IgG and IgE are responses which also characterize immune regulation.

Immune elimination, on the other hand, involves a number of non-specific biological amplification systems including inflammatory changes which appear to be controlled and orchestrated by specifically-primed T cells. These cells contribute to the protective effector mechanisms in the mucosal layer by enhancing and augmenting the process of immune regulation. Inflammatory changes include the recruitment and the activation of various leucocytes and the generation of a battery of mediators which are thought to be important in protection against reinfection with helminths.

The lamina propria is rich in lymphoid cells, some 60% of which are recognized as $CD3^+$ (T cells). There are more resting B cells than plasma cells (1×10^{10} per mm^3 compared to 2×10^5 per mm^3). The gut mucosa also contains dendritic macrophage cells which reside mainly beneath the epithelium at the tip of the villi, and since they bear MHC Class II molecules, they are assumed to be involved in antigen presentation.

10.2. GUT-ASSOCIATED LYMPHOID TISSUE (GALT)

The structure of the lymphoid compartment in the small intestine is unique to this organ. This distinct GALT is composed of three basic elements in addition to the draining mesenteric lymph nodes: (I) a number of lymphoid aggregated follicles called Peyer's patches; (II) diffused lymphoid cells which are present in large numbers within the lamina propria; (III) intra-epithelial lymphocytes present between the villar columnar epithelial cells. These three components are involved in the process of antigen recognition, processing, presentation and subsequent effector immune mechanisms including the initiation of the inflammatory reaction.

The disruption to the integrity of the epithelial (mucosal) layer by various agents including food antigens, helminthic parasites, bacteria and their toxins, as well as drugs results in antigenic stimulation and subsequent mobilization of antibody-producing cells (plasma cells). In man, the predominant secreted immunoglobulin from lamina propria B cells is IgA (80%) although IgM (17%), IgG and IgE (3%) plasma cells are also present (Figure 10.2). Thus large numbers of lymphoblasts are generated, both in the Peyer's patches and draining mesenteric lymph nodes, committed mainly to IgA antibody synthesis (i.e. with IgA expressed on their surface membrane) and some IgM. The production of copious amounts of IgA antibody, protects the epithelial layer from further penetration by antigen into the lamina propria. Once there,

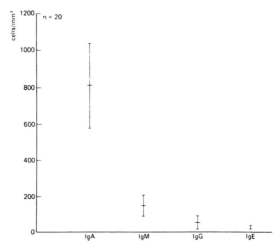

Figure 10.2. Relative frequency of plasma cells of various immunoglobulin classes within the lamina propria of human intestinal mucosal tissue (reproduced from Kilby *et al.*, 1976, by kind permission).

committed effector B cells appear to be unable to return to blood circulation. However, B cells committed to IgA synthesis migrate to the mesenteric lymph nodes to secrete antibody which subsequently reaches the intestinal capillaries via circulation. Passage to the intestinal lumen (Figure 10.3) is accomplished by active transportation through enterocytes and in some species (notably rodents) through the hepato-biliary pump. T cells, on the other hand, are found mainly in intra-epithelial sites (Ferguson, 1977) or in the lamina propria and reach the latter by travelling through the mesenteric lymph nodes into the thoracic lymph duct and then via peripheral circulation to reach the mucosa.

Immunocytochemical studies have revealed differences between the sub-classes of T cells present in the lamina propria and the epithelium. Helper-inducer T cells (CD4+) appear to predominate in the lamina propria while suppressor-cytotoxic T cells (CD8+) form the majority of cells found within the epithelium. Studies using human foetal ileum have suggested that the CD4:CD8 ratio may not be totally antigen-dependent (Spencer *et al.*, 1986a).

Peyer's patches

This aggregated lymphoid tissue, found in the lamina propria of the small intestine is composed of clusters of lymphoid cells which are unencapsulated (Figure 10.4). Studies by Spencer *et al.* (1986b) have shown that these clusters appear in human foetal ileum within the first 14 to 16 weeks of life, where both B and T cells are seen. These two cell types begin to segregate separately into two distinct structures to form the Peyer's patches. Their germinal centres

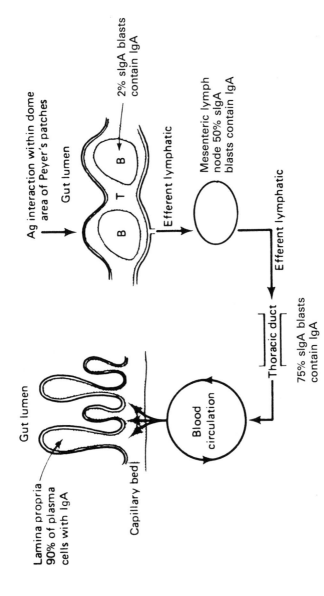

Figure 10.3. Diagramatic presentation of the maturation journey of B blast cells from Peyer's patches through the mesenteric lymph node and thoracic lymph duct to the lamina propria where the majority of plasma cells produce IgA (reproduced from Parrott, 1976 with kind permission of author and publishers).

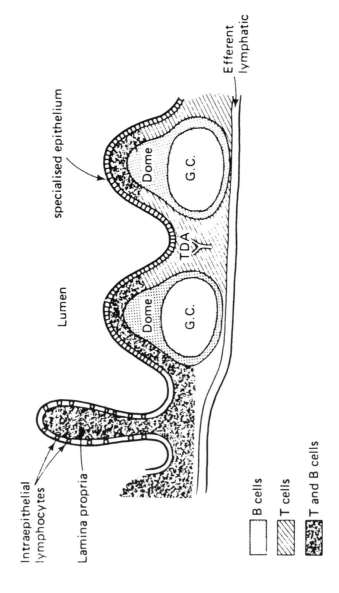

Figure 10.4. Schematic representation of the structure of a murine Peyer's patch: G.C. = germinal centres (B-cell rich); TDA = thymus-dependent area (T-cell rich) (reproduced from Parrott, 1976 by kind permission of publishers).

contain mostly B cells while the thymus-dependent areas (the inter-follicular spaces) are rich with T cells. However, mixtures of T cells and B cells are also found throughout the villar spaces.

Peyer's patches are clearly observed in the intestinal tissue of rodents as prominent swellings apparent even on the serosal aspect of the gut wall. Their counterparts in human adult tissue are rather more difficult to observe. Peyers's patches are separated from the intestinal lumen by a distinct dome-shaped overlaying epithelial layer of M (microfold) cells which are cuboidal with a distinctive ultrastructure and a specialized role in antigen presentation. In contrast to the columnar epithelial cells of the intestine, M cells have few micro-villi and a very poorly developed glycocalyx, but they appear to have active endoplasmic reticulum and vesicular transport system. Lymphoid cells within the Peyer's patches found adjacent to the base of the M cells may be introduced to the antigen via the M cells. It is still not fully clear whether dome M cells are the only source of antigenic presentation to lymphocytes in the Peyer's patches. This is generally accepted since M cells have a high expression of MHC Class II (HLA-DR) in the crypt epithelium directly adjacent to Peyer's patches. This view is further supported by the fact that Peyer's patches have no afferent lymphatics.

Intra-epithelial lymphocytes

Lymphocytes are divided according to the type of receptors for antigens expressed on their membrane surface. B lymphocytes express receptors for the Fc portion of Ig and, once induced by T cells, manufacture specific antibodies. T lymphocytes are sub-divided into CD4$^+$ helper/inducer cells, which express receptors specific for antigenic fragments juxtaposed to MHC Class II (HLA-DR) molecules, and CD8$^+$ suppressor/cytotoxic cells which recognize the combination of antigen and MHC Class I molecules. The vast majority of mature T lymphocytes (i.e CD3$^+$) in the peripheral blood and lymphoid organs recognize antigen via alpha/beta (α/β) chain receptors. Recent studies on human and murine T cells have identified a novel gene rearrangement which encoded for a third T-cell-specific chain called gamma (γ) with a second variable chain called delta (δ; Saito *et al.*, 1984). T cells that express CD3 and bear the $\gamma\delta$ receptors are relatively infrequent in lymphoid tissue (1–3%) when compared with those that express $\alpha\beta$ which represent the majority of CD3 cells (Janeway, 1988). Most T lymphocytes located on epithelial surfaces have been shown to express CD3 $\gamma\delta$ and as such may be important in surveillance of mucosal surfaces. Normal murine intestinal intra-epithelial lymphocytes (IEL) found interdigitating between columnar epithelial cells are CD3$^-$, 90% of which are CD8$^+$ (i.e. CD4$^-$; only 10% are CD4$^+$ with a CD8$^+$:CD4$^+$ ratio of approximately 3:1 in man, mouse and rat). IEL are therefore a distinct set of CD3 lymphocytes which express the $\gamma\delta$ heterodimers of antigen recognition units (Goodman and Lefrancois, 1988; Bonneville *et al.*, 1988). Recent studies have further suggested that, at least in normal mice, IEL may have two distinct lineages, one population which is thymus-dependent (CD3$^+$) which operates via αβ T-cell receptor and another subset which is thymus-independent (CD3$^-$) which appear to utilize the γδ T-cell receptor (Viney *et al.*, 1989).

A population of these IEL are granular (gEL) and were studied for their possible relationship to mucosal mast cells. The latter cells are found in sub-epithelial sites and are considered, both *in vitro* and *in vivo*, to be under T-cell control. Guy-Grand *et al.* (1978) proposed that murine gEL may be novel precursors of mucosal mast cells. This view remains controversial especially since mice genetically deficient in mast cell precursors (W/W^v) express a normal population of gEL in their mucosal tissue. It is now generally accepted that MMC and gEL have separate ontological pathways and may be derived from distinct progenitors. However, in ovine intestinal mucosa, gEL are considered to be partially degranulated MMC (Huntley *et al.*, 1984).

Mesenteric lymph nodes

The efferent lymphatics within the small intestine drain into regional lymph nodes located in the mesenteries connecting the various parts of the intestine, which in turn drain into the thoracic lymph duct. These relatively large typical lymphoid nodes are the receiving station for dividing cells from the Peyer's patches and are the store for many of the effector T and B blasts which are antigen-specific. Lymphocytes obtained from mesenteric lymph nodes of sensitized or infected animals have been used as the main source of immune competent cells for adoptive transfer of immunity in syngeneic host/parasite systems. *In vivo*-activated lymphocytes recirculate via the thoracic duct and blood stream before homing preferentially to the intestinal mucosa to execute their specific role in defense.

10.3. IgA SECRETION

The secretion of immunoglobulin, particularly IgA, is a predominant feature of the gut immune response. This class of mucosal IgA antibody is antigenically distinct from that present in the circulation and is found in a secretory form (11S) which is a dimer joined by a secretory piece (s.p.). In addition, secretory IgA (SIgA) contains another additional distinct polypeptide chain called the J-chain (which is also found in secretory IgM). The latter polypeptide is synthesized by lymphocytes, while the secretory piece is a glycoprotein manufactured by the Golgi apparatus of gut enterocytes and added to the dimeric IgA molecule during uptake of IgA and transportation across the epithelial cells into the gut lumen. The secretory piece may be important in the protection of IgA against destruction by both pancreatic and bacterial products. Secretory IgA has the main function of exerting immune exclusion by preventing, in combination with non-specific (innate) defense mechanisms, further entry by antigens into the lamina propria and reducing the ability of bacteria and other antigens to stick to the intestinal mucosal wall. IgA, unlike both IgG or IgM, does not activate complement via the classical pathway, but aggregated human IgA is able to do so through the alternative pathway.

10.4. CYTOKINES AND MUCOSAL IMMUNITY

Immunoregulatory mechanisms operating at the gut mucosal interphase involve the interplay between T cells, their products and other immune cells, especially B cells and macrophages. Antigen processing and presentation via either M cells lining the dome of Peyer's patches or dendritic cells (found within mucosa or in lymphoid follicles) leads to the activation of T lymphocytes. This is followed by the induction of B-cell proliferation and differentiation and specific-antibody production, which is mainly IgA. The release of putative cytokines (e.g. IL-2, IL-3, IL-4 and IL-6) is thought to play an important role in the amplification of the local immune response and the subsequent recruitment and activation of secondary inflammatory (effector) cells.

Peyer's patches contain precursors for B cells, which are committed to IgA production, and also T-helper cells specific for IgA B cells. IL-2 is secreted from Peyer's patches T cells when stimulated (e.g. experimentally with Con-A) and appears to have an effect on increasing the production of IgA from committed plasma cells. IFN-gamma has a synergistic effect on the action of IL-2. However, IL-5 appears to have a direct and selective *in vitro* effect on Peyer's patches B cells (but not spleen) committed only to IgA production (i.e. not IgG) by enhancing IgA synthesis. IL-5 does not induce the proliferation of IgA secreting cells, *in vitro*, but appears to increase their frequency in Peyer's patches, *in vivo*. Whether this is achieved through directed locomotion (migration), isotype switching or other mechanisms is not yet clear. However, it is thought that IL-5 may be involved in isotype switching from IgM to IgA. Recent studies suggested that IL-6 may be more efficient than IL-5 in enhancing IgA synthesis by Peyer's patches IgA B cells and may exert a synergistic effect with IL-5. IL-4 may also have an enhancing influence on the action of IL-5, but this aspect requires further confirmation. The role of IL-3 as a pluripotential cytokine involved in the growth and differentiation of mucosal mast cells, *in vitro*, is the subject of a separate chapter (Chapter 3) in this book.

10.5. THE INFLAMMATORY RESPONSE TO INTESTINAL HELMINTHS

Inflammation is the response of living tissue to injury and is manifested by changes in the vasculature, increased permeability (exudation) and a number of leucocyte events, the aim of which is to repair and heal the affected site.

In the case of intestinal helminthiases, these inflammatory changes may be elicited by the tissue-damaging activities or toxic effects of the worms themselves. However, there is a large body of experimental evidence suggesting that much of the inflammation is a consequence of the immune response to the parasites (reviewed in Wakelin, 1978).

Studies using a number of host-parasite systems where intestinal inflammation is a prominent feature of the host response have consistently underlined the likely protective role of inflammatory changes in the expulsion of worms from the gut. Primary infections of rats with *Nippostrongylus brasiliensis* (Chapter 8, Figure 8.7) and mice with *Trichinella spiralis* (Chapter 6, Figure 6.2) exhibit a 'spontaneous cure' response leading to the expulsion of intact worms from the host intestine following the development of a mucosal inflammatory response. Subsequent infections show classical secondary immune-response kinetics with accelerated rejection of the intestinal worm burden. This type of response may, depending on factors such as host species and strain, number of priming infections as well as the size of the infective dose, take place over several hours rather than days. This process, known as 'rapid expulsion' (RE) is akin to an immediate hypersensitivity response (reviewed in Wakelin, 1978; Miller, 1984). Sheep rendered hyperimmune to the abomasal-dwelling nematodes, *Haemonchus contortus* (Chapter 8, Figure 8.7) and *Ostertagia circumcincta*, by multiple high-dose infections develop a stage-specific 'self-cure' response to further larval challenge. This mechanism may also be mediated by mucosal inflammatory changes (Miller, 1984).

The characteristic inflammatory changes occurring in the intestine during infection with the majority of helminths include: inflammatory cell infiltration into the mucosa, particularly mast cells and eosinophils; mucosal oedema with leakage of plasma into the lumen; villous atrophy; mucus hypersecretion; disruption of epithelial integrity; and increased gut motility. These immunopathological changes, while detrimental to normal intestinal functioning in the short term, are ultimately beneficial to the host since they, almost certainly, contribute to the removal of the worm burden from the gut. It should be noted that the inflammatory response of the intestine acts in an essentially non-specific fashion. For example, mice immune to *T. spiralis* will expel both *T. spiralis* and *N. brasiliensis* when given a concurrent challenge with these two species. There is no evidence of cross immunity between the two species, and the removal of *N. brasiliensis* appears to be as a result of the non-specific effects of inflammation induced by the immune response to *T. spiralis* (Kennedy, 1980).

A number of other species combinations have been studied and have shown similar responses to that described for *T. spiralis* and *N. brasiliensis* (reviewed in Kazacos, 1975). However, there are other worm combinations where previous or concurrent infection inhibits or delays the immune expulsion of another species; this is thought to be due to immune depression of the host exerted by the initial infection (Jenkins, 1975; Jenkins and Behnke, 1977; Behnke *et al.*, 1978).

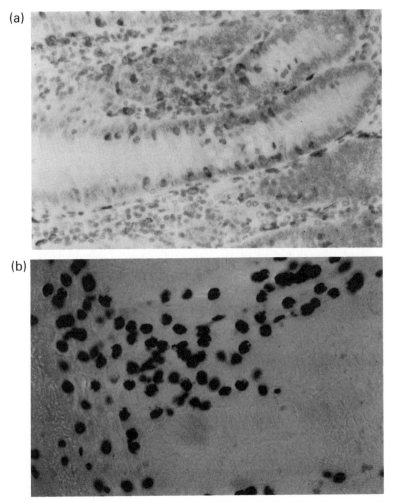

Figure 10.5. (a) Mucosal mast cells (stained with Alcian blue) and (b) tissue eosinophils (stained with immunoperoxidase) in villi of rat intestine during a secondary *N. brasiliensis* infection (photomicrographs by courtesy of Dr. H.R.P. Miller).

T-cell dependence of the inflammatory response

Intestinal helminth infections in mammals are usually associated with mastocytosis and eosinophilia in the intestinal mucosa (Figure 10.5), goblet cell hyperplasia and the development of an IgE response. The central role of the T lymphocyte in initiating and regulating these responses has now been well established in several host-parasite systems in laboratory rodents (reviewed in Miller, 1984). In the mouse, CD4-positive T cells (i.e. of the helper/inducer phenotype) from parasite-immune animals can adoptively transfer immunity to *T. spiralis* (Grencis *et al.*, 1985) and are required for the spontaneous cure response as well as the development of intestinal mast cell hyperplasia and an IgE response during primary infection with *N. brasiliensis* nematodes (Katona *et al.*, 1988). Accelerated rejection of *T. spiralis* by normal rats can also be transferred from immune rats by CD4$^+$ T cells (T$_h$), and eosinophilia and resistance to infection with the peritoneal-dwelling cestode, *Mesocestoides corti*, can be transferred by an *M. corti*-antigen-specific T-cell line of T$_h$ phenotype (Lammas *et al.*, 1987).

It is important to stress that there is no evidence that T cells 'themselves mediate worm expulsion directly. Rather, it is more likely that the lymphokines produced by these cells following specific interaction with worm antigens promote the growth and activation of other cell types, such as mast cells, eosinophils and goblet cells which, in turn, mediate inflammatory responses detrimental to worm establishment and survival. Recombinant human interleukins-3 and -5 (IL-3 and IL-5) have both been shown to enhance antibody-dependent cytotoxicity by human eosinophils (Lopez *et al.*, 1987, 1988). Repeated administration of IL-3 to nude mice (Table 10.1) restores their ability to expel *Strongyloides ratti* and to mount an intestinal mucosal mastocytosis (Abe and Nawa, 1988). These latter *in vitro* and *in vivo* observations underline the role of the T cell in orchestrating the immune response to intestinal helminth infections.

Table 10.1. Effect of IL-3 treatment on the recovery of adult worms and the intestinal mast cell response in *S. ratti*-infected nu/nu mice (modified from Abe and Nawa, 1988).

Treatment	No. of adult worms recovered from small intestine	Caecum
1. 29 000 BU IL-3 2 × daily i.p. Days 4–13 post-infection	0 ± 0	6 ± 2
2. Saline 2 × daily i.p. Days 4–13 post-infection	74 ± 13 (P<0.001)	27 ± 5 (P<0.05)

Recent evidence has implicated the T cell more directly in intestinal pathology with possible relevance to intestinal helminthiasis. Specific *in vitro* activation of mucosal T cells by Pokeweed mitogen in foetal human gut explants rapidly results in epithelial cell hyperplasia and villous atrophy (MacDonald and Spencer, 1988). This phenomenon was prevented by treatment of the gut explant cultures with cyclosporin A, a potent inhibitor of T-cell activation, prior to stimulation of T lymphocytes. In the absence of myeloid cells in these tissue preparations, the phenomenon was considered to be T-cell dependent (MacDonald and Spencer, 1988). This result may suggest a direct role for the T cell in the villous atrophy which accompanies many intestinal helminth infections. The T-cell dependence of villous atrophy and crypt cell hyperplasia observed during *N. brasiliensis* infection in rats has previously been suggested by experiments in which T-cell-depleted rats did not show such intestinal pathology associated with worm infection (Ferguson and Jarrett, 1975). It is possible that lymphokines produced by activated mucosal T cells have a direct effect on the proliferation of epithelial cells and/or mesenchymal cells in the lamina propria.

Role of mast cells and IgE in intestinal inflammation and mucosal defense

An absolute requirement for mast cells during the expulsion of intestinal nematodes in rodents has yet to be definitively established. Nevertheless, there is strong circumstantial evidence that these cells may play an important role in the patho-physiological changes in the gut leading to worm expulsion. Mucosal mast cells (MMC) which, in the rat, are functionally and biochemically distinct from connective tissue mast cells (CTMC; covered in greater detail in Chapter 3) have been shown to be functionally active during the expulsion phase of primary infections with *N. brasiliensis* and *T. spiralis* (Woodbury *et al.*, 1984). In these experiments, MMC activation was detected by the presence of the MMC-specific enzyme, rat mast cell protease II (RMCPII) in the serum of infected rats (Figure 10.6). Mast cell activation has also been demonstrated in sheep immune to *H. contortus* and *Ostertagia circumcincta* following intra-abomasal and oral challenge, respectively, with these nematodes (Huntley *et al.*, 1987). Release of a sheep mast cell protease (SMCP) was detectable in blood and lymph fluid following parasite challenge of immune sheep. Although levels of SMCP measured were very low, this was attributed to the presence of potent protease inhibitor(s) in ovine blood and lymph fluid which interfered with the SMCP assay system.

Mast cells (and basophils) bear surface-bound IgE on their high-affinity $Fc_\epsilon R$ receptors. Depending on species, other isotypes also bind to mast cells with lower affinities (see Chapter 3). Very high levels of IgE are produced following infection with intestinal nematodes, such as *N. brasiliensis*, where serum IgE concentrations can exceed 200 $\mu g\ ml^{-1}$ compared to normal levels of less than 1 $\mu g\ ml^{-1}$ (reviewed in Block *et al.*, 1972), and much of this IgE is parasite non-specific. Arguments against the involvement of mast cells/IgE in expulsion of primary *N. brasiliensis* infections have been based on intestinal mastocytosis and the serum IgE response following the onset of worm expulsion. However, it is known that shortly after the arrival of worms in the gut, the existing MMC in the lamina propria discharge and become histologically undetectable. This initial activation is thought to be mediated by a worm-derived degranulating factor (Block *et al.*, 1972). The claim that rises in serum IgE follow parasite expulsion may also not hold for a number

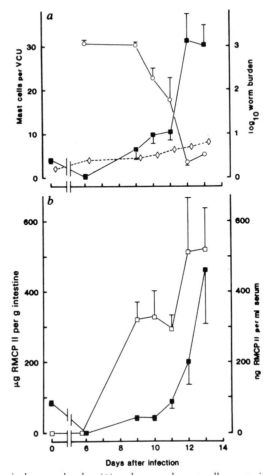

Figure 10.6. *a*) Intestinal worm burden (○) and mucosal mast cell counts in the jejunum (■) and ileum (◇) of rat infected with *N. brasiliensis*.
b) Concentration of RMCPII in the serum (□) and jejunum (■) of naive (day 0) and *N. brasiliensis*-infected rats (data represent mean ± s.e.m.; (after Woodbury *et al.*, 1984). Reprinted from *Nature*, 312, 450–52. Copyright © 1984 Macmillan Magazines Ltd.

of reasons. Firstly, the earliest IgE produced may be sequestrated onto the surface of mast cells (or basophils) and be undetectable in the serum. Secondly, parasite-specific IgE has been detected using passive cutaneous anaphylaxis (PCA) reactions and, since worm-specifc IgE constitutes only a small part of the total IgE present at the early stages of the IgE response, the sensitization of skin mast cells in the PCA test with non-specific IgE may explain the lack of a detectable PCA response to worm antigens. Thirdly, skin and peritoneal mast cells are sensitized with anaphylactic antibody (presumably IgE) prior to the rejection of a primary *N. brasiliensis* infection (Block *et al.*, 1972), and parasite-specific IgE can be detected in the mesenteric lymph nodes, at the onset of worm expulsion, several days prior to elevation in serum IgE (Allan and Mayrhofer, 1981).

More evidence in favour of the protective role of IgE was provided when rats treated with anti-epsilon antisera showed diminished resistance to *T. spiralis* infection (Dessein *et al.*, 1981) (Figure 10.7). On balance, it seems likely that mast cell/IgE-mediated mucosal anaphylaxis plays a role in the elimination of intestinal worms although the extent of the involvement is, as yet, unclear. The presence of low-affinity IgE receptors (Fc$_\epsilon$RII) on a sub-population of T cells and B cells, macrophages, eosinophils and platelets further implicates the IgE isotype in immune responses against helminths. The latter three cell types have been shown to be capable of killing IgE-coated helminths *in vitro* (reviewed in Capron and Capron, 1987). However, it is not possible to extrapolate from these *in vitro* results since there is no evidence that such a mechanism may also operate *in vivo* against adult worms in the gut lumen. IgE is very susceptible to proteolysis and appears to be rapidly degraded within the intestine (Kolmannskog, 1987). Helminths which breach or invade the intestinal mucosa may, however, become susceptible to IgE-dependent, inflammatory, cell-mediated cytotoxicity.

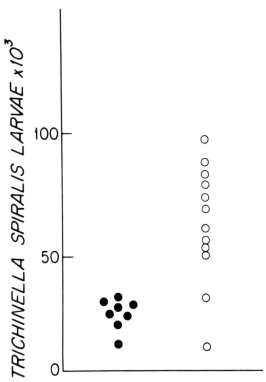

Figure 10.7. Susceptibility of untreated (●) or IgE suppressed (○) rats to *T. spiralis* infection as measured by the recovery of muscle larvae from individual rats (after Dessein *et al.*, 1981 by copyright permission of the Rockefeller University Press).

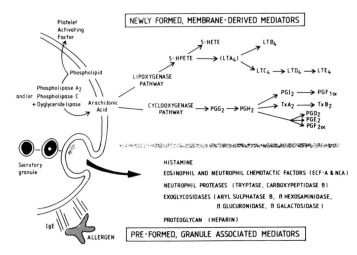

Figure 10.8. Schematic representation of the activation of the mast cell and the release of preformed and *de novo*-generated mediators following transmembrane signal (e.g IgE/antigen complex). Abbreviations: 5-HETE = 5-hydroxyeicosatetraenoic acid; 5-HPETE = 5 hydroperoxyeico-satetraenoic acid; LTB$_4$ = leukotriene B$_4$ (chemotactic factor); LTC$_4$, LTD$_4$, LTE$_4$ = leukotrienes comprizing SRS-A sulfidopeptides; PGD$_2$, PGE$_2$, PGF$_2\alpha$ = prostaglandins; PGI$_2$ = prostacyclins; TxA$_2$, TxB$_2$ = thromboxanes (figure by courtesy of Professor A.B. Kay).

Following interaction of parasite antigens with specific IgE on the surface of mast cells, an array of biologically-active mediators are released (Figure 10.8). In addition to the IgE/allergen trigger, a growing list (Table 10.2) of other mast cell degranulation stimuli (reviewed in Befus *et al.*, 1987) has been

Table 10.2. Activators of mast cell histamine release from various mediator cells.

Agents	*Target cells*
IgE and allergen	All mast cell + basophils
Anaphylatoxins: C3a and C5a	Rat PMC (Johnson *et al.*, 1975). Rat IMMC; human IMC (Befus *et al.*, 1988)
Vasoactive intestinal peptide, somatostatin, bradykinin, neurotensin	Rat PMC (no effect on rat IMMC) (Shanahan *et al.*, 1984)
Substance P	Rat PMC; rat IMMC (Denburg and Bienenstock, 1983)
Eosinophil peroxidase and hydrogen peroxide and halide	Rat PMC (Henderson *et al.*, 1980)
Human eosinophil major basic protein	Human basophils
Chymase (RMCPI)	RAT PMC (Schick *et al.*, 1984)
Interleukin 1	Human basophils; human mast cells (adenoidal; Subramanian and Bray, 1987)
Antigen-specific T-cell factor	Mouse skin mast cells (not known if intestinal MMC respond; Askenase *et al.*, 1983)

described including complement components (C3a and C5a), neuropeptides, eosinophil peroxidase, eosinophil major basic protein, an antigen-specific, T-cell-derived factor (acts on CTMC but its effect on MMC is unknown) and worm-derived degranulating agents. The activation of mast cells by any of these stimuli results in the release and generation of a battery of mediators (Figure 10.8). These are associated with the cell in two forms: Pre-packaged mediators (e.g. histamine) which are stored within the granule matrix and released within seconds of physiological stimulation of the cell and newly-generated, membrane-associated mediators involving the metabolism of membrane phospholipids and the generation of potent mediators of inflammation such as leukotrienes (LT), platelet-activating factor (PAF) or prostaglandins (PG).

Role of eosinophils in intestinal inflammation and mucosal defense

An increase in the numbers of peripheral blood and tissue eosinophils is a major hallmark of helminth infections, especially of those species with tissue-invasive stages (reviewed in Befus and Bienenstock, 1982). Intestine-dwelling worms, such as *T. spiralis* and *N. brasiliensis*, provoke a marked T-dependent eosinophilia in the lamina propria and submucosa which is temporally-associated with the expulsion phase of these infections. Accumulation of eosinophils in the intestinal mucosa may arise as a result of chemotactic factors, such as PAF and LTB_4 released following mast cell/IgE-worm antigen interaction, T-cell-mediated eosinophilopoiesis either locally and/or in the bone marrow, and chemoattractants released from the worms themselves.

A vast amount of data has been generated on the cytotoxic properties of these cells against opsonized helminth targets, *in vitro* (reviewed in Butterworth, 1984). Eosinophils have been shown to degranulate and release their stored mediators following contact with non-phagocytosable surfaces, such as worms or their larvae (reviewed in McLaren, 1980). Eosinophil granule peroxidase (EPO) may play a role in immunity to *T. spiralis* (Castro *et al.*, 1974), and EPO has been shown to be released onto the surfaces of *T. spiralis* and *N. brasiliensis in vitro* (McLaren *et al.*, 1977). Eosinophil major basic protein (MBP) is highly toxic to nematodes *in vitro* (Wassom and Gleich, 1979). It is important to stress here that the relevance of such *in vitro* phenomena to intestinal helminthiases is questionable. Eosinophils are sometimes found in the parasitized intestinal epithelium, however, luminal-dwelling helminths are unlikely to come into contact with sufficient numbers of eosinophils to suffer significant eosinophil-mediated damage. Increased epithelial permeability may allow the passage of eosinophil granule contents and membrane-derived lipid mediators into the lumen with potentially worm-damaging results. Helminths which penetrate the epithelial layer will, of course, be vulnerable to contact with mucosal eosinophils.

In addition to preformed granule contents, perturbation of eosinophil IgG and IgE receptors leads to the generation of pro-inflammatory leukotrienes and PAF. Murine intestinal mucosal eosinophils have recently been shown to bear surface IgA, and the frequency of IgA-positive eosinophils was increased considerably following infection with the cestode, *Hymenolepis diminuta* (van der Vorst *et al.*, 1988). Given the predominance of this isotype at mucosal surfaces, IgA-dependent activation of eosinophils (which has yet to be demonstrated) is of potential importance in intestinal defense against helminths. The importance of IgE-dependent activation of eosinophils was suggested in an elegant study in which peritoneal eosinophils from *N. brasiliensis*-infected rats showed markedly enhanced IgE-dependent killing of schistosomula of *Schistosoma japonicum*, *in vitro*, associated with increased expression of $Fc_\epsilon RII$ on these cells (Kojima *et al.*, 1985). The enhanced cytotoxic capacity of eosinophils commenced at the onset of expulsion of a primary *N. brasiliensis* infection when peripheral blood eosinophil counts started to rise.

While there is abundant evidence of the helminthicidal activities of eosinophils *in vitro*, there is, as yet, only limited data to support a protective role for these cells in intestinal helminthiases. However, evidence of eosinophil activation has been suggested by the measurement of increased concentrations of phospholipase B following nematode infections of rodents (Larsh *et al.*, 1974; Goven, 1979) and the immune expulsion of *Trichostrongylus colubriformis* from guinea pigs was partially delayed by treatment with anti-eosinophil serum (Gleich *et al.*, 1979).

Role of mucus in intestinal defense

The generation of mucin glycoprotein by goblet cells is a normal feature of gastro-intestinal function. Mucus hyper-secretion, however, has been observed in nematode infections in ruminants and in *N. brasiliensis* and *T. spiralis* infections in laboratory rodents, and a protective role for mucus has been suggested in these infections (Miller, 1984). Rats infected with *N. brasiliensis* have increased numbers of intestinal goblet cells and proliferation of both goblet cells and mucosal mast cells which may be under similar immunological control. Both responses can be transferred by adoptive immunization with T cells and serum from immune rats (Miller, 1980), and there is evidence that the release of goblet cell mucus in passively-sensitized rats following intraduodenal antigen challenge is IgE-dependent (Lake *et al.*, 1980; Levy and Frondoza, 1983). Mucus release during intestinal worm infection may, therefore, be triggered by mediators released by MMC or eosinophils following cross linking of surface-bound, parasite-specific IgE. Alternatively goblet cells may have receptors for IgE. Mucus secretion from goblet cells is stimulated by a variety of secretagogues and mechanisms including cholinergic stimulation (Specian and Neutra, 1982), immune complexes (Walker *et al.*, 1977) and C3a, which is a potent stimulator of mucus secretion acting independently of mast cell activation (Marom *et al.*, 1985).

Intestinal mucus may serve a protective function by excluding or trapping worms in immune animals (Lee and Ogilvie, 1981; Miller *et al.*, 1981) by preventing intimate contact with the mucosa thus preventing their establishment. The possible involvement of anti-worm antibody and also complement in mucus trapping of nematodes has been suggested by the observation that a specific component (presumably IgA antibody) and a non-specific component (possibly complement, since nematode cuticles are capable of activating complement) were involved in retention of nematode larvae in the mucus layer following their incubation in immune serum (Lee and Ogilvie, 1982). Acquired resistance to *Strongyloides ratti* in rats results in severe damage to the parasites and immune-damaged worms invariably possess mucus plugs around their oral orifice (Figure 10.9). Thus mucus may also interfere with parasite food intake (Moqbel and McLaren, 1979).

Role of inflammatory mediators in intestinal defense mechanisms

Preformed and newly-generated mediators are released from inflammatory cells known to infiltrate the helminth-infected gut. The exact role of each

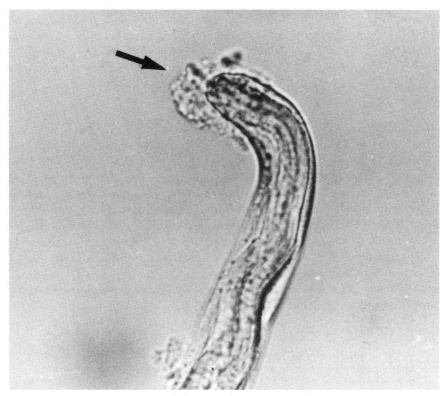

Figure 10.9. Anterior end of an adult *S. ratti* worm recovered 20 days post-infection exhibiting a mucous oral plug (arrowed).

mediator in enhancing the inflammatory process and the extent of their contribution to worm damage and expulsion is, as yet, unclear. However, experimental evidence suggests that a number of these mediators have the potential to participate in events leading to the elimination of worms from the gut. It is unlikely that a single mediator is responsible for the final trigger which leads to worm expulsion. It is more probable that a combination of mediators is involved, both directly by their spasmogenic and vasoactive properties, and indirectly by recruiting and activating other cells in the inflamed tissue resulting in pathophysiological changes in the intestinal environment which are deleterious to parasite survival. The biological properties and cellular sources of the main mediators concerned are listed in Table 10.3 and will now be considered in more detail.

Newly-generated mediators

These include the leukotrienes (LT) and prostaglandins (PG) which are synthesized *de novo* by the metabolism of arachidonic acid which is activated by lipoxygenase and cyclooxygenase enzymes, respectively (Figure 10.10). Arachidonic acid is derived from membrane phospholipids via the activity of phospholipase A_2 following transmembrane signalling or perturbation (Figure 10.10). The known pharmacological actions of leukotrienes and prostaglandins are summarized in Table 10.4. Platelet-activating factor (PAF) is another potent membrane-derived mediator formed from membrane phospholipids by phospholipase A_2 as a consequence of cell activation.

Leukotrienes

LTC_4, D_4, and E_4 comprize what was previously identified as slow-reacting substance of anaphylaxis (SRS-A). These SRS-A LTs exhibit potent biological activities (reviewed in Lewis and Austen, 1984) at nanomolar concentrations and are known to cause sustained smooth muscle contraction, increase vascular permeability and, together with hydroxyeicosatetraenoic acids (HETEs), enhance the production of mucus from goblet cells *in vitro*. A direct anti-worm activity of SRS-A LTs has been observed in gastrointestinal mucus of sheep resistant to *T. colubriformis* which exhibited a potent inhibitory activity against larval migration *in vitro*. This activity, which was absent in mucus obtained from non-immune sheep, appeared to have physicochemical properties similar to LTC_4 (Douch *et al.*, 1983).

LTB_4, on the other hand, is a potent chemotactic factor for human neutrophils and, to a lesser extent, eosinophils (Nagy *et al.*, 1982). Rat neutrophils, on the other hand, do not exhibit a chemotactic response to LTB_4, although they do show aggregation, degranulation and chemokinesis *in vitro* in response to this lipid (Kreisle *et al.*, 1985). LTB_4 has also been shown to cause *in vivo* vasodilatation of capillaries and increase vascular permeability thus promoting plasma leakage. In addition, LTB_4 treatment of

Table 10.3. Pathological changes associated with helminth-induced intestinal inflammation. The possible mediators involved and their cell source(s).

Inflammatory event and relevant mediators	*Possible cell source(s)*
Cell infiltration and activation:	
Leukotriene B$_4$	Epithelial cells, neutrophils, macrophages, mast cells
HETEs	Mast cells, neutrophils, epithelial cells
Platelet activating factor (PAF)	Mast cells, neutrophils, eosinophils, platelets, macrophages, endothelial cells
Histamine	Mast cells, basophils
Eosinophil chemotactic factor of anaphylaxis (ECF-A)	Mast cells
Neutrophil chemotactic activity (NCA)	Mast cells, mononuclear cells, T cells
Mucosal oedema:	
Histamine	Mast cells, basophils
SRS-A leukotrienes (LTC$_4$, LTD$_4$)	Mast cells, eosinophils
Prostaglandins (PGD$_2$, PGF$_{2\alpha}$)	Macrophages, platelets, (mast cells)
PAF	Mast cells, neutrophils, eosinophils, platelets, macrophages, endothelial cells
Mucus hypersecretion:	
SRS-A (LTC$_4$, LTD$_4$)	Mast cells, eosinophils
Histamine	Mast cells, basophils
HETEs	Mast cells, neutrophils, epithelial cells
Epithelial shedding:	
Proteolytic enzymes (RMCPII)	Mast cells (mucosal)
Neuropeptides	-
Products of oxidative metabolism (H$_2$O$_2$)	Neutrophils, macrophages
Major basic protein (MBP)	Eosinophils
PAF	Mast cells, neutrophils, eosinophils, platelets, macrophages, endothelial cells
Increased gut motility (peristalsis):	
Histamine	Mast cells, basophils
SRS-A (LTC$_4$, LTC$_4$)	Mast cells, eosinophils
PGF$_2\alpha$	Mast cells
PAF	Mast cells, neutrophils, eosinophils, platelets, macrophages, endothelial cells
5-Hydroxytryptamine (serotonin)	Mast cells, platelets, intestinal enterochromaffin cells

human granulocytes enhances their C3b receptors (Nagy *et al.*, 1982) and increases the cytotoxic capacity of granulocytes against schistosomula of *Schistosoma mansoni in vitro* (Moqbel *et al.*, 1983a, 1983b). Elevated levels of LTB$_4$ have been measured in human intestinal mucosa from patients with inflammatory bowel disease which is associated with a marked mucosal neutrophil infiltrate (Sharon and Stenson, 1984).

Leukotrienes (both SRS-A and LTB$_4$) are released *in vivo* in rats undergoing systemic anaphylaxis induced by *N. brasiliensis* antigen (Moqbel *et al.*, 1986) and also during rapid expulsion of *T. spiralis* (Moqbel *et al.*, 1987).

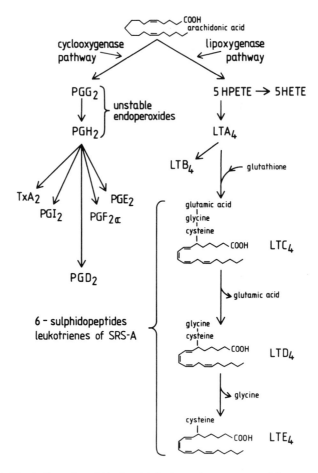

Figure 10.10. Metabolism of arachidonic acid via the cyclooxygenase and lipoxygenase pathways (figure courtesy of Dr. O. Cromwell).

Leukotrienes were present in intestinal luminal washings and plasma, and their generation was temporally-associated with MMC activation as indicated by RMCPII release (Figure 10.11). The possible participation of eosinophils, which are also present in large numbers in the intestinal mucosa in both these model systems in the generation of SRS-A LTs, could not be excluded and requires further investigation. Given the known potent biological activities of the LTs, the release of these mediators during gut anaphylaxis and rapid expulsion of nematodes from immune rats suggest that they may be implicated in causing inflammatory changes in the intestine.

The cellular sources of SRS-A LTs in helminth-induced intestinal inflammation are most likely to be MMC and eosinophils. Isolated MMC from the intestines of *N. brasiliensis*-infected rats generate substantial quantities of

Table 10.4. Known pharmacological actions of lypoxygenase and cyclooxygenase metabolites of arachidonic acid.

Lipoxygenase metabolites	Pharmacological action
$LTC_4 > LTD_4 > LTE_4$ (in order of potency)	Contraction of intestinal smooth muscle in most species; rat ileum does not respond
LTC_4, LTD_4	Vasoconstriction
LTC_4, LTD_4, LTB_4	Increased capillary permeability
LTC_4, LTD_4	Mucous secretion by goblet cells
LTC_4, LTD_4	Paralysis and loss of motor coordination of adult and larval nematodes, *in vitro*
LTB_4	Leucocyte chemotaxis and aggregation
5-HETEs	Increased fluid transport across epithelium
Cyclo-oxygenase metabolites	
PGD_2	Vasodilation, contractile for smooth muscle and potentiates SRS-A-mediated contraction; chemotactic for eosinophils
PGE_2, $PG1_2$	Potent vasodilators
PGE_1, $PG1_2$	Modulate mast cell histamine release
$PGF_2\alpha$	Increases peristalsis
TxA_2	Vasoconstriction

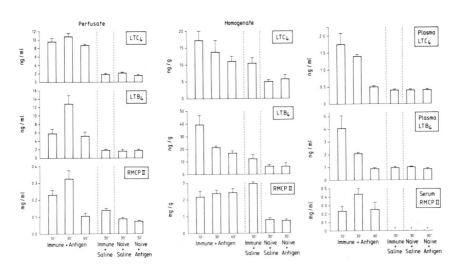

Figure 10.11. Levels of immunoreactive LTC_4, LTB_4 and RMCPII of luminal perfusate (per ml), intestinal homogenate (per gm wet weight) and blood of immune and naive rats following systemic anaphylaxis induced by i.v. challenge with *N. brasiliensis* antigens (or saline as control; reproduced with permission from Moqbel *et al.*, 1986 and Journal of Immunology).

LTC$_4$ upon activation with worm antigen (Heavey *et al.*, 1988). Human eosinophils also generate large amounts of LTC$_4$ following interaction with IgG-coated surfaces (Shaw *et al.*, 1985) and have recently been shown to release LTC$_4$ during contact with *S. mansoni* schistosomula opsonized with specific IgE (Moqbel *et al.*, 1990). The major sources of LTB$_4$ are likely to be MMC (Heavey *et al.*, 1988) and neutrophils (Fitzharris *et al.*, 1987). Intestinal epithelial cells, which are the cells most exposed to high concentrations of worm antigens, may also be involved in the generation of lipid mediators. Epithelial cells from canine trachea can generate LTB$_4$ (Nadel and Holtzman, 1984). In contrast, human tracheal epithelial cells have the capacity to convert arachidonic acid into a number of its 15-lipoxygenase products (Hunter *et al.*, 1985).

PLATELET-ACTIVATING FACTOR (PAF)

PAF or 1-0-alkyl-2-acetyl-sn-glyceryl-3-phosphorylcholine was first described in terms of its biological actions on platelets but has since been found to have multiple effects and must now be considered as a broad-spectrum mediator of inflammatory reactions (reviewed in Barnes *et al.*, 1988). Like LTs and PGs, PAF is formed by the action of phospholipase A$_2$ on membrane phospholipids following activation of a wide range of cell types (Table 10.3). *In vitro*, PAF is the most potent eosinophil chemotactic agent yet described (Wardlaw *et al.*, 1986) and enhances both IgG- and IgE-dependent killing of schistosomula of *S. mansoni* by human eosinophils (Moqbel *et al.*, 1987; Walsh *et al.*, 1989).

In vivo, PAF is the most potent gastrointestinal ulcerogenic agent yet described in the rat, nanogram quantities per animal being capable of causing extensive small intestinal necrosis, mucosal haemorrhage and epithelial shedding from tips of villi (Wallace and Whittle, 1986). Infusion of PAF in several species mimics the signs and symptoms of acute systemic anaphylaxis which is inhibitable by PAF-receptor antagonists (Barnes *et al.*, 1988). Other *in vivo* effects of PAF include extravasation of plasma fluid, constriction of smooth muscle and intravascular and perivascular leucocyte accumulation.

PAF has recently been shown to be released into the intestinal lumen (in nanogram quantities per ml of intestinal lavage fluid) of *N. brasiliensis*-immune rats undergoing systemic anaphylaxis following intravenous challenge with worm antigen (Moqbel *et al.*, 1989). PAF was not present in intestinal contents of immune rats given saline instead of antigen, and PAF release in shocked rats was associated with MMC activation (Figure 10.12). The main intestinal lesions observed were epithelial shedding, hypersecretion of mucus, mucosal haemorrhaging and plasma leakage into the lumen. Given the known effects of PAF on the rat gastrointestinal tract, PAF is implicated in the intestinal pathology of *N. brasiliensis* model system. By extrapolation, PAF may also be involved in similar but more moderate lesions associated with other intestinal worm infections. Experiments with PAF antagonists should reveal the extent of the role played by this mediator in the latter case.

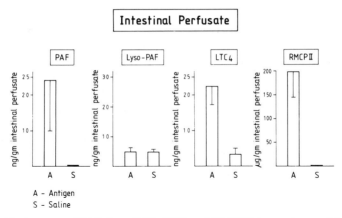

Figure 10.12. Amounts of PAF, Lyso-PAF, LTC$_4$ and RMCPII (mean ± s.e.m.) in intestinal perfusates of *N. brasiliensis*-primed rats, 15 min after i.v. injection of either worm antigen (A) or saline (S). (Lyso-PAF is the biologically-inactive precursor and metabolite of PAF.)

PAF is synthesized by an ever-growing list of appropriately-stimulated cells (Table 10.3) of which eosinophils, MMC and possibly vascular endothelial cells are of most relevance in intestinal helminthiases. Eosinophils generate large amounts of PAF following interaction with IgG-coated surfaces (Champion *et al.*, 1988). It is not yet known if MMC generate PAF, but murine bone marrow-derived, cultured mast cells considered to be *in vitro* analogues of MMC, generate PAF upon IgE-mediated activation (Mencia-Huerta *et al.*, 1983). Furthermore, vascular endothelial cells synthesize PAF upon stimulation with histamine, LTs and interleukin-1 (Barnes *et al.*, 1988).

PROSTAGLANDINS

This family of mediators which include the prostaglandins (PGD,E,G,H and F), prostacyclins (PGI) and thromboxanes (TxA and B), are the products of the metabolism of arachidonic acid by the action of cyclooxygenase enzymes (Figure 10.10; Table 10.4). The role of these mediators in immune reaction to helminths is unclear. In general, PGE$_1$, PGE$_2$ and PGI$_2$ are considered to have anti-inflammatory activities, and both inhibit histamine release from rat peritoneal mast cells. PGF$_2\alpha$ and PGD$_2$ are pro-inflammatory; the former increases peristalsis and the latter causes constriction of smooth muscle and, in common with PGE$_2$, potentiates the activities of the SRS-A LTs (Bach, 1982). Conflicting results regarding the involvement of PGs in expulsion of *N. brasiliensis* from the rat have been obtained.

Inhibition of PG formation by cyclooxygenase blockage using aspirin delays the expulsion of *N. brasiliensis* (Kelly *et al.*, 1974). However, intraduodenal injection of PGE$_1$ or PGE$_2$ did not affect the survival of worms in the gut (Kassai *et al.*, 1989). Cyclooxygenase blockade by indomethacin causes accelerated development of intestinal mast responsiveness to worm antigen in

N. brasiliensis-infected rats, possibly by inhibiting PGE_1 formation which modulates mast cell activity (Befus *et al.*, 1982). The main cellular source of PGs in intestinal inflammation are likely to be the MMC which synthesize moderate amounts of PGD_2 upon IgE-mediated activation, although in considerably lesser quantities than LTC_4 (Heavey *et al.*, 1988).

Preformed mediators

These are stored in cytoplasmic granules and lysosomes of mast cells, granulocytes, macrophages, NK cells and platelets and are released following ligand-receptor interactions on the cell surface.

HISTAMINE AND SEROTONIN (5-HYDROXYTROPTAMINE; 5-HT)

Histamine has been extensively studied and has a multitude of pro-inflammatory (Bach, 1982) and immunoregulatory effects. These include stimulation of mucus secretion, vasodilation and plasma leakage, stimulation of gut motility, chemokinesis of granulocytes, moderate enhancement of eosinophil-mediated cytotoxicity, and activation of both suppressor and contrasuppressor lymphocytes (Siegel *et al.*, 1982). 5-HT stimulates water and sodium secretion in the rodent small intestine and smooth muscle contraction in the same site and also acts as a neurotransmitter.

Conflicting results concerning the possible involvement of histamine and 5-HT in immunity to intestinal helminths have been obtained. Antihistamines and anti-5-HT agents have been reported to have no inhibitory effect on RE of *T. spiralis* from immune rats (Bell *et al.*, 1982). Conversely, expulsion of *T. colubriformis* from guinea pigs was inhibited by depletion of these biogenic amines (Rothwell *et al.*, 1974). In the sheep, histamine levels in blood and intestinal contents bore no relationship to resistance to trichostrongyle infections, whereas SRS-A-like activity in intestinal mucus correlated with immunity to these worms (Douch *et al.*, 1984).

The main cellular sources of histamine during intestinal helminthiasis are likely to be MMC and basophils. 5-HT is derived from rodent, but not human mast cells, platelets, intestinal entero-chromaffin cells and at neuronal terminals.

MAST CELL PROTEASES

The amounts, types of enzyme and substrate specificities of granule proteases depend on species, location and mast cell type (MMC *versus* CTMC). The predominant granule protease of MMC is perhaps of most relevance in intestinal helminth infections and, of these, the chymotryptic enzyme, RMCPII in the rat MMC, has been most extensively studied (Miller, 1984). This serine protease acts on type IV basement membrane collagen and is very efficient at removing epithelial cells from intestinal mucosa *in vitro* (Miller, 1984). Release of RMCPII has been shown to correlate with the leakage of plasma proteins into the intestinal lumen during *N. brasiliensis* antigen-mediated systemic anaphylaxis in primed rats which is accompanied by massive epithelial shedding. The enzyme may also be responsible for creating gaps in the epithelium observed during primary *N. brasiliensis* infection thereby causing local permeability changes (Miller, 1984). Such leak lesions would allow the passage of the non-secretory immunoglobulin isotypes IgG and IgE, plasma proteins such as complement, and inflammatory mediators into the intestinal lumen with potential worm-damaging effects.

LYSOSOMAL ENZYMES

Lysosomal enzymes are released during phagocytosis or when cells are exposed to surfaces too large to ingest, the latter applying more to eosinophils than neutrophils. Neutrophil lysosomes contain a number of potentially tissue-damaging enzymes including lysozyme, elastase, myeloperoxidase and collagenase. Eosinophil granules contain relatively large amounts of peroxidase which, together with a combination of hydrogen peroxide and halides, is toxic for schistosomula *in vitro* (reviewed in Gleich and Adolphson, 1986).

The above list of preformed and newly-generated mediators is not exhaustive but probably includes those which, according to current knowledge, are most likely to play a role in intestinal defense mechanisms against parasites. The list is constantly growing and will eventually include interleukins and other cytokines yet to be described, with direct pro-inflammatory properties.

The hypothesis that mast cells and eosinophils and their products play a role in generating non-specific inflammatory changes likely to create an unfavourable environment for intestinal worms is an attractive one. The hypothesis is supported by the efficacy of corticosteroids in inhibiting worm expulsion. Corticosteroid treatment abolishes the mucosal mast cell response, inhibits eosinophil infiltration and goblet cell hyperplasia and prevents the generation of LTs, PGs and PAF by inhibiting phospholipase A_2 activation. While the effects of corticosteroids on lymphocyte responsiveness cannot be ruled out, their potent anti-inflammatory effects must surely support a role for intestinal inflammation in expulsion of helminths.

10.6. CONCLUSIONS

Studies on the mechanisms of protection against intestinal helminths, using laboratory host/parasite models, have revealed a complex jigsaw puzzle which as yet has not been fully and clearly assembled. The exposure of the intestinal mucosal tissue to antigenic onslaught triggers a series of intricate and multi-factorial events which, at least in murine systems, lead to acquired resistance to reinfection. There is strong evidence that these reactions may be under the influence of complex genetic and environmental factors (see Chapter 6 and Wakelin, 1985). These immunological events include the initial antigen-specific T-helper-lymphocyte response which in turn induces the synthesis and generation of antigen-specific immunoglobulins (especially secretory IgA and IgE) from committed B cells.

The role of T cells appears to be pivotal in immunity against intestinal nematode infections with T cells of the helper phenotype having the ability to adoptively transfer protection against reinfection (Nawa and Miller, 1978; Grencis *et al.*, 1985; Lammas *et al.*, 1987). These cells may be recruited from the lymphatic tissue but may also be generated locally in the intestine in response to antigenic challenge (Korenaga *et al.*, 1989).

The release of a number of T-cell products and other cytokines may, on the one hand, amplify the T-cell response and, on the other, signal to myeloid progenitor cells both in the bone marrow or possibly locally in the relevant tissue to recruit and/or differentiate into inflammatory cells. The development of intestinal inflammation, unlike the early antigen-specific, T-dependent steps, are non-MHC-linked and appear to be the important component of the protective response against gut parasites. The release of an array of cytotoxic and pro-inflammatory mediators from the cellular infiltrate is an important accompaniment of the interaction between the parasite and the cells found within the epithelial mucosal interphase. The exact role of each mediator in enhancing the inflammatory process and the extent of their contribution to worm damage and expulsion is not yet clear. It is likely that the action of a combination of mediators, acting in concert with selective cytokines, results in the generation of pathophysiological changes which are inimical to the survival and/or establishment of the parasite.

The advances currently being made in the molecular basis of the inflammatory process will help to identify more precisely the sequence of events and mechanisms associated with immune responses against helminthic infections. Such advances augur well towards more effective preventive measures and therapies for these important diseases in man and domestic animals.

ACKNOWLEDGEMENTS

The authors wish to thank Christine Bensted and Jennifer Mitchell for excellent secretarial help with this manuscript.

REFERENCES

Abe, T. and Nawa, Y. (1988), Worm expulsion and mucosal mast cell response induced by repetitive IL-3 administration in *Strongyloides ratti*-infected nude mice, *Immunology*, **63**, 181–5.

Allan, W. and Mayrhofer, G. (1981), The distribution and traffic of specific homocytotropic antibody-synthesising cells in *Nippostrongylus brasiliensis*-infested rats, *Australian Journal of Experimental Biology and Medical Science*, **59**, 723–37.

Askenase, P.W., Van Loveren, H., Rosenstein, R.W. and Ptak, W. (1983),

Immunologic specificity of antigen-binding T cell-derived factors that transfer mast cell dependent, immediate hypersensitivity-like reactions. *Monographs in Allergy*, **18**, 249–55.

Bach, M.K. (1982), Mediators of anaphylaxis and inflammation, *Annual Reviews in Microbiology*, **36**, 371–413.

Barnes, P.J., Chung, K.F. and Page, C. (1988), Platelet-activating factor as a mediator of allergic disease, *Journal of Allergy and Clinical Immunology*, **81**, 919–34.

Befus, A.D. and Bienenstock, J. (1982), Factors involved in symbiosis and host resistance at the mucosa—parasite interface, *Progress in Allergy*, **31**: 76–177.

Befus, A.D., Pearce, F.L. and Bienenstock, J. (1987), Intestinal mast cells in pathology and host resistance, in Brostoff, J. and Challacombe S.J. (Eds.), *Food Allergy and Intolerance*, Saunders.

Befus, A.D., Johnston, N., Berman, L. and Bienenstock, J. (1982), Relationship between tissue sensitization and IgE antibody production in rats infected with the nematode, *Nippostrongylus brasiliensis*, *International Archives of Allergy and Applied Immunology*, **67**, 213–8.

Befus, A.D., Fujimaki, H., Lee, T.D.G. and Swieter, M. (1988), Mast cell polymorphisms—present concepts, future directions, *Dig. Dis. Sci.*, **33**, Supplement, 165–245.

Behnke, J.M., Wakelin, D. and Wilson, M.M. (1978), *Trichinella spiralis*: delayed rejection in mice concurrently infected with *Nematospiroides dubius*, *Experimental Parasitology*, **46**, 121–30.

Bell, R.G., McGregor, D.D. and Adams, L.S. (1982), Studies on the inhibition of rapid expulsion of *Trichinella spiralis* in rats, *International Archives of Allergy and Applied Immunology*, **69**, 73–80.

Block, K.J., Cygan, R.W., and Waltin, J. (1972), The IgE system and parasitism: role of mast cells, IgE, and other antibodies in host response to primary infection with *Nippostrongylus brasiliensis*, in Ishizaka, K. and Dayton, J. (Eds.), *The Biological Role of the Immunoglobulin E System*, p.119, Washington: US Government printing.

Bonneville, M., Janeway, C.A. Jr., Ito, K., Haser, W., Ishida, I., Nakaushi, N. and Tonegawa, S. (1988), Intestinal intra-epithelial lymphocytes are a distinct set of T cells, *Nature*, **336**, 479–81.

Brandtzaeg, P., Valnes, K., Scott, H., Rognum, T.O., Bijerke, K. and Baklien, K. (1985), The human gastrointestinal secretory immune system in health and disease, *Scandinavian Journal of Gastroenterology*, **20**, Supplement 114, 17–38.

Butterworth, A.E. (1984), Cell-mediated damage to helminths, *Advances in Parasitology*, **23**, 143–235.

Capron, M. and Capron, A. (1987), The IgE receptor of human eosinophils, in Kay, A.B. (Ed.), *Allergy and Inflammation*, p. 151, London: Academic Press.

Castro, G.A., Roy, S.A. and Stockstill, R.D. (1974), *Trichinella spiralis*: peroxidase activity in isolated cells from rat intestine, *Experimental Parasitology*, **36**, 307–15.

Champion, A., Wardlaw, A.J., Moqbel, R., Cromwell, O., Shepherd, D. and Kay, A.B. (1988), IgE-dependent generation of PAF by normal and low-density human eosinophils, *Journal of Allergy and Clinical Immunology*, **81**, 207 (Abstract 157).

Denburg, J.A. and Bienenstock, J. (1983), Mast cell subpopulations and their response to substance P, in Skrabanek,P. and Powell D. (Eds.), *Substance P*, pp. 145–6, Dublin Book Press.

Dessein, A.J., Parker, W.L., James, S.L. and David, J.R. (1981), IgE antibody and resistance to infection. I. Selective suppression of the IgE antibody response in rats diminishes the resistance and the eosinophil response to *Trichinella spiralis* infection, *Journal of Experimental Medicine*, **153**, 423–6.

Douch, P.G.F., Harrison, G.B.L., Buchanan, L.L. and Greer, K.S. (1983), *In vitro* bioassay of sheep gastrointestinal mucus for nematode paralysing activity mediated by a substance with some properties characteristic of SRS-A, *International Journal of Parasitology*, **13**, 207–12.

Douch, P.G.C., Harrison, G.B.L., Buchanan, L.L. and Brunsdon, R.V. (1984), Relationship of histamine in tissues and anti-parasitic substances in gastrointestinal mucus to the development of resistance to trichostrongyle infections in young sheep, *Veterinary Parasitology*, **16**, 273–88.

Ferguson, A. (1977), Intraepithelial lymphocytes of the small intestine, *Gut*, **18**, 921–37.

Ferguson, A. and Jarrett, E.E.E. (1975), Hypersensitivity reactions in the small intestine. 1. Thymus dependence of experimental 'partial villous atrophy', *Gut*, **16**, 114–7.

Fitzharris, P., Cromwell, O., Moqbel, R., Hartnell, A., Walsh, G.M., Harvey, C. and Kay A.B. (1987), Leukotriene B_4 generation by human neutrophils following IgG-dependent stimulation, *Immunology*, **61**, 449–55.

Gleich, G.J. and Adolphson, C.R. (1986), The eosinophilic leukocyte: structure and function, *Advances in Immunology*, **39**, 177–253.

Gleich, G.J., Olsen, G.M., and Herlich, H. (1979), The effect of antiserum to eosinophils on susceptibility and acquired immunity of the guinea pig to *Trichostrongylus colubriformis*, *Immunology*, **37**, 873–80.

Gleich, G.J., Loegering, D.A., Mann, K.G. and Maldonado, J.E. (1976), Comparative properties of the Charcot-Leyden crystal protein and the major basic protein from human eosinophils, *Journal of Clinical Investigation*, **57**, 633–40.

Goodman, T. and Lefrancois, L. (1988), Expression of the T cell receptor on intestinal CD8+ intraepithelial lymphocytes, *Nature*, **333**, 855–8.

Goven, A.J. (1979), The phospholipase B content of the intestines of sensitized rats challenged with varied larvae doses of *Nippostrongylus brasiliensis*, *International Journal of Parasitology*, **9**, 345–9.

Grencis, R.K., Riedlinger, J. and Wakelin, D. (1985), L3T4-positive T lymphoblasts are responsible for transfer of immunity to *Trichinella spiralis* in mice, *Immunology*, **56**, 213–8.

Guy-Grand, D., Griscelli, C. and Vassalli, P. (1978), The mouse gut T lymphocyte, a novel type of T cell. Nature, origin and traffic in mice in normal and graft-versus-host conditions, *Journal of Experimental Medicine*, **148**, 1661–77.

Heavey, D.J., Ernst, P.B., Stevens, R.L., Befus, A.D., Bienenstock, J. and Austen, K.F. (1988), Generation of leukotriene C_4, leukotriene B_4, and prostaglandin D_2 by immunologically activated rat intestinal mucosal mast cells, *Journal of Immunology*, **140**, 1953–7.

Henderson, W.R., Chi, E.Y. and Klebanoff, S.J. (1980), Eosinophil peroxidase-produced mast cell secretion, *Journal of Experimental Medicine*, **152**, 265–79.

Hunter, J.A., Finkbeiner, W.F., Nadel, J.A., Goetzl, E.J. and Holtzman, M.J. (1985), Predominant generation of 15- lipoxygenase metabolites or arachidonic acid by epithelial cells from human trachea, *Proceeedings of the National Academy of Sciences* (USA), **82**, 4633–7.

Huntley, J.F., Newlands, G. and Miller, H.R.P. (1984), Isolation and characterisation of globular leucocytes; their derivation from mucosal mast cells in parasitized sheep, *Parasite Immunology*, **6**, 371–90.

Huntley, J.F., Gibson, S., Brown, D., Smith, W.D., Jackson, F. and Miller, H.R.P. (1987), Systemic release of a mast cell proteinase following nematode infections in sheep, *Parasite Immunology*, **9**, 603–14.

Janeway, C.A. Jr. (1988), Frontiers of the immune system, *Nature*, **333**, 804–6.

Jenkins, D.C. (1975), The influence of *Nematospiroides dubius* on subsequent *Nippostrongylus brasiliensis* infections in mice, *Parasitology*, **71**, 349–55.

Jenkins, S.N. and Behnke J.M. (1977), Impairment of primary expulsion of *Trichuris muris* in mice concurrently infected with *Nematospiroides dubius*, *Parasitology*, **75**, 71–8.

Johnson, A.R., Hugli, T.E. and Muller-Eberhard, H.J. (1975), Release of histamine from rat mast cells by the complement peptides C3a and C5a, *Immunology*, **28**, 1067–80.

Kassai, T., Redl, P., Jecsai, G., Balla, E. and Harangozo, E. (1989), Studies on the involvement of prostaglandins and their precursors in the rejection of *Nippostrongylus brasiliensis* from the rat, *International Journal of Parasitology*, **10**, 115–20.

Katona, I.M., Urban, J.F. Jr., and Finkelman, F.D. (1988), The role of L3T4[+] and Lyt-2[+] T cells in the IgE response and immunity to *Nippostrongylus brasiliensis*, *Journal of Immunology*, **140**, 3206–11.

Kazacos, K.R. (1975), Increased resistance in the rat to *Nippostrongylus brasiliensis* following immunisation against *Trichinella spiralis*, *Veterinary Parasitology*, **1**, 165–74.

Kelly, J.D., Dineen, J.K., Goodridge, B.S. and Smith, I.D. (1974), Expulsion of *Nippostrongylus brasiliensis* from the intestine of rats. Role of prostaglandins and pharmacologically active amines (histamine and 5-hydroxy-tryptamine) in worm expulsion, *International Archives of Allergy and Applied Immunology*, **47**, 458–65.

Kennedy, M.W. (1980), Immunologically mediated, non-specific interactions between the intestinal phases of *Trichinella spiralis* and *Nippostrongylus brasiliensis* in the mouse, *Parasitology*, **80**, 61–72.

Kilby, A., Walker-Smith, J.A. and Wood, C.B.S. (1976), Studies on the immunoglobulin containing cells and intra-epithelial lymphocytes in the small intestinal mucosa of infants with the post-gastroenteritis syndrome. *Australian Paediatric Journal*, **12**, 241.

Kojima, S., Yamamoto, N., Kanozawa, T., and Ovary, Z. (1985), Monoclonal IgE-dependent eosinophil cytotoxicity to haptenated schistosomula of *Schistosoma japonicum*: enhancement of the cytotoxicity and expression of Fc receptors for IgE by *Nippostrongylus brasiliensis* infection, *Journal of Immunology*, **134**, 2719–22.

Kolmannskog, S. (1987), Similarities between IgE in human faeces and a chymotrypsin-digest of an IgE myeloma protein, *International Archives of Allergy and Applied Immunology*, **82**, 100–7.

Korenaga, M., Wang, C.H., Bell, R.G., Zhu, D. and Ahmed, A. (1989), Intestinal immunity to *Trichinella spiralis* is a property of OX8[−] OX22[−] T-helper cells that are generated in the intestine, *Immunology*, **66**, 588–94.

Kreisle, R.A., Parker, C.W., Griffin, G.L., Senior, R.M. and Stenson, W.F. (1985), Studies of leukotrience B$_4$-specific binding and function in rat polymorphonuclear leukocytes: absence of a chemotactic response, *Journal of Immunology*, **134**, 3356–63.

Lake, A.M., Bloch, K.J., Sinclair, K.J., and Walker, W.A. (1980), Anaphylactic release of intestinal goblet cell mucus, *Immunology*, **39**, 173–8.

Lammas, D.A., Mitchell, L.A., and Wakelin, D. (1987), Adoptive transfer of enhanced eosinophilia and resistance to infection in mice by an *in vitro* generated T-cell line specific for *Mesocestoides corti* larval antigen, *Parasite Immunology*, **9**, 591–601.

Larsh, J.E. Jr., Ottolenghi, A. and Weatherly, N.F. (1974), *Trichinella spiralis*: phospholipase in challenged mice and rats, *Experimental Parasitology*, **36**, 299–306.

Lee, G.B. and Ogilvie, B.M. (1981), The mucus layer in intestinal nematode infections, in Ogra, P.L. and Bienenstock, J. (Eds). *The Mucosal Immune System in Health and Disease* pp.175–83, *Proceedings 81st Ross Conference on Pediatric Research*, Ross Laboratories. Ohio: Columbus.

Lee, G.B. and Ogilvie, B.M. (1982), The intestinal mucus barrier to parasites and bacteria, in Chantler, E.N., Elder, J.B. and Elstein, M. (Eds.), *Mucus in Health and Disease II. Advances in Experimental Medicine and Biology*, **Vol. 144**, pp.247–8, New York: Plenum.

Levy, D.A. and Frondoza, C. (1983), Immunity to intestinal parasites: role of mast cells and goblet cells, *Federation Proceedings*, **42**, 1750–5.

Lewis, R.A. and Austen, K.F. (1984), The biologically active leukotrienes biosynthesis, metabolism, receptors, functions and pharmacology, *Journal of Clinical Investigations*, **73**, 889–97.

Lopez, A.F., Sanderson, C.J., Gamble, J.R., Campbell, H.D., Young, I.G. and Vadas, M.A. (1988), Recombinant human interleukin 5 is a selective activator of human eosinophil function, *Journal of Experimental Medicine*, **167**, 219–24.

Lopez, A.F., To, L-B., Yang, Y-C., Gamble, J.R., Shannon, M.F., Burns, G.F., Dyson, P.G., Juttner, C.A., Clark, S. and Vadas, M.A. (1987), Stimulation of proliferation, differentiation and function of human cells by primate IL-3, *Proceedings of the National Academy of Sciences* (USA), **84**, 2761–5.

MacDonald, T.T. and Spencer, J.J. (1988), Evidence that activated mucosal T cells play a role in the pathogenesis of enteropathy in human small intestine, *Journal of Experimental Medicine*, **167**, 1341–9.

Marom, Z, Shelhamer, J., Berger, M., Frank, M. and Kaliner, M. (1985), Anaphylatoxin (3a) enhances mucous glycoprotein release from human airways *in vitro*, *Journal of Experimental Medicine*, **161**, 657–68.

McLaren, D.J. (1980), *Schistosoma mansoni*: the parasite surface in relation to host immunity, *Tropical Medicine Research Studies I*, Chichester: John Wiley.

McLaren, D.J., Mackenzie, D.D. and Ramalho-Pinto, F.J. (1977), Ultrastructural observations on the *in vitro* interaction between rat eosinophils and some parasitic helminths (*Schistosoma mansoni, Trichinella spiralis* and *Nippostrongylus brasiliensis*), *Clinical Experimental Immunology*, **30**, 105–18.

Mencia-Huerta, J-M., Razin, E., Ringel, E.W., Corey, E.J., Hoover, D., Austen, K.F. and Lewis, R.A. (1983), Immunologic and ionophore-induced generation of leukotriene B4 from mouse bone marrow-derived mast cells, *Journal of Immunology*, **130**, 1885.

Miller, H.R.P. (1980), The structure, origin and function of mucosal mast cells, a brief review, *Biologie Cellulaire.*, **39**, 229–32.

Miller, H.R.P. (1984), The protective mucosal response against gastrointestinal nematodes in ruminants and laboratory animals, *Veterinary Immunology and Immunopathology*, **6**, 167–259.

Miller, H.R.P., Huntley, J.F. and Wallace, G.R. (1981), Immune exclusion and mucus trapping during the rapid expulsion of *Nippostrongylus brasiliensis* from primed rats, *Immunology*, **44**, 419–29.

Moqbel, R. and McLaren, D.J. (1979), *Strongyloides ratti*: Studies on the structure and characteristics of normal and immune-damaged worms, *Experimental Parasitology*, **49**, 139–52.

Moqbel, R., MacDonald, A.J., Cromwell O. and Kay, A.B. (1990), Release of leukotriene C₄ (LTC₄) from human eosinophils following adherence to IgE- and IgG-coated schistosomula of *Schistosoma mansoni*, *Immunology*, **69**, 435–442.

Moqbel R., MacDonald, A.J., Kay, A.B. and Miller H.R.P. (1989), Platelet activating factor (PAF) release during intestinal anaphylaxis in rats, *FASEB Journal*, **3**, A1337 (Abstract 6454).

Moqbel, R., Sass-Kuhn, S.P., Goetzl, E.J. and Kay, A.B. (1983a), Enhancement of neutrophil- and eosinophil-mediated complement-dependent killing of schistosomula of *Schistosoma mansoni in vitro* by leukotriene B₄, *Clinical Experimental Immunology*, **52**, 519–27.

Moqbel, R., Nagy, L., Lee, T.H., Sass-Kuhn, S.P. and Kay, A.B. (1983b), Effect of lipoxygenase products of arachidonic acid on complement receptor enhancement and killing of *Schistosomula mansoni, in vitro*, in Piper, P.J. (Ed.), *Leukotrienes and Other Lipoxygenase Products*, pp.241–7, Chichester: Research Studies Press.

Moqbel, R., Wakelin, D., MacDonald, A.J., King, S.J., Grencis, R.K. and Kay, A.B. (1987), Release of leukotrienes during rapid expulsion of *Trichinella spiralis* from immune rats, *Immunology*, **60**, 425–30.

Moqbel, R., King, S.J., MacDonald, A.J., Miller, H.R.P., Cromwell, O., Shaw, R.J. and Kay, A.B. (1986), Enteral and systemic release of leukotrienes during anaphylaxis of *Nippostrongylus brasiliensis*-primed rats, *Journal of Immunology*, **137**, 296–301.

Nadel, J.A. and Holtzman, M.J. (1984), Regulation of airway responsiveness and secretion: role of inflammation, in Kay, A.B., Austen, K.F. and Lichtenstein, L.M. (Eds.), *Asthma: Physiology, Immunopharmacology and Treatment*, pp. 129–53, London: Academic Press.

Nagy, L., Lee, T.H., Goetzl, E.J., Pickett, W.C. and Kay, A.B. (1982), Complement receptor enhancement and chemotaxis of human neutrophils and eosinophils by leukotrienes and other lipoxygenase products, *Clinical Experimental Immunology*, **47**, 541–7.

Nawa, Y. and Miller, H.R.P. (1978), Protection against *Nippostrongylus brasiliensis* by adoptive immunization with immune thoracic duct lymphocytes, *Cellular Immunology*, **37**, 51–60.

Parrott, D.M.V. (1976), The gut as a lymphoid organ. *Clinics in Gastroenterology*, **5**, 211–28.

Rothwell, T.L.W., Jones, W.O. and Love, R.J. (1974), Studies on the role of histamine and 5-hydroxytryptamine in immunity against the nematode, *Trichostrongylus colubriformis*. II. Inhibition of worm expulsion from guinea pigs by treatment with reserpine, *International Archives of Allergy and Applied Immunology* **47**, 875–86.

Saito, H., Kranz, D.M., Takagaki, Y., Hayday, A.C., Eisen, H.N. and Tonegawa S. (1984), Complete primary structure of a heterodimeric T-cell receptor deduced from cDNA sequences, *Nature*, **309**, 757–62.

Schick, B., Austen, K.F. and Schwartz, L.B. (1984), Activation of rat serosal mast cells by chymase, an endogenous secretory granule protease, *Journal of Immunology*, **132**, 2571–6.

Shanahan, F., Denburg, J.A., Bienenstock, J. and Befus, A.D. (1984), Mast cell heterogeneity, *Canadian Journal of Physiology and Pharmacology*, **62**, 734–7.

Sharon, P. and Stenson, W.F. (1984), Enhanced synthesis of leukotriene B$_4$ by colonic mucosa in inflammatory bowel disease, *Gastroenterology*, **86**, 453–60.

Shaw, R.J., Walsh, G.M., Cromwell, O., Moqbel, R., Spry, C.J.F. and Kay, A.B. (1985), Activated human eosinophil generate SRS-A leukotrienes following IgG-dependent stimulation, *Nature*, **316**, 150–2.

Siegel, J.N., Schwartz, A., Askenase, P.W. and Gershon, R.K. (1982), T cell suppression and contrasuppression induced by histamine H2 and H1 receptor agonist, respectively, *Proceedings of the National Academy of Sciences* (USA), **79**, 5052–6.

Specian, R.D. and Neutra, M. (1982), Regulation of intestinal goblet cell secretion. 1. The role of parasympathetic stimulation, *American Journal of Physiology*, **242**, 370–9.

Spencer, J., Finn, T. and Isaacson, P.G. (1986b), Human Peyer's patches: an immunohistochemical study, *Gut*, **27**, 153–7.

Spencer, J., Dillon, S.B., Isaacson, P.G. and MacDonald, T.T. (1986a), T cell subclasses in fetal human ileum, *Clinical Experimental Immunology*, **65**, 553–8.

Subramanian, N. and Bray, M.A. (1987), Interleukin-1 releases histamine from human basophils and mast cells *in vitro*, *Journal of Immunology*, **138**, 271–5.

van der Vorst, E., Dhont, H., Cesbron, J.Y., Capron, M., Dessaint, J.P., and Capron, A. (1988), The influence of an *Hymenolepsis diminuta* infection on IgE and IgA bound to mouse intestinal eosinophils, *International Archives of Allergy and Applied Immunology*, **87**, 281–5.

Viney, J.L., MacDonald, T.T. and Kilshaw, P.J. (1989), T cell receptor expression in intestinal intra-epithelial lymphocyte subpopulations of normal and athymic mice, *Immunology*, **66**, 583–7.

Wakelin, D. (1978), Immunity to intestinal parasites, *Nature*, **273**, 617–20.

Wakelin, D. (1985), Genetic control of immunity to helminth infections, *Parasitology Today*, **1**, 17–23.

Walker, W.A., Wu, M., and Bloch, K.J. (1977), Stimulation by immune complexes of mucus release from goblet cells of the rat small intestine, *Science*, **197**, 370–2.

Wallace, J.L. and Whittle, B.J.R. (1986), Profile of gastrointestinal damage induced by platelet-activating factor, *Prostaglandins*, **32**, 137–41.

Walsh, G.M., Moqbel, R., Nagakura, T., Iikura, Y. and Kay, A.B. (1989), Enhancement of the expression of the eosinophil IgE receptor (Fc R2) and its function by platelet-activating factor, *Journal of Lipid Mediators*, (in press).

Wardlaw, A.J., Moqbel, R., Cromwell, O. and Kay, A.B. (1986), Platelet activating factor: potent chemotactic and chemokinetic factor for human eosinophils, *Journal of Clinical Investigations*, **78**, 1701–6.

Wassom, D.L. and Gleich, G.J. (1979), Damage to *Trichinella spiralis* newborn larvae by eosinophil major basic protein, *American Journal of Tropical Medicine and Hygiene*, **28**, 860–3.

Woodbury, R.G., Miller, H.R.P., Huntley, J.F., Newlands, H.G.F.J., Palliser, A.C. and Wakelin, D. (1984), Mucosal mast cells are functionally active during spontaneous expulsion of intestinal nematode infections in rat, *Nature*, **312**, 450–2.

11. Intestinal pathology

Gilbert A. Castro

11.1. INTRODUCTION

The alimentary canal provides specific stimuli that are conducive to the infectivity and sustenance of parasites. On the other hand, parasites, once established, can cause the enteric environment to undergo drastic change. Evoked changes may contribute to the pathogenesis of disease or may represent adaptations that allow the host to maintain homeostasis. In this chapter the changes that parasites produce in the gastrointestinal (GI) tract are reviewed. A major aim is to employ tenets or precepts from the disciplines of pathology, physiology and immunology to present an integrated picture of the host-parasite interaction. Relative to this it is worth noting that changes in tissue morphology are reflective of underlying biochemical and physiological alterations (Figure 11.1). Furthermore, smooth muscle and epithelial tissues that perform functions unique to the GI tract are regulated in their behaviour by several physiological systems. Either or both the effector and modulatory systems may be perturbed during infection.

To fairly describe enteric parasitism it must be envisioned as a kaleidoscope of physiological events initiated when the parasite enters the GI tract and modulated by its subsequent growth and development. Although changes evoked are most prominent during infection, physiological vestiges of the interaction may be evident long after the parasite has completed its life cycle and has been eliminated from the host. By integrating and summing the effects of parasites on their hosts as a function of time, it should be possible to rationally assess the impact of observed changes on the host-parasite interaction. It might well be concluded that physiological changes, which entail a transient period of disease, represent a response on the part of the host to effectively shift the symbiotic relationship (Schmidt and Roberts, 1985) from one of parasitism to commensalism.

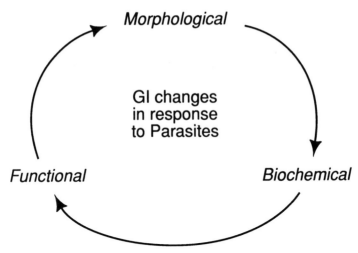

Figure 11.1. Inseparable relationships between biochemical, functional and morphological changes during parasitism.

11.2. LUMINAL EVENTS DURING INFECTION

The lumen and mucosal surface of the GI tract are constantly changing in geometry, primarily because of the contractile behaviour of the tunica muscularis and of smooth muscle fibres in the mucosa and submucosa. Smooth muscle function is regulated by intrinsic (nervous, paracrine and myogenic) as well as extrinsic (nervous and endocrine) factors. Parasites infecting via the enteral route have adapted to changes in motility patterns and to the presence of gastric acid, bile and enzymes. In fact, these latter factors which retard development of some micro-organisms (Mims, 1976) are often required by protozoa and helminths to successfuly complete the infectious process (Rogers, 1962).

In addition to the environmental factors already noted, enteric parasites find the luminal surface of the GI tract covered by an unstirred layer of fluid—unstirred in the sense that it does not readily mix with the bulk of the intestinal contents. Physiologically this layer represents a mucoid 'barrier' across which solutes, such as nutrients, diffuse before being assimilated. Also, the unstirred layer has been referred to as an 'antiseptic paint' containing, in addition to mucus, antibodies that defend against noxious agents.

Mucus

Evidence from several sources supports the hypothesis that intestinal mucus is important in defense against parasites. Ackert *et al.*, (1939) were early proponents of such a hypothesis, suggesting that greater resistance to *Ascaridia galli* in old chickens compared with young ones was due to more goblet cells and greater mucus production in the aged host. This suggestion was supported by experiments showing a parasitocidal effect of mucus (Frick and Ackert, 1948). Dobson (1967) demonstrated that sheep-derived mucus inhibited the respiration of the enteric nematode *Oesophagostomum columbianum* and attributed this, in part, to antibodies. He later demonstrated (Dobson and Bawdin, 1974) that the increased susceptibility of protein deficient sheep to infection with *O. columbianum* was associated with a decrease in mucus production.

Lee and Ogilvie (1982) concluded that the failure of preadult stages of *Trichinella spiralis* (Chapter 6, Figure 6.2) to establish in the intestine of immune rats was due to entrapment of worms in mucus, an event dependent on a heat-sensitive component in serum. The mucus-trapping theory of immunity, although attractive, remains debatable since Bell *et al.*, (1984) observed that rejection of infective *Trichinella* larvae occurred without mucus trapping in rats transiently immunized by intestinal infection alone. The importance of mucus trapping in excluding L_1 larvae from the intestine is further questioned by experiments showing that rejected and presumably 'trapped' worms passing from the small intestine of an immune rat into the surgically anastomosed intestine of a non-immune host readily infected the latter (Hessel *et al.*, 1982). As in murine trichinosis the entrapment of *Nippostrongylus brasiliensis* (Chapter 8, Figure 8.7) in immune hosts is an established phenomenon, but the relationship between mucus and functional immunity remains unclear (Miller *et al.*, 1981; Miller, 1987).

The factors that control mucous cell proliferation, a characteristic feature of catarrhal inflammation induced by enteric parasites, are not fully understood. Miller and Nawa (1979) showed, in primary nippostrongylosis in the rat, that goblet cells increased significantly from basal numbers at day 8 post-inoculation (PI) to a maximum level at day 14 PI (Figure 11.2). When thoracic duct lymphocytes were collected at day 10 PI and transferred to naive recipients that were concurrently infected, the maximal goblet cell response was seen at day 8 PI, i.e. there was an earlier onset of goblet cell hyperplasia in infected recipients of sensitized lymphocytes. These experiments are important in demonstrating that goblet cell proliferation, hence epithelial development, is partially controlled by immunological elements. The mechanism by which lymphocytes enhance mucus production and secretion could involve direct effects on epithelium. Alternatively, indirect pathways may be involved that include other cell types. For example mucosal mast cells, which also undergo lymphocyte-dependent proliferation during nippostrongylosis, may be involved in regulating mucus production through the release of histamine (Wells, 1962). An intriguing parasite model that may prove beneficial in determining factors that control goblet cell proliferation and mucus secretion involves the rat infected with the non-intestinal strobilocercus stage of the tapeworm, *Taenia taeniaformis*. These hosts show an unexplained but profound goblet cell hyperplasia in gastric and duodenal mucosa (Cook *et al.*, 1981).

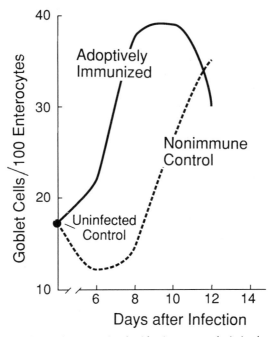

Figure 11.2. Goblet cell hyperplasia associated with nippostrongylosis in the rat (non-immune control) and the earlier onset of peak expression resulting from adoptive immunization with immune lymphocytes [original data from Miller and Nawa (1979) as summarized by Castro (1989a)].

IgA

Relative to the association of antibodies with the unstirred layer, one of the most well-defined components of the mucosal immune system is secretory IgA. The secretion of this immunoglobulin is depicted as a reaction against antigens in free form or as components of pathogens, blocking their entrance into the mucosa. Other antibody isotypes such as IgG, IgM and IgE are present in the lumen. These types are normally found in serum but may be translocated to the lumen through leaks in the epithelial barrier. Leakage may be greatly enhanced as a result of local anaphylactic reactions (Murray, 1972).

IgA has been implicated in the exclusion of bacteria, viruses, and a variety of non-infectious antigens (McGhee and Mestecky, 1983) from the mucosa. IgG_1 appears to function in controlling the entry of soluble antigens from the gut lumen by interfering with their uptake at the level of the mucus coat (Walker *et al.*, 1974). Despite evidence that secretory antibodies offer protection to micro-organisms, there are only a few examples demonstrating protection against parasites. Treatment of onchospheres of *T. taeniaformis* with IgA *in vitro* blunted their infectivity for mice (Lloyd and Soulsby, 1978), and the exposure of trophozoites of *Giardia muris* to phagocytic cells plus IgA decreased their infectiveness (Kaplan *et al.*, 1985). *Giardia* is not invasive and its attachment to enterocytes purportedly is mediated in part by a ligand-carbohydrate interaction. A lectin on the surface of *Giardia* apparently becomes expressed when the parasite is exposed to the proteolytic enzyme trypsin. This lectin allows the parasite to bind to a mannose-6-phosphate receptor on the enterocyte surface (Lev *et al.*, 1986). If *Giardia* depends on specific membrane-binding sites for attachment, then antibodies directed against those surface receptors might interfere with the infective process. The plausibility of such a defense mechanism against enteric protozoa draws strength from observations that invasiveness of *Eimeria* species is blocked by monoclonal antibodies to sporozoites or their host cells (Augustine and Danforth, 1985, 1987; Whitmire *et al.*, 1988).

11.3. FACTORS INFLUENCING PATHOGENESIS

Inasmuch as enteric parasites do not produce toxins analogous to bacterial enterotoxins, their pathogenicity relates to their capacity to attach to and/or penetrate the gut mucosa. Whereas conditions that influence those processes are poorly understood, several factors influence pathogenesis once an infection is established. These include the type of parasite involved, intensity of infection, duration of infection, immune state of the host, and unique characteristics of the injured tissue or organ.

Parasite type bears on size, degree of multiplication *in situ*, and level of invasiveness. Tissue invaders generally induce more severe reactions than lumen dwellers or organisms that interact superficially with the mucosa. The number of organisms present in the GI tract often dictate the degree of pathology associated with infections. A few helminths may have little effect on the host whereas heavy infections arising from a single inoculum, repeated exposure, or auto-infections could have more dire consequences. Mechanical obstruction of the small bowel lumen may occur on the presence of a bolus of large parasites, as in human ascariasis and taeniasis. In human strongyloidiasis, ancylostomiasis and schistosomiasis, the degree of symptoms are related partly to the overall worm burden and partly to the long duration of infection. As will be explained below, the immune system may attenuate pathologic changes by preventing reinfection, or it may augment the disease process due to involvement in inflammation. Since the GI tract responds in a similar manner to a variety of noxious stimuli (Sprinz, 1962), commonalities in enteric pathology caused by a variety of parasites should be considered the rule rather than the exception.

Pathologic changes may be localized or more intense in the intestine, but physiological processes foreign to the GI tract may be affected. Pathology may be initiated by the active nature of the parasite or by proliferative changes that they induce in host tissues. Examples of pathology induced by the direct action of parasites include blockage of the absorptive surface of the small bowel caused by overgrowth with *G. lamblia*, mechanical blockage associated with heavy worm burdens as in ascariasis, the absorption of vitamin B_{12} by the fish tapeworm, *Diphyllobothrium latum*, trauma and blood loss due to hookworms, and lysis of tissue by *Entamoeba histolytica*. Proliferative changes, which encompass various forms and degrees of acute and chronic inflammation, are induced by numerous species.

Inflammation may be evoked non-specifically or it may be immunologically mediated. Eliciting antigens may be derived from somatic tissues of parasites or from their excretions and secretions. Hypersensitivity states play important roles in mediating inflammatory lesions, including granuloma formation and fibrosis (Sections 5.10 and 6.4). The latter is illustrated in the tissue response to eggs of *Schistosoma* species. Mucosal inflammation is generally accompanied by disturbances in digestive, absorptive, secretory, and motor functions (Castro, 1976). These conditions may contribute to local symptoms of diarrhoea, steatorrhoea or abdominal pain and to systemic symptoms such as anorexia, weight loss and cachexia (Johnson, 1965). Systemic signs of infection are often reflected in haematologic changes such as anaemia and eosinophilia (Olson and Schultz, 1963).

Hookworm disease is an example of an infection that causes both local and systemic changes. The initial lesion is traumatic in nature. Cells and tissues are forced out of their normal spatial relationships as adult worms employ hard mouth parts to attach to the mucosal surface. Trauma leads to a local inflammation. Blood loss, attributed to the feeding habits of the worms, results in anaemia of the iron-deficiency type.

The histological appearance of lesions caused by *E. histolytica* supports the notion that this protozoan invades tissues through a lytic process. This view is supported by information on enzyme synthesis. *Entamoeba* synthesizes amylase, acid phosphatase, hyaluronidase as well as enzymes with gelatinase, pepsin and trypsin activities (Bragg and Reeves, 1962; Naegraith, 1963). Lushbaugh *et al.* (1984) identified an enzyme similar to cathepsin B in amoebic extracts that was correlated with virulence. This enzyme, released also by intact trophozoites, was cytotoxic to tissue culture cells but to date has not been shown to be cytotoxic for intestinal tissue of experimental animals.

Stirewalt (1963) catalogued a number of helminth-derived secretions having the potential to lyse tissues. However, these lytic factors, from both larval and adult stages, have not been investigated thoroughly enough to determine their role in pathogenesis. It is usually assumed that lytic secretions aid parasites and are harmful to the host, allowing translocation of parasites from one site to another and leading to inflammation and associated sequelae. On the other hand, micro-organisms, such as bacteria, secrete various enzymes that are unrelated to pathogenicity (Mims, 1976). In fact, lytic factors, such as proteinases, collagenases, and lipases that allow micro-organisms to deciminate, may expose them more readily to host defense mechanisms. Based on this consideration, it is possible that lytic enzymes secreted by protozoan and metazoan parasites may have little or no role in pathogenicity.

Table 11.1 summarizes some local effects caused by GI parasites which are discussed in this chapter. In rare instances pathology may be caused by the

Table 11.1. Local effects of gastrointestinal parasites on the host.

Effects	Exemplary etiologic agent	Selected reference
Traumatic damage	Hookworm	Kalkofen, 1974
Lytic necrosis	*Entamoeba histolytica*	Lushbaugh *et al.*, 1984
Tissue proliferation		Miller, 1984
Inflammation	Numerous species	Leid and Williams, 1979
Granuloma formation	*Schistosoma mansoni*	Domingo and Warren, 1969
Altered villus architecture	*Nippostrongylus brasiliensis*	Symons and Fairbairn, 1962
Altered smooth muscle		Stephenson *et al.*, 1980
architecture	*Ascaris lumbricoides*	
Mechanical obstruction	*Ascaris lumbricoides*	Blumenthal and Schultz, 1975
Intestinal enzyme		Castro, 1981
deficiency	Numerous species	
Impaired absorption	*Giardia lamblia*	Hoskins *et al.*, 1967
Altered response to GI		Dembinski *et al.*, 1979a,b
hormones	*Trichinella spiralis*	
Intestinal secretion	*Trichinella spiralis*	Castro *et al.*, 1979
	Capillaria phillipinesis	Whalen *et al.*, 1969
Competition for nutrients	*Diphyllobothrium latum*	von Bonsdorf, 1956
Altered motility	Numerous species	Castro, 1988
Hypersensitivity	Numerous species	Wakelin, 1984

accidental wandering of enteric parasites into ectopic sites, as in the haematogenous spread of *E. histolytica* from its primary site in the GI tract, the migration of *Ascaris* into the common bile duct, pancreatic duct or appendix, and the nocturnal migration of pinworms into the genital tract of female hosts. Where the life cycle of some 'enteric' helminths involve a transient period of development in the lungs, there may be respiratory tract involvement. It is noteworthy that pathologic responses elicited by parasites are not unique to these organisms, albeit high titres of IgE and eosinophilia are considered hallmarks, if not pathognomonic, of certain infections.

11.4. ANATOMICAL INTERRELATIONSHIPS

Inflammation

Essentially any stimulus that causes local damage to tissue will evoke an inflammatory response. Also, reinfection with parasites may be associated with inflammation mediated by immunological reactions. The immunology of the GI mucosa is extremely complex. Major components of the mucosal immune system include macrophages, lymphocytes, plasma cells, mast cells, and granulocytes. An appreciation of the interactions among these cell types relative to their maturation, proliferation, and functions can be derived from recent authoritative reviews (Leid and Williams, 1979; Elson *et al.*, 1986; Miller, 1987). See also Chapters 3, 4 and 5.

Macrophage as a prototype of cell involvement

In order to illustrate the complexity of interactions among immunological elements within the limited context of this chapter, the macrophage is selected for special attention. Macrophages in general are important in the presentation of antigens to lymphocytes (Chapter 4). The specific role of gut macrophages in antigen processing is unknown. Mayrhofer *et al.* (1983) described macrophage-like cells bearing Ia antigens in the lamina propria of Peyer's patches and mesenteric lymph nodes and proposed an antigen-processing role for them. Monocyte-derived macrophages that have been activated by antigen secrete interleukin 1 (IL1). This paracrine substance activates T lymphocytes leading to the expression of IL2 surface receptors. These activated T cells also secrete IL2 which is a proliferative agent. IL2 stimulates subsets of other lymphocytes with appropriate surface receptors. This leads to the proliferation of T helper, T suppressor, and T cytotoxic cells. Concomitantly, the activated T cells release an activating factor which causes macrophages to continue the production of IL1, giving rise to a mutual activation process.

In addition to antigen processing, gut macrophages may act as effector cells. This potential is extrapolated from knowledge of other macrophage populations. Lung macrophages can kill infective larvae of *N. brasiliensis* in the presence of the C3 component of complement (Egwang *et al.*, 1985). Wing and Remington (1978) reported that mouse peritoneal macrophages activated by bacteria (*Listeria monocytogenes*) or protozoa (*Toxoplasma gondii*) adversely affect infective larvae of *Trichinella spiralis*. The finding of micro-organisms in phagocytic cells of the gut mucosa (Takeuchi and Sprinz, 1967) and the phagocytosis by macrophages of sporozoa such as *T. gondii* (McLeod and Remington, 1977) and of amastigote stages of haemoflagellates (Williams *et al.*, 1976; Nussenzweig, 1982) support the premise that macrophages are important effector cells in host resistance.

The fact that the monocytes and macrophages are activated non-specifically and are the source of a variety of mediators deleterious to host tissues as well as to micro-organisms, make them prime candidates for initiating pathophysiologic processes. These mediators include vasoactive and permeability modulators (prostaglandin, leukotrienes, prostacyclin, and thromboxane), chemotactic agents, proteases (collagenase, elastase, plasminogen-activating factor and acid hydrolase), procoagulating factor, fibroblast growth factor, and oxidants (Carrico *et al.*, 1986; Halliwell, 1987).

Lamina propria, epithelium and smooth muscle

Considerable evidence is available emphasizing the impact of inflammatory-immunological elements in the lamina propria on epithelial and smooth muscle architecture in the GI tract. Symons and Fairbairn (1962) reported villous atrophy and crypt hyperplasia in the rat associated with intestinal inflammation caused by *N. brasiliensis* (Figure 11.3). Other parasitic infections in which these changes are evident include trichostrongylosis (Barker, 1973), hookworm disease (Sheehy *et al.*, 1962), and trichinosis (Castro *et al.*, 1967). In addition to the changes in villus structure, epithelial cell morphology *per se* is affected (Castro *et al.*, 1967). Symons (1965) reported that *Nippostrongylus* stimulates epithelial cell turnover. Resulting differentiated enterocytes are characterized by a decrease in number and height of microvilli (Symons 1976a).

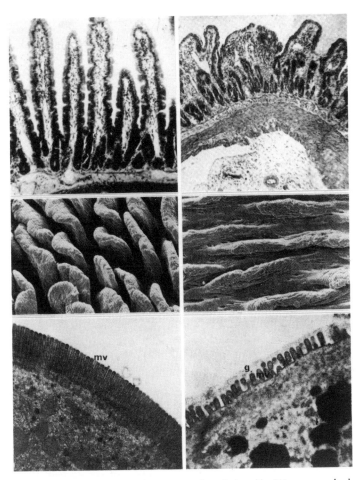

Figure 11.3. Morphological changes in the jejunum of rats induced by *Nippostrongylus brasiliensis*. Uninfected control tissues on left side of plate and infected tissues on the right. Top: Light microscopic view illustrating shortening of villi and elongation of crypts; original magnification × 100 (from Symons and Fairbairn, 1962). Middle: Scanning electron micrograph depicting atrophic villi; original magnification × 75 (from Symons, 1976a). Bottom: Electron micrograph emphasizing the reduction in brush-border height and surface area caused by infection; mv = microvilli; g = glycocalyx; L = lipid inclusion; original magnification × 30 000 (from Symons *et al.*, 1971).

In addition to these morphological changes, parasitism also induces biochemical changes in the epithelial cell brush border. Isolated brush border membranes (BBM) from rats infected with *Nippostrongylus* and *Trichinella* show a reduced capacity to bind the lectin wheat germ agglutinin (WGA; Castro and Harari, 1982; Harari and Castro, 1988). In trichinosis this reduction is maximally expressed as early as two weeks after inoculation of hosts with L_1 larvae and remains so for at least 12 weeks thereafter (Figure 11.4). The failure of BBM to bind WGA at normal levels is due to a decrease in sialic acid residues (Harari and Castro, 1983) that, along with N-acetylglucosamine, specifically bind WGA. In another study of BBM from trichinized rats, the methylation of phospholipids was quantitatively altered as compared with BBM from uninfected hosts (Harari and Castro, 1985).

Manson-Smith *et al.* (1979), working with *Trichinella*-infected mice, and Ferguson and Jarret (1975), studying rats infected with *Nippostrongylus*, demonstrated that villous atrophy and crypt hyperplasia were influenced by T lymphocytes (Figure 11.5). Already mentioned was the report by Miller and Nawa (1979) that lymphocytes influenced goblet cell hyperplasia in nippostrongylosis. These examples emphasize the capacity of immunological elements to alter villus architecture.

The relationship between proliferative or inflammatory changes in the lamina propria and changes in smooth muscle structure is as impressive as that described for epithelial tissue. An increase in small bowel smooth muscle mass in rats with nippostrongylosis was described by Symons (1957). Analogous changes have been reported in trichinized rats and guinea pigs (Lin and Olson, 1970; Castro *et al.*, 1976), in rats infected with acanthocephalans

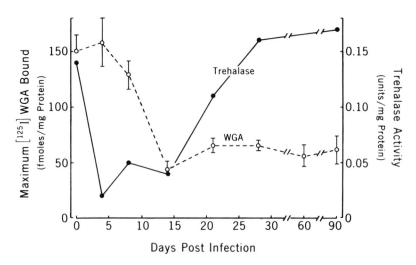

Figure 11.4. Epithelial disaccharidase (trehalase) activity and maximum binding of wheat germ agglutinin (WGA, lectin) associated with brush-border membranes from the rat intestine as a function of time after infection with *Trichinella spiralis* (from Castro and Harari, 1982).

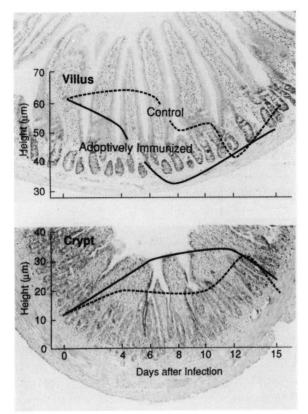

Figure 11.5. Influence of the intestinal phase of infection with *Trichinella spiralis* on villous atrophy and crypt elongation in mice that were thymectomized, irradiated and repleted with T lymphocytes and bone marrow cells [original data from Manson-Smith *et al.*, (1979) as summarized by Castro (1989a)].

(Crompton and Singhi, 1982), and in horses (Clayton and Duncan, 1977) and pigs (Stephenson *et al.*, 1980) with ascaris infections. Whereas the basis for the change in smooth muscle mass is subject to speculation, the accelerated response in guinea pigs reinfected with *Trichinella* (Lin and Olson, 1970) suggests that immune factors are involved.

11.5. FUNCTIONAL INTERRELATIONSHIPS

Lamina propria, epithelium and smooth muscle

Based on the premise that morphological changes are accompanied by underlying physiological and biochemical changes, the villous and smooth muscle alterations described above foretell of concomitant functional changes. Current considerations of the mucosal immune system implicate epithelium

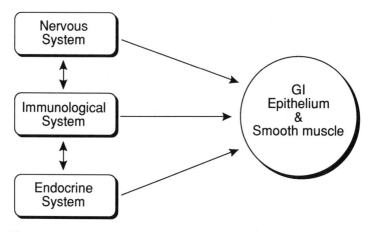

Figure 11.6. Immunophysiological pathways of communication in the gut.

and smooth muscle in the GI tract as effector tissues of immune responses. Those effectors may communicate directly with more 'classically' recognized immunological elements via paracrine secretions or indirectly via neural and endocrine pathways. Such communications are depicted conceptually in Figure 11.6. Whereas the involvement of the central nervous system in the regulation of immune phenomena is not a new idea, the role of the enteric nervous system is a fresh consideration supportable by recently acquired data (Castro *et al.*, 1987; Bienenstock *et al.*, 1988). Such data are summarized in Figure 11.7. Here immediate hypersensitivity in the jejunum of *Trichinella*-sensitized rats is depicted as transducing antigenic signals into chloride secretion by epithelium. From this example it is easy to visualize a similar scheme involving

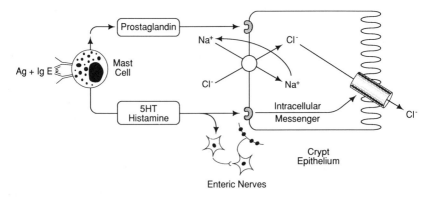

Figure 11.7. Immunophysiological pathway involved in epithelial Cl− secretion in jejunum from rats sensitized by infection to *Trichinella spiralis* and challenged with *Trichinella*-derived antigen. Mast cell-derived 5-hydroxytryptamine (5-HT), histamine, and prostaglandin may act directly on epithelial cells or indirectly through enteric neural pathways (from Castro *et al.*, 1987).

smooth muscle as the target tissue and with altered motility as a functional endpoint of the anaphylactic reaction.

Allergic responses of the type presented in Figure 11.7 highlight several points that are significant to an integrative view of host-parasite interactions. Emphasized is (a) the requirement for antigen processing, antibody production and cell-antibody interactions, (b) how responses initiated in the lamina propria can be transduced immediately to effect changes in the luminal compartment, and (c) a mechanism by which a stimulus applied to one point in the intestinal mucosa can be transduced to be expressed panmucosally. The latter point is important in explaining how only a few, small, highly localized parasites may induce inflammatory lesions that are amplified over large areas. Such a phenomenon was pondered by Larsh (1963) in his studies of intestinal inflammation induced by *T. spiralis* in mice.

Gastrointestinal motility

Alterations in GI motility caused by parasites are reflected in numerous measurements of intraluminal pressure, gastric emptying, intestinal transit and myoelectric activity in intact hosts, and of contractile and fluid-propelling properties of isolated tissues or gut segments. Biochemical changes have been detected through the use of various methods to monitor muscle function while tissues have been manipulated pharmacologically. There have been relatively few direct studies of smooth muscle biochemistry in infected hosts. Table 11.2 is a compendium of experiments in which various techniques were used to assess smooth muscle function during parasitism. More detailed information regarding the design of these experiments and the specific types of results obtained are discussed in a recent review (Castro, 1988).

When interpreting motility changes such as those listed in Table 11.2, one should be mindful that alterations vary drastically with the feeding state. As noted below infected hosts eat less than normal. Therefore, it is important to dissociate those changes that might result from altered feeding patterns from those caused directly by the parasite.

Immunological reactions may intervene to regulate smooth muscle function in appropriately sensitized and challenged hosts. Substances such as acetylcholine, substance P, vasoactive intestinal peptide, histamine, serotonin, prostaglandins and leukotrienes released by inflammatory or immunologically competent cells may modulate smooth muscle physiology either directly or through neural pathways (Burks, 1981). Positive Schultz-Dale reactions (*in vitro* anaphylaxis) which can be elicited in isolated intestines from parasitized hosts provide a rational basis for considering smooth muscle as a target tissue under immunological control. Indeed, Palmer and Castro (1986) reported the *in vivo* equivalent of the Schultz-Dale reaction in rats sensitized to *T. spiralis* and challenged by reinfection. Altered motility induced in the intact host by specific antigenic challenge was detected through measurements of myoelectric activity (Figures 11.8 and 11.9).

Table 11.2. Changes in gastrointestinal motility induced by parasites[a].

Organ	Method of detection	Parasite	Host
Oesophagus	Manometry	*Trypanosoma cruzi*	Human
Stomach	Manometry	*Trypanosoma cruzi*[b]	Human
	Electromyography	*Trichostrongylus axei*	Sheep
		Chabertia ovina	Sheep
		Haemonchus contortus	Sheep
		Eimeria magna	Rabbit
	Transit measurement	*Eimeria neischulzi*	Rat
Small Intestine	Manometry	*Trypanosoma cruzi*	Human
		Trichinella spiralis	Guinea Pig
	Electromyography	*Strongylus vulgaris*	Pony
		Trichostrongylus axei[c]	Sheep
		Chabertia ovina	Sheep
		Haemonchus contortus	Sheep
		Hookworms	Dog
		Trichinella spiralis	Dog
		Trichinella spiralis	Rat
		Eimeria neischulzi	Rat
		Eimeria magna	Rabbit
	Transit measurement	*Hymenolepis nana*	Mouse
		Trichinella spiralis	Mouse
		Trichinella spiralis	Rat
		Nippostrongylus brasiliensis	Rat
	Propulsion measurement	*Trichinella spiralis*	Guinea Pig
	Length-tension relationships	*Nippostrongylus brasiliensis*	Rat
		Trichinella spiralis	Rat
Caecum-Colon-Rectum	Manometry	*Trypanosoma cruzi*	Human
	Electromyography	*Eimeria magna*	Rabbit
Gall Bladder	Manometry	*Trypanosoma cruzi*[b]	Human

[a] Summarized from Castro, 1988
[b] Meneghelli, 1985
[c] Berry et al., 1986

Parasite-induced alterations in GI motility are interpreted generally in relation to the pathogenesis of disease or as an adaptive response to defects in digestion, absorption and secretion. Motility changes can be related to symptoms such as vomiting, abdominal pain, and diarrhoea. Whereas alterations in contractile behaviour alone are inadequate to cause diarrhoea,

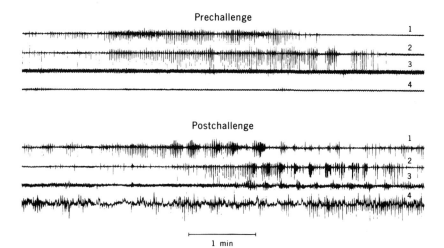

Figure 11.8. Tracings of intestinal myoelectric activity in an immunized rat before and after challenge with *Trichinella spiralis*. Patterns of slow waves and action potentials were recorded from four monopolar electrodes sutured to the antimesenteric, seromuscular surface of the small intestine at ~15 cm intervals beginning 5 cm distal to the pyloris (electrode 1) and progressing toward the caecum. Challenge with *T. spiralis* was carried out by adminstering L_1 larvae directly into the duodenum via a duodenocutaneous catheter. Electromyograph, recorded before and for ~6 min after challenge, illustrates slow waves (slow depolorization and repolarizations) and spike potentials (rapid depolarizations and repolarizations evident as upward and downward deflections) superimposed on slow waves. In prechallenge state, the phasic spike activity is evident at electrode sites 1 and 2 and appeared later (not shown) at sites 3 and 4. In postchallenge state, intense spiking occurred simultaneously at all electrode sites (from Palmer and Castro, 1986).

propulsive forces may be an important contributing factor when they occur in conjunction with secretion or other factors that contribute to the enteropooling of fluid (Figure 11.10). Increased aboral propulsion may prevent adequate mixing and contact of fluid with the absorbing surface.

Whereas changes in GI motility may possibly benefit the host by clearing enteric parasites from the lumen or superficial mucosa through aborally directed propulsive forces, there is little evidence to support this idea. In one study, however, the intensity of infection with *Hymenolepis nana* in mice was increased by slowing small bowel transit through the administration of opium (Larsh, 1947). Despite this example and a few related ones (Castro, 1988), the evidence implicating altered GI motility as a defense system against infection by parasites is ambiguous at best.

Gastrointestinal secretion

Endocrine-controlled enzyme secretion

Parasites induce changes in gastric, pancreatic and intestinal secretions. Secretion of hydrochloric acid (HCl) and pepsinogen by gastric mucosa

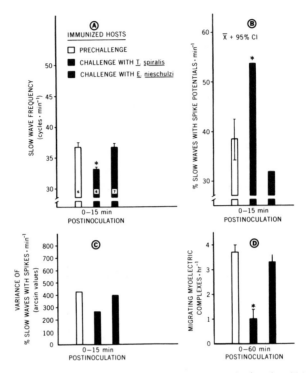

Figure 11.9. Small intestinal myoelectric parameters in rats immunized against *Trichinella spiralis* and challenged with *Eimeria nieschulzi* or *T. spiralis*. Measurements were made after intraduodenal inoculation with saline (prechallenge) or saline containing *E. nieschulze* sporozoites. Number of animals in each group is designated at base of bars in *panel A*. Asterisks indicate significant difference ($P < 0.05$) compared with prechallenge values. *A*: values are means and SE. *B*: values are percentages expressed as mean and 95% confidence interval (CI) for prechallenge group. Means for other groups excluded from 95% CI are significantly different from prechallenge group ($P < 0.05$). *C*: values are based on arcsines and are variances calculated for mean percentages shown in *panel B*. *D*: values are means and SE (from Palmer and Castro, 1986).

impacts on cavital or luminal digestion. Effective digestion of proteins in the stomach is dependent on the conversion of pepsinogen to pepsin through the action of HCl. In this regard gastric hypoacidity occurs in sheep infected with *Trichostrongylus colubriformis*, *Ostertagia circumcincta* or *Haemonchus contortus* and in cattle harbouring *Ostertagia ostertagi* (Ross and Dow, 1965; Anderson *et al.*, 1976a; Dakkak *et al.*, 1982). In the study of ovine ostertagiasis, the decreased acidity was attributed to gastritis and was associated with abnormally high serum gastrin levels (Anderson *et al.*, 1976b). Rats harbouring liver stages of *T. taeniaformis* develop a hyperplastic gastric mucosa that is deficient in acid secretion and also display hypergastrinaemia (Cook *et al.*, 1981). Gastrin is a GI hormone important in regulating gastric acid secretion and, additionally, in promoting growth of specific GI tissues. Hypergastrinaemia in ovine infection with *Ostertagia* and murine taenaisis can

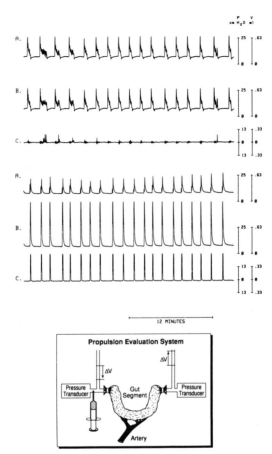

Figure 11.10. Propulsive wave forms produced by a segment of jejunum from an uninfected guinea pig (top Tracing A,B,C) and from a guinea pig infected with *Trichinella spiralis* for 10 days (lower Tracings A,B,C). Changes in pressure and volume expelled at the oral and aboral ends of fluid-filled segments were measured with the propulsion evaluation system illustrated in the bottom diagram. Propulsive behaviour is measured continuously as a regular pattern of changes in pressure (P, cm H_2O) and fluid flow (V, ml) at the oral (A) and aboral (B) ends of the segment. The net changes in pressure or volume flow are measured in Trace C (A minus B), i.e. a greater aboral value would yield an upward deflection whereas a greater oral value would yield a downward deflection in Trace C. Quantitative data is collected automatically by on-line computer analysis of tracings and is reported as the maximum pressure per complex, the duration of each complex, the interval between complexes and the maximum rate of fluid expulsion during each complex. An obvious change caused by infection is net aboral propulsion of fluid (from Alizadeh *et al.*, 1987).

be explained by a similar mechanism. Since HCl, which suppresses gastrin release in the normal physiological feedback system, is present only at low levels, the hormone is secreted in excess.

Physiologically, acid and pepsin secretion is stimulated by gastrin. Such stimulation is suppressed by secretin in the canine host. Secretin is a GI

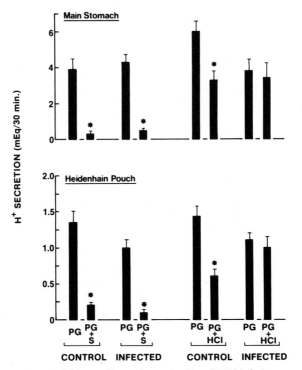

Figure 11.11 Maximum inhibitory effects of secretin (S) and HCl infusions on pentagastrin (PG; 1μg/kg-hr)-induced acid secretion from the main stomach and Heidenhain pouch before infection and during the first week of infection with *Trichinella spiralis* in the dog. Results obtained with a secretin dose of 4 unit/kg-hr and an HCl infusion from PG stimulation alone. Asterisk indicates significant difference (P < 0.05) compared with PG stimulation alone for respective group.

hormone which is released from the duodenum under stimulation by HCl. In dogs infected with intestinal stages of *T. spiralis*, the modulatory effects of secretin are blunted because of impaired release of the hormone from the duodenal mucosa (Dembinski *et al.*, 1979a) in response to HCl-stimulation (Figure 11.11). The decreased release of secretin in response to HCl was evident also in the reduced secretion of bicarbonate by the pancreas (Dembinski *et al.*, 1979b). Secretin is a potent pancreatic bicarbonate secretogogue.

Cholecystokinin (CCK) release from duodenal mucosa in response to a fatty acid stimulus is also reduced during the intestinal phase of primary but not secondary trichinosis (Dembinski *et al.*, 1979b). Impaired CCK output was reflected in the reduction of pancreatic enzyme secretion in trichinized dogs. Uninfected dogs readily secreted pancreatic enzymes when fatty acids were placed in their duodenum.

Dembinski *et al.* (1979b) proposed that suppressed hormone release from the duodenal mucosa, as summarized in Figure 11.12, was due to parasite-induced mucosal inflammation. Since inflammation underlies many physiologi-

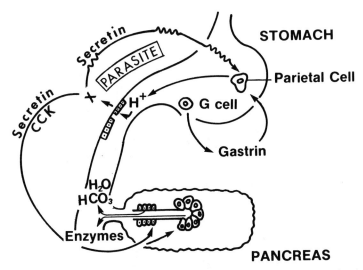

Figure 11.12. Influence of an enteric trichinosis on secretin and CCK release in the dog. The stimulatory effects of secretin and CCK on pancreatic secretion and the inhibitory effect of secretin on gastrin-stimulated acid secretion are reduced in infected hosts when HCl or fat is infused intraduodenally to release endogenous hormones as compared with exogenously infused hormones (results from Dembinski *et al.*, 1979a, 1979b, as summarized by Castro, 1981).

cal malfunctions associated with enterocytes during intestinal trichinosis (Castro, 1981), it is possible that epithelial endocrine cells are affected also. Accordingly, inflammation could lead to a decrease in the number of functional CCK-secreting cells in the crypts, a decrease in the number or avidity of membrane receptors for luminal secretagogues or to a reduction in hormone synthesis or release.

Hormonal malregulation may explain pancreatic enzyme deficiencies observed in humans with giardiasis (Okado *et al.*, 1983). On the other hand, *Giardia* may produce enzyme inhibitors as does another lumen-dwelling parasite, *Hymenolepis diminuta* (Pappas and Read, 1972). Despite such a possibility, physiological significance of such inhibitors remains to be resolved.

Diarrhoea

Since many enteric parasite infections are characterized by diarrhoea (World Health Organization, 1980), it is important to understand fluid movement in the GI tract in order to appreciate the potential cause of this common symptom. It is within this context that intestinal secretion of solutes and fluid will be considered. By definition diarrhoea is an increase in the frequency of bowel movements or in the volume or fluidity of the stool. Diarrhoea may be caused by the osmotic retention of water in the gut lumen, exudative lesions, disorders of intestinal secretion, or decreased contact time between fluid in the gut lumen and the absorptive surface. The role of these mechanisms in contributing to diarrhoea in parasitic infections will be discussed further.

OSMOTIC DIARRHOEA

Osmotic retention of water in the gut lumen is a passive phenomenon in that the fluid movement is dependent on the porosity of the mucous membrane (hydraulic conductivity) and the difference in hydrostatic pressure between the interstitial fluid compartment and the luminal compartment. It should be noted that transmembrane pressure differences, due to asymmetrical distribution of osmotically active solutes, are equivalent to hydrostatic pressure differences in terms of forces that contribute to volume flow (Byrne and Schultz, 1988). Since the mucous membrane of the gut is a complex biological membrane, an increase in fluid movement between blood and lumen may be affected by altered capillary wall structure, a loss of epithelial cells, an increase in size of pores in the apical or basolateral membranes, or an increase in the leakiness of tight junctions. Despite changes in porosity, net fluid movement from blood to gut lumen and vice versa still occurs only in response to a pressure gradient. Therefore, osmotic diarrhoeas in general can occur without a change in membrane leakiness. The driving force is a blood to lumen pressure gradient formed by the presence of excessive osmotically active molecules in the lumen. Such molecules may accumulate in parasitic diseases in which membrane digestion and absorptive functions are reduced. Additionally, selective biochemical or physiological defects may contribute to increased osmotic pressure in the gut lumen. Ganguly *et al.* (1987) reported ion transport disturbances in mice infected with *Giardia lamblia*. They found through studies of ion transport by vesicles of small intestinal microvilli that Na^+ and Cl^- absorption was decreased in association with an increase in the concentration of calmodulin in the microvillar core. Although the subcellular mechanism is unclear, calmodulin is known to inhibit Na^+ and Cl^- absorption and enhance Ca^{++} absorption.

EXUDATIVE AND PROTEIN LOSING DIARRHOEA

Inflammation or ulceration can contribute to an increase in faecal volume, especially in the colon which has relatively poor absorptive properties and where evacuation occurs quickly. These types of lesions allow the exudation of serum proteins, mucus and blood-derived cells. Amoebic dysentery is an example of exudative diarrhoea.

Because inflammation is common to parasite infections of the gut, it is not unusual to find increased mucosal permeability associated with the transepithelial movement of macromolecules such as proteins. Selected examples of parasite-induced protein-losing enteropathies are sheep infected with *O. circumcincta* and rats infected with *N. brasiliensis* (Halliday and Mulligan, 1968; Murray, 1972). Leakage of serum proteins occurred through capillary endothelium and the zona occludens of mucosal epithelial cells during nippostrongylosis. Although this effect may be interpreted as an adaptive change in that it allows the egress of antibodies that can act against parasites, a considerable amount of plasma protein is lost in the process.

In addition to blood-to-lumen movement, intraluminal antigenic molecules, due to a local lumen-to-tissue concentration gradient, may diffuse through the leaky mucosal membranes and stimulate the host mucosal immune system. Relative to this, Bloch *et al.* (1979) reported the enhanced intestinal absorption of bovine serum albumin from rats with nippostrongylosis.

SECRETORY DIARRHOEA

Bacterial enterotoxins associated with Asiatic cholera, staphylococcal and clostridial food poisoning and infections with various strains of *Escherichia coli* are the prototype agents that cause secretory diarrhoea. These agents induce secretion by increasing intraepithelial cAMP which mediates active chloride secretion. The blood-to-lumen active Cl^- secretion is accompanied by passive sodium movement and is followed by water movement. Agents such as GI hormones (vasoactive intestinal peptide, secretin), luminally derived long chain fatty acids and hydroxy bile acids stimulate the cAMP system and induce secretion. As noted by a WHO scientific work group (World Health Organization, 1980) secretory type diarrhoeas caused by parasites have not been recognized. Evidence is accumulating, however, suggesting that parasites may evoke this type of diarrhoea. The secretagogue action of fatty acids and bile acid is significant in that steatorrhoea is a symptom of giardiasis, and this infection is often associated with fat and bile acid malabsorption. In ascariasis, intestinal obstruction may cause secretory diarrhoea (Phillips, 1972) although the biochemical steps leading to enhanced secretion are not known. *Strongyloides* infection has been linked to the intestinal secretion of chloride, sodium and water (Kane *et al.*, 1984). Although the underlying malfunction of altered ion transport is not known, diarrhoea induced by the nematode *Capillaria philippinensis* in humans may be through a mechanism related to that seen in cholera patients (Whalen *et al.*, 1969).

The use of a perfusion technique involving the dilution or concentration of a radioactive, non-absorbable marker to estimate net transmural fluid movement in the small intestine demonstrated a decrease in net fluid absorption and sometimes net secretion in *T. spiralis*-infected rats (Castro *et al.*, 1979). These effects had their onset several days after primary infection but within minutes after a secondary infection. Subsequent studies using an *in vitro* system have provided strong evidence that Cl^- secretion is the driving force for epithelial fluid secretion and that inflammatory mediators such as histamine, serotonin and prostaglandin, released during intestinal anaphylaxis, are the physiological secretagogues (Russell and Castro 1985; Russell 1986; Castro *et al.*, 1987; Harari *et al.*, 1987). These results are detailed in Figure 11.7. In reference to serotonin-induced Cl^- secretion following *Trichinella*-induced anaphylaxis, it is interesting that the anion secretory response elicited by lysates of *E. histolytica* in rabbit ileum and rat colon were mimicked by serotonin. Furthermore this amine was identified as a component of the amoeba-derived lysate (McGowan *et al.*, 1983).

It is now evident that the local immune system can serve as a specific trigger for eliciting common physiological responses. Therefore, it is probable that antigens of non-parasite origin gaining access to appropriately sensitized tissue may evoke responses similar to those stimulated by parasite-derived antigens. Furthermore, mucosal tissues distant to the parasite's habitat become reactive to antigenic stimulation. For example, the colon of rats infected with *N. brasiliensis* and *T. spiralis* elicits an ionic secretory response when challenged with worm-derived metabolic antigens (Baird *et al.*, 1985; Russell and Castro, 1989), despite the fact that these nematodes do not infect the colon. Recently derived knowledge about antigen-induced ion and fluid secretion (see Castro 1989b, 1989c for reviews) should have a profound influence on how one envisions the potential effects of parasites on gut function.

11.6. NUTRITIONAL DISTURBANCES

Based on recent reviews (Solomons and Keusch, 1981; Beisel, 1982; Russell and Castro, 1987; Stephenson, 1987), it is evident that the role of nutrition in host-parasite interactions in the GI tract remains poorly understood. Figure 11.13 illustrates generalizations that have been deduced from a variety of observations dealing with nutrition and immunity. The interactions depicted provide a means whereby various physiological and parasitological processes can be related in a logical manner. It is generally assumed that severe malnutrition reduces resistance to infectious agents by decreasing humoral and cellular resistance mechanisms. Infection in turn may impair nutrient assimilation because of impaired digestive and absorptive processes, parasite acquisition of nutrients or anorexia. The difficulty of addressing nutrition and parasitism is evident from the sentiment expressed by Layrisse and Vargas (1975) that the literature on the topic suffered pathetically from all the confounding variables that came into play. Judging from a recent review of this topic (Stephenson, 1987), the related literature remains vague and tentative.

Adaptations

Compensatory absorptive mechanisms

The assimilation of high molecular weight nutrients involves sequential digestion by enzymes acting within the gut lumen followed by membrane digestion and absorption. Despite reports that parasites produce anti-enzymes and that enteric infections lead to reductions in the release of luminally active enzymes, deficiencies in brush-border enzymes and to malabsorption (Russell and Castro, 1987), the consequences of these factors on the host are not readily predictable.

The reason for this is that the GI tract has great adaptive potential and in many cases the host is able to adjust to the effects of parasitism and thereby maintain homeostasis. Experimental findings that reflect the adaptive capability of the host are worth considering.

Symons (1976a) presents objective evidence questioning the general import-

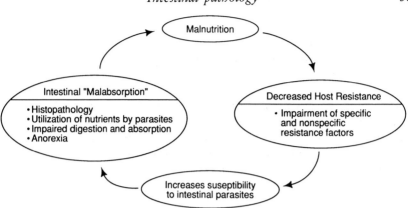

Figure 11.13. Working hypothesis of the interrelationships between host nutrition and parasitism.

ance of digestive and absorptive deficiencies in the pathogenesis of enteric parasitism. Much of that evidence is derived from a systematic study of rats infected with *N. brasiliensis*. Some of that evidence from Symons' pioneering studies are presented in Table 11.3.

Relative to the assimilation of egg albumin labelled with radioactive iodine, this protein was digested and absorbed at a slower rate from the gut of hosts with nippostrongylosis. Quantitative estimates were made from measurements of unassimilated protein 1 h after a 2 ml bolus of a 5% albumin solution was

Table 11.3. Digestion and absorption of nutrients by rats during the intestinal phase of nippostrongylosis.

	Host	
Parameter	Uninfected	Infected
	% of protein meal digested and absorbed	
1 hr[a]	50	31.1
17 hr[b]	78.5 (n = 18)	77.5 (n = 15)
	Glucose absorbed	
Jejunum[c] (mM hr^{-1}g^{-1} dry wt mucosa)	3.92 ± 0.61 (n = 12)	0.88 ± 0.29 (n = 12)
Ileum[d] (mM hr^{-1}g^{-1} dry wt mucosa)	1.06 ± 1.30 (n = 6)	2.67 ± 1.30 (n = 6)
Entire small intestine[c] (mg/40 min)	181 ± 42 (n = 11)	200 ± 20 (n = 10)

[a] Symons (1960a)
[b] Symons and Jones (1970)
[c] Symons (1960b)
[d] Symons (1961)

delivered by stomach tube (Symons, 1960a). In analogous studies in which protein assimilation was measured 17 h after gastric intubation, infection had no significant influence on the percentage of the test meal digested and absorbed (Symons and Jones, 1970). In the latter experiment, a carbon 14-labelled *Chlorella* protein was used as a marker in the dietary protein that comprized 9–21% of the administered meal.

Further examination of data on glucose absorption (Table 11.3) indicates that during the intestinal phase of nippostrongylosis, glucose uptake by jejunum was significantly reduced compared with that in jejunum from uninfected hosts (Symons, 1961). These results were obtained using an *in vivo* perfusion method to examine surgically 'isolated' segments of small bowel. In other studies employing the Cori method (Wilson, 1962) to test the glucose absorptive capacity of the entire small intestine, the rate of glucose uptake was judged similar in infected and uninfected hosts (Symons, 1960b).

Inasmuch as glucose absorption was greater in the ileum of infected rats (Table 11.3), Symons hypothesized that the failure to observe malabsorption when the entire intestine was taken into consideration despite jejunal defects, was because of an ileal compensatory absorptive mechanism (Symons, 1961). This hypothesis was supported by results of experiments employing the same host-parasite system performed by Schofield (1977) who used everted sac preparations (Wilson, 1962) to make measurements of glucose absorption. Conceivably, compensation occurs as a result of infection-induced damage to the jejunum. Under normal circumstances the glucose absorptive capacity in the rat is greater in the jejunum than in the ileum (Baker *et al.*, 1960). However, it is well known that when the jejunum is damaged, the ileum is capable of adapting structurally and functionally to carry out greater than normal absorption (Dowling and Riecken, 1973). The magnitude of the adaptive capability of the ileum has been clearly demonstrated in jejunal-bypass operations. Under these conditions glucose absorption almost doubles to reach a level equal to about 70% of that in the jejunum (Dowling, 1973).

Nolla *et al.*, (1985) reported results of studies on the rat-*Nippostrongylus* system that were at variance with those reported by Symons (1961). Using a triple segment perfusion system in anaesthetized rats, no ileal compensatory increase in glucose absorption was observed during infection. In fact, malabsorption of glucose was noted in uninfected as well as infected regions of intestine. Despite this report (Nolla *et al.*, 1985), Symons' (1961) comments about physiological compensation as an adaptive mechanism remains realistic in that it is consistent with the known adaptability of the small intestine. Therefore, compensatory mechanisms command consideration in interpretations of the role of localized or regionalized malabsorptive defects in the pathogenesis of enteric infections.

Physiological reserves

Another factor that is often overlooked in the interpretation of regional malabsorption in disease conditions is the extremely large functional reserve capacities in the GI tract. An example of this capacity has been presented in a recent consideration of hookworm-induced malabsorption (Castro *et al.*, 1989). That example is appropriate for reconsideration here because of its general value in the interpretation of data in a number of host-parasite interactions. The hexose absorbing capabilities of the small intestine was quantified by Crane (1975) from data reported by Holdsworth and Dawson (1964) on glucose and fructose absorption in the human intestine. By measuring the absorptive capacity of a 30 cm segment of small bowel in normal humans, the total capacity for absorption of a mixture of glucose and fructose was determined to be 10 211 g/day. This uptake rate over a 24 h period is comparable to more than 22 pounds of sugar and over 50 000 calories.

$$\text{Measured: Glucose} = \frac{0.4 \text{ g}}{\text{min} \times 30 \text{ cm}}$$

$$\text{Fructose} = 0.9 \times \text{glucose}$$

$$\text{Calculated: Glucose} = \frac{0.4}{\text{min} \times 30 \text{ cm}} \times \frac{1440 \text{ min}}{\text{day}} \times 280 \text{ cm} = 5374 \text{ g/day}$$

$$\text{Fructose} = 5374 \times 0.9 = 4837 \text{ g/day}$$

THUS

Total Daily Capacity = 10 211 g > 22 lb > 50 000 cal

In addition to the uptake of elemental dietary carbohydrates by specific brush-border membrane carriers, Crane (1975) described a route for the direct translocation of monosaccharides by disaccharidases without involvement of normal carrier molecules. This hydrolase-related system provides an alternate and additive pathway for sugar transport when carriers for monosaccharides are saturated with substrates. The products of any disaccharide substrate of a brush-border enzyme can be translocated. Furthermore, the products of translocation are additive when more than one disaccharide is present in the gut lumen. From these considerations it is evident that the absorptive capacity for carbohydrates is, as Crane (1975) stated, 'ten times more than would be needed for even the most unreasonable individual caloric requirements'.

The examples presented indicate that the regulation of sugar assimilation is not associated with digestive and absorptive events at the level of the brush border. Physiological control is exerted by receptors in the proximal small bowel involved in a negative feedback mechanism which affects the motility of the stomach. Carbohydrates, fats, and proteins entering the intestine from the stomach trigger responses that slow gastric emptying (Hunt and Knox, 1968). Receptors identified to date respond to osmotic pressure, acid or lipid. The intestine adapts dramatically to maintain homeostasis, even after a 75% proximal resection (Dowling and Riecken, 1974). The remaining intestine undergoes hypertrophy with an increase in absorptive surface. Also, the rate of passage of nutrients through the gut decreases (Nemeth *et al.*, 1981), allowing more time for absorption.

Other mechanisms in the GI tract indicate that minor reductions in digestive enzymes should have little or no effect on nutrient assimilation. The enzymatic secretion from the pancreas is so great that 85–90% of exocrine pancreatic function has to be lost before homeostatic mechanisms break down and pancreatic insufficiency is expressed (Dworken, 1974). The overabundance of digestive enzymes, coupled with the enormous absorptive capacity and the adaptive potential of the intestine may explain why malabsorption measured over short periods *in vitro* or for only specific regions within the gut do not interfere significantly with nutrient assimilation by the entire intestine.

Regional defects without compensation

While compensatory mechanisms command consideration in interpretations of the role of localized malabsorptive defects on the pathogenesis of enteric infections, it should also be noted that the host's intestine may not be able to adapt to compensate for malabsorption of nutrients whose uptake is restricted to a specific region, such as vitamin B_{12} which is absorbed in the ileum. 'Isolated' damage to the ileum could lead to vitamin B_{12} deficiency. Interestingly, vitamin B_{12} is one of the few dietary nutrients whose impaired uptake can lead to an identifiable deficiency in parasitized hosts. A small percentage of human infections with the fish tapeworm, *Diphyllobothrium latum*, involves B_{12} deficiency expressed as a pernicious-like anaemia and caused by the absorption of the vitamin by the parasite (von Bonsdorff, 1956). *Diphyllobothrium* can break the B_{12}-intrinsic factor complex in the gut and absorb the free vitamin.

Malabsorption, metabolic changes and reduced food intake as contributors to weight loss

Malabsorption along with altered metabolism and decreased food intake are believed to be the major factors contributing to weight loss or poor weight gain in hosts infected with enteric parasites. Upon evaluating the relative importance of these three factors, the preponderance of evidence favours the conclusion that the primary cause of weight loss is decreased food consumption (Symons, 1976a, 1976b, 1989). An argument favouring this conclusion stems from the observation that parasites which live in the abomasum, small intestine, distal ileum and large bowel of ruminants all cause weight loss. Direct evidence in support of Symon's conclusion comes from paired feeding experiments involving parasitized hosts and from studies of infected, parenterally fed hosts (Castro et al., 1979). Rats infected with enteric stages of the nematode *T. spiralis* and placed on an oral liquid diet in which glucose was the major caloric source, lost over 20% of their starting body weight during the first week of infection. In contrast, uninfected controls gained ~7% of their starting weight. If the infected rats were fed by infusion of the liquid diet intraduodenally so that their food intake was isocaloric with

Table 11.4. Body weights of rats on a stock or liquid diet as influenced by intestinal infection with *Trichinella spiralis*[a].

Feeding regimen	Mean (±S.E.) body weight (g)[b]		%change
	Initial	Final	
Stock diet			
Control (n = 5)	224 ± 5	279 ± 8[c]	+14.0 ± 0.8
Infected (n = 5)	240 ± 1	188 ± 5[c,d]	−21.8 ± 1.7[d]
Liquid diet (orally)			
Control (n = 6)	224 ± 4	237 ± 4[c]	+ 6.8 ± 2.0
Infected (n = 6)	230 ± 4	181 ± 7[c]	−21.3 ± 2.8[d]
Liquid diet (introduodenally)			
Control (n = 6)	210 ± 3	214 ± 3	+ 2.7 ± 2.4
Infected (n = 6)	210 ± 6	208 ± 4	− 0.7 ± 3.6
Liquid diet (intravenously)			
Control (n = 5)	208 ± 2	219 ± 6	+ 5.4 ±10.5
Infected (n = 5)	201 ± 4	215 ± 7	+ 6.8 ± 5.9

[a] From Castro *et al.*, 1979.
[b] Initial body weight was taken when the specific feeding regimen was started; final weight was taken 7 or 8 days later.
[c] Indicates significant difference ($P < 0.06$) compared with initial weight.
[d] Indicates significant difference as compared with weight of respective control.

uninfected rats, then body weight paralleled that of uninfected counterparts (Table 11.4). This occurs despite evidence of a reduced capacity of infected animals to absorb glucose and indicates that reduced food intake, rather than decreased absorption or altered metabolism, accounts for the failure of parasitized animals to gain weight.

The perceived importance of reduced food intake in contributing to weight loss has focused attention on anorexia and the physiology of appetite control in this process. Future advances in the understanding of how malnutrition and associated sequelae (Figure 11.13) can be controlled may lie in the ability to control appetite, with the proviso that adequate nutrients are available for assimilation.

11.7. CONCLUSIONS

This chapter is organized primarily to impart an appreciation of how parasites affect structure-function relationships and regulatory processes involved in propulsion of food through the GI tract, nutrient assimilation, and ion and fluid transport. It should be evident that pathological changes in these processes induced by different parasites are often similar. This similarity is due to the fact that the changes relate more to the host response than to the specific actions of parasites *per se*. For example, toxins with specific modes of action which play a role in bacterial infections are of no known consequence in infections with enteric protozoa or helminths. Since the pathogenicity of enteric parasites depends predominantly on direct contact with the mucosa, inflammation, either non-specifically induced or immunologically mediated, plays a pivotal role in causing anatomical, biochemical, and physiological lesions in epithelium and smooth muscle. Despite a growing knowledge of gut alterations caused by parasites, it is still difficult to determine their significance to the health and well being of the host. In order to intervene effectively on the side of the host, an observer must decide cautiously and with all faculties whether a parasite-induced change is harmful (contributing to the pathogenesis of disease), beneficial (contributing to adaptation of the host to the presence of the parasite), or without effect. It is argued, therefore, that the most accurate decisions will rest on the most thorough understanding of the integrative aspects of enteric parasitism.

REFERENCES

Ackert, J.E., Edgar, S.A. and Frick, L.P. (1939), Goblet cells and age resistance of animals to parasitism, *Transactions of the American Microscopic Society*, **58**, 81–9.

Alizadeh, H., Castro, G.A. and Weems, W.A. (1987), Intrinsic jejunal propulsion in guinea pigs during parasitism with *Trichinella Spiralis*, *Gastroenterology*, **93**, 784–90.

Anderson, N., Blake, R. and Titchen, D.A., (1976a), Effects of a series of infection of *Ostertagia circumcincta* on gastric secretion of sheep. *Parasitology*, **72**, 1–2.

Anderson, N., Hansky, J. and Titchen, D.A., (1976b), Hypergastrinemia during parasitic gastritis in sheep, *Journal of Physiology* (London), **256**, 51P–52P.

Augustine, P.C. and Danforth, H.D. (1985), Effects of hybridoma antibodies on invasion of cultured cells by sporozoites of *Eimeria*, *Avian Diseases*, **29**, 1212–23.

Augustine, P.C. and Danforth, H.D., (1987), Use of monoclonal antibodies to study the invasion of cells by *Eimeria* sporozoites, in Agabian, N., Goodman, H. and Nogueira, N. (Eds.), *Molecular Strategies of Parasitic Invasion*, pp. 511–20, New York: Allan R. Liss.

Baird, A.W., Cuthbert, A.W. and Pearce, F.L. (1985), Immediate hypersensitivity reactions in epithelia from rats infected with *Nippostrongylus brasiliensis*, *British Journal of Pharmacology*, **85**, 787–95.

Baker, R.D., Searle, G.W. and Nunn, A.S. (1960), Glucose and sorbose absorption at various levels of rat small intestine, *American Journal of Physiology*, **200**, 301–4.

Barker, I.K. (1973), Scanning electromicroscopy of the duodenal mucosa of lambs infected with *Trichostrongylus colubriformis*, *Parasitology*, **67**, 307–14.

Beisel, W.R. (1982), Synergism and antagonism of parasitic diseases and malnutrition, *Reviews in Infectious Diseases*, **4**, 746–50.

Bell, R.G., Adams, L.S. and Ogden, R.W. (1984), Intestinal mucus trapping in the rapid expulsion of *Trichinella spiralis* by rats, Induction and expression analyzed by quantitative worm recovery, *Infection and Immunity*, **45**, 267–72.

Berry, C.R., Merrit, A.M., Burrows, C.F., Campbell, M. and Drudge, J.H. (1986), Evaluation of the myoelectrical activity of the equine ileum infected with *Strongylus vulgaris* larvae, *American Journal of Veterinary Research*, **47**, 27–30.

Bienenstock, J., Perdue, M., Stanisz, A. and Stead, R. (1988), Neurohumoral regulation of gastrointestinal immunity, *Gastroenterology*, **93**, 1431–4.

Bloch, K.J., Bloch, D.B., Stearns, M. and Walker, W.A. (1979), Intestinal uptake of macromolecules. VI. Uptake of protein antigen *in vivo* in normal rats and in rats with *Nippostrongylus brasiliensis* or subjected to mild systemic anaphylaxis, *Gastroenterology*, **77**, 1039–44.

Blumenthal, D.S. and Schultz, M.G. (1975), Incidence of intestinal obstruction in children infected with *Ascaris lumbricoides, American Journal of Tropical Medicine and Hygiene*, **24**, 801–5.

Bragg, P.D. and Reeves, R.E. (1962), Studies on the carbohydrate metabolism of a gram-negative anaerobe (*Bacteroides symbiosus*) used in the culture of *Entamoeba histolytica, Journal of Bacteriology*, **83**, 76–84.

Byrne, J.H. and Schultz, S.G. (1988), *An Introduction to Membrane Transport and Bioelectricity*, pp.39–49. New York: Raven Press.

Burks, T.F. (1981), Actions of drugs on gastrointestinal motility, in Johnson L.R., (Ed.), *Physiology of the Gastrointestinal Tract*, **1st ed., Vol. 2**, pp.495–516, New York: Raven Press.

Carrico, C.J., Meakins, J.L., Marshall, J.C., Fry, D. and Maier, R.V. (1986), Multiple-Organ-Failure Syndrome, *Archives of Surgery*, **121**, 196–208.

Castro, G.A. (1976), in Van den Bossche, H. (Ed.), *Biochemistry of Parasites and Host–Parasite Relationships*, pp.343–58, Amsterdam: North Holland Publishing Company.

Castro, G.A. (1981), Physiology of the gastrointestinal tract in the parasitized hosts. in Johnson, L.R., *Physiology of the Gastrointestinal Tract*, **1st ed., Vol. 1**, pp.1381–406, New York: Raven Press.

Castro, G. (1982), Immunological regulation of epithelial function, *American Journal of Physiology*, **243**, G321–G329.

Castro, G.A. (1988), Parasitic infections and gastrointestinal motility, in Wood, J.D. (Ed.), *Handbook of Physiology: Gastrointestinal System*, Section 6, Vol. 1, pp. 1133–52, Bethesda: The American Physiological Society.

Castro, G.A. (1989a), Immunophysiology of enteric parasitism, *Parasitology Today*, **5**, 11–19.

Castro, G.A. (1989b), Intestinal Neuroimmune interactions, in Singer, M.V. and Goebell, H. (Eds.), *Nerves and the Gastrointestinal Tract, Falk Symposium No. 50*; pp. 287–97, Lancaster: MTP Press.

Castro, G.A. (1989c), Gut Immunophysiology: common functions for a common mucosal immune system, *News in Physiological Sciences*, **4**, 59–64.

Castro, G.A. and Harari, Y. (1982), Intestinal epithelial membrane changes in rats immune to *Trichinella spiralis, Molecular and Biochemical Parasitology*, **6**, 191–204.

Castro, G.A., Olson, L.J. and Baker, R.D. (1967), Glucose malabsorption and intestinal histopathology in *Trichinella spiralis*-infected guinea pigs, *Journal of Parasitology*, **53**, 595–612.

Castro, G.A., Hessel, J.J. and Whalen G. (1979), Altered intestinal fluid movement in response to *Trichinella spiralis* in immunized rats, *Parasite Immunology*, **1**, 259–66.

Castro, G.A., Harari, Y. and Russell, D.A. (1987), Mediators of anaphylaxis-induced ion transport changes in small intestine, *American Journal of Physiology*, **253**, G540–G548.

Castro, G.A., Behnke, J.M. and Weisbrodt, N.W. (1989), Hookworm infection and malabsorption: Where do we stand? in Schad, G.A. and Warren, K.S. (Eds.), *Hookworm Disease: Current Status and New Directions*, London: Taylor and Francis Ltd., in press.

Castro, G.A., Copeland, E.M., Dudrick, S.J. and Ramaswamy, K. (1979), Enteral and parenteral feeding to evaluate malabsorption in intestinal parasitism, *American Journal of Tropical Medicine and Hygiene*, **28**, 500–7.

Castro, G.A., Badial-Aceves, F., Smith, J.W., Dudrick, S.J. and Weisbrodt, N.W. (1976), Altered small bowel propulsion associated with parasitism, *Gastroenterology*, **71**, 620–5.

Clayton, H.M. and Duncan, J.L. (1977), Experimental *Parascaris equorum* infection of foals, *Research in Veterinary Sciences*, **23**, 109–14.

Cook, R.W., Williams, J.F. and Lichtenberger, L. (1981), Hyperplastic gastropathy in the rat due to *Taenia taeniaformis* infection: parabiotic transfer and hypergastrinemia, *Gastroenterology*, **80**, 728–34.

Crane, R.K. (1975), in Jones, A. and Hodge, J. (Eds.), *Physiological Effects of Food Carbonhydrates*, pp.2–19, Washington, DC: American Chemical Society.

Crompton, D.W.T. (1984), Influence of parasitic infection on food intake. *Federation Proceedings*, **43**, 239–45.

Crompton, D.W.T. and Singhi, A. (1982), Intestinal responses of rats to *Moniliformis* (Acanthocephala), *Parasitology*, **84**: 1XX (abstract).

Dakkak, A., Fioramonti, J. and Buena, L. (1982), *Haemonchus contortus*: Abomasal transmural potential difference and permeability changes associated with experimental infection in sheep, *Experimental Parasitology*, **53**, 209–16.

Dembinski, A.B., Johnson, L.R. and Castro, G.A. (1979a), Influence of parasitism on secretin-inhibited gastric secretion, *American Journal of Tropical Medicine and Hygiene*, **28**, 854–9.

Dembinski, A.B., Johnson, L.R. and Castro, G.A. (1979b), Influence of enteric parasitism on hormone-regulated pancreatic secretion in dogs, *American Journal of Physiology*, **237**, R232–8.

Dobson, C. (1967), Changes in the protein of the serum and intestinal mucus of sheep with reference to the histology of the gut and immunological response to *Oesophagostomum columbianum* infections, *Parasitology*, **57**, 201–19.

Dobson, C. and Bawdin, R.J. (1974), Studies on the immunity of sheep to *Oesophagostomum columbianum*: Effects of low protein diet in resistance to infection and cellular reactions in the gut, *Parasitology*, **69**, 239–55.

Domingo, E.O. and Warren, K.S. (1969), Pathology and pathophysiology of the small intestine in murine schistosomiasis mansoni, including a review of the literature, *Gastroenterology*, **56**, 231–40.

Dowling, R.H. (1973), The influence of luminal nutrition on interstinal adaptation after small bowel resection and by-pass, in Dowling, R.H and Riecken, E.O. (Eds.), *Intestinal Adaptation*, pp.35–46, Stuttgart: F.K. Schattauer Verlag.

Dowling, R.H. and Riecken, E.O. (Eds.), (1973), *Intestinal Adaptation*, Stuttgart: F.K. Schattauer Verlag, 271 p.

Dworken, H.J. (1974), *The Alimentary Tract*, pp.251–75, Philadelphia: W.B. Saunders Co.

Egwang, T.G., Befus, A.D. and Gauldie, J. (1985), Activation of alveolar macrophages following infection with the parasite nematode *Nippostrongylus brasiliensis*, *Immunology*, **54**, 581–8.

Elson, C.O., Kagnoff, M.F., Fiocchi, C., Befus, A.D., and Targan, S. (1986), Intestinal immunity and inflammation: recent progress, *Gastroenterology*, **91**, 746–68.

Ferguson, A. and Jarrett, E.E. (1975), Hypersensitivity reactions in the small intestine. I. Thymus dependence of experimental villous atrophy, *Gut*, **16**, 114–7.

Frick, L.P. and Ackert, J.E. (1948), Further studies on duodenal mucus as a factor in age resistance of chickens to parasitism, *Journal of Parasitology*, **34**, 192–201.

Ganguly, N.K., Garg, U.C., Mahajan, R.C., Kanwaj, S.S., Rai, N. and Walia, B.N.S. (1987), Intestinal brush border calmodulin: Key role in the regulation of

NaCl transport in *Giardia lamblia* infected mice, *Biochemistry International*, **14**, 249–56.

Halliday, G.J. and Mulligan, W. (1968), Parasitic hypoalbuminemia: studies on type II ostertagiasis in cattle, *Research in Veterinary Science*, **9**, 224–30.

Halliwell, B. (1987), Oxidents and human disease: some new concepts, *FASEB Journal*, **1**, 358–64.

Harari, Y. and Castro, G.A. (1983), Sialic acid deficiency in lectin-resistant intestinal brush border membranes from rats following the intestinal phase of trichinellosis, *Molecular and Biochemical Parasitology*, **9**, 73–81.

Harari, Y. and Castro, G.A. (1985), Phosphatidyl ethanolamine methylation in intestinal brush border membrane from rats resistant to *Trichinella spiralis*, *Molecular and Biochemical Parasitology*, **15**, 317–26.

Harari, Y. and Castro, G.A. (1988), Evaluation of a possible functional relationship between chemical structure of intestinal brush border and immunity *Trichinella spiralis* in the rat, *Journal of Parasitology*, **74**, 244–288.

Harari, Y., Russell, D.A. and Castro, G.A. (1987), Anaphylaxis mediated Cl$^-$ secretion and parasite rejection in rat intestine, *Journal of Immunology*, **138**, 1250–5.

Hessell, J., Ramaswamy, K. and Castro, G.A. (1982), Reduced hexose transport by enterocytes associated with rapid, noninjurious rejection of *Trichinella spiralis* from immune rats, *Journal of Parasitology*, **68**, 202–7.

Holdsworth, C.D. and Dawson, A.M. (1964), The absorption of monosaccharides in man, *Clinical Science*, **27**, 371–9.

Hoskins, L.C., Winawer, S.J., Broitman, S.A. (1967), Clinical giardiasis and intestinal malabsorption, *Gastroenterology*, **53**, 265–79.

Hunt, J.N. and Knox, M.T. (1968), Regulation of gastric emptying. in Code, C.F. (Ed.), *Handbook of Physiology, Section 6, Alimentary Canal*, **Vol. 4**, 1917–35, Washington: American Physiological Society.

Johnson, C.V. (1965), Malabsorption syndromes, clinical and theoretical considerations, *Postgraduate Medicine*, **37**, 667–76.

Kalkofen, U.P. (1974), Intestinal trauma resulting from the feeding activities of *Ancylostoma caninum*, *American Journal of Tropical Medicine and Hygiene*, **23**, 1046–53.

Kane, M.G., Luby, J.P. and Krejs, G.J. (1984), Intestinal secretion as a cause of hypolcalemia and cardiac arrest in a patient with strongyloidiasis, *Digestive Diseases and Sciences*, **29**, 768–72.

Kaplan, B.S., Uni, S., Aikawa, M., Mahmoud, A.A.F. (1985), Effector mechanism of host resistance in murine giardiasis: specific IgG and IgA cell-mediated toxicity, *Journal of Immunology*, **134**, 1975–81.

Larsh, J.E. (1947), The relationship in white mice of intestinal emptying time and natural resistance to *Hymenolepis*, *Journal of Parasitology*, **33**, 79–84.

Larsh, J.E., Jr. (1963), Experimental trichinosis, in Dawes, B. (Ed.), *Advances in Parasitology*, pp.213–86, New York: Academic Press.

Layrisse, R. and Vargas, A. (1975), Nutrition and intestinal parasitic infection, *Progress in Food and Nutrition Science*, **1**, 645–667.

Lee, G.B. and Ogilvie, B.M. (1982), The intestinal mucus layer in *Trichinella spiralis* infected rats, in Strober, W., Hanson, L.A. and Sell, K.W. (Eds.), *Recent Advances in Mucosal Immunity*, pp.319–29, New York: Raven Press.

Leid, R.W. and Williams, J.F. (1979), Helminth parasites and the host inflammatory system, in Florkin, M. and Sheer, B.T. (Eds.), *Chemical Zoology*, **Vol. XI**, pp.229–71, New York: Academic Press.

Lev, B., Ward, H., Kousch, G.T. and Pereira, M.E.A. (1986), Lectin activation of *Giardia lamblia* by host protease: A novel host-parasite interaction, *Science*, **232**, 71–3.

Lin, T.-M. and Olson, L.J. (1970), Pathophysiology of reinfection with *Trichinella spiralis* in guinea pigs during the intestinal phase, *Journal of Parasitology*, **56**, 529–39.

Lloyd, S. and Soulsby, E.J.L. (1978), The role of IgA immunoglobulins in the passive transfer of protection to *Taenia taeniaformis* in the mouse, *Immunology*, **34**, 939–45.

Lushbaugh, W.B., Hofbauer, A.F. and Pittman, F.E. (1984), Proteinase activities of *Entamoeba histolytica* cytotoxin, *Gastroenterology*, **87**, 17–27.

Manson-Smith, D.F., Bruce, R.G. and Parrott, D.M.V. (1979), Villous atrophy and expulsion of intestinal *Trichinella spiralis* are mediated by T cells, *Cellular Immunology*, **47**, 285–92.

Mayrhofer, G., Pugh, C.W. and Barclay, A.N. (1983), The distribution, ontogeny and origin in the rat of Ia-positive cells with dendritic morphology and of Ia antigen in epithelia, with special reference to the intestine, *European Journal of Immunology*, **13**, 112–22.

McGhee, J.R. and Mestecky, J. (1983), The secretory immune system, *Annals of the New York Academy of Sciences*. **409**, 896.

McGowan, K., Kane, A., Asarkof, N., Wicks, J., Guerina, V., Kellum, J., Baron, S., Gintzler, A.R., and Donowitz, M. (1983), *Entamoeba histolytica* causes intestinal secretion: Role of serotonin, *Science*, **221**, 762–4.

McLeod, R. and Remington, J.S. (1977), Influence of infection with *Toxoplasma* on macrophage function and role of macrophages in resistance to Toxoplasma, *American Journal of Tropical Medicine and Hygiene*, **26**, 170–86.

Meneghelli, U.G. (1985), Chaga's disease: A model of denervation in the study of digestive tract motility, *Brazilian Journal of Medical and Biological Research*, **18**, 255–64.

Miller, H.R.P. (1984), The protective mucosal response against gastro-intestinal nematodes in ruminants and laboratory animals, *Veterinary Immunology and Immunopathology*, **6**, 167–259.

Miller, H.R.P. (1987), Gastrointestinal mucus, a medium for survival and for elimination of parasitic nematodes and protozoa, *Parasitology*, **94**, S77–100.

Miller, H.R.P. and Nawa, Y. (1979), *Nippostrongylus brasiliensis*: Intestinal goblet cell response in adaptively immunized rats, *Experimental Parasitology*, **47**, 81–90.

Miller, H.R.P., Huntley, J.F. and Wallace, G.R. (1981), Immune exclusion and mucus trapping during the rapid expulsion of *Nippostrongylus brasiliensis* from primed rats, *Immunology*, **44**, 419–29.

Mims, C.A. (1976), *The Pathogenesis of Infectious Disease*, London: Academic Press, 246 p.

Murray, M. (1972), Immediate hypersensitivity effector mechanisms II. *In vivo* reactions, in Soulsby, E.J.L. (Ed.), *Immunity to Animal Parasites*, pp.155–90, New York: Academic Press.

Naegraith, B.G. (1963), Pathogenesis and pathogenic mechanisms in protozoal diseases with special reference to amebiasis and malaria, in Garnham, P.C.C., Pierce, A.E. and Roitt, I. (Eds.), *Immunity to Protozoa; A Symposium of The British Society for Immunology*, pp.48–65, Oxford: Blackwell Scientific Publications.

Nemeth, P.R., Kwee, D.J. and Weisbrodt, N.W. (1981), Adaptation of intestinal muscle in continuity after jejunoileal bypass in the rat. *American Journal of Physiology*, **241** (*Gastrointestinal Liver Physiology*, **4**), G259–63.

Nolla, H., Bristol, J.R. and Mayberry, L.F. (1985), *Nippostrongylus brasiliensis*: Malabsorption in experimentally infected rats, *Experimental Parasitology*, **59**, 180–4.

Nussenzweig, R.S. (1982), Parasitic disease as a cause of immunosuppression, *The New England Journal of Medicine*, **306**, 423–4.

Okado, M., Ri, S., Omae, T., Fuchigami, T. and Kohrogi, N. (1983), The BTPABA pancreatic function test in giardiasis, *Postgraduate Medical Journal*, **59**, 79–82.

Olson, L.J. and Schultz, C.W. (1963), Nematode induced hypersensitivity reaction in guinea pigs: onset of eosinophilia and positive Schultz-Dale reactions following graded infections with *Toxocara conis*, *Annals of the New York Academy of Science*, **113**, 440–55.

Palmer, J.M. and Castro, G.A. (1986), Anamnestic stimulus-specific myoelectric responses associated with intestinal immunity in the rat, *American Journal of Physiology*, **250**, G266–73.

Pappas, P.W. and Read, C.P. (1972), Inactivation of α and β-chymotrypsin by intact *Hymenolepis diminuta*, (Cestoda), *Biological Bulletin*, **143**, 605–16.

Phillips, S.F. (1972), Diarrhoea: A current view of the pathophysiology, *Gastroenterology*, **63**, 495–518.

Rogers, W.P. (1962), *The Nature of Parasitism*, New York: Academic Press, 287 p.

Ross, J.G. and Dow, C. (1965), The course and development of the abomasal lesions in calves experimentally infected with the nematode *Ostertagia ostertagi*, *British Veterinary Journal*, **121**, 228–33.

Russell, D.A. (1986), Mast cells in the regulation of intestinal electrolyte transport, *American Journal of Physiology*, **14**, G253–62.

Russell, D.A. and Castro, G.A. (1985), Anaphylactic-like reaction of small intestinal epithelium in parasitized guinea pigs, *Immunology*, **54**, 573–9.

Russell, D.A. and Castro, G.A. (1987), Physiology of the gastrointestinal tract in the parasitized host, in Johnson, L.R. (Ed.), *Physiology of the Gastrointestinal Tract*, 2nd ed., **Vol. 2**, pp.1749–80, New York: Raven Press.

Russell, D.A. and Castro, G.A. (1989), Immunological regulation of colonic ion transport, *American Journal of Physiology*, **256**, G396–G403.

Schmidt, G.D. and Roberts, L.S. (1985), *Foundations of Parasitology*, St. Louis: Times Mirror/Mosby College Publishing 775 p.

Schofield, A.M. (1977), Intestinal absorption of hexose in rats infected with *Nippostrongylus brasiliensis*, *International Journal of Parasitology*, **7**, 159–65.

Sheehy, T.W., Meroney, W.H., Cox, R.S., Jr. and Soler, J.E. (1962), Hookworm disease and malabsorption, *Gastroenterology*, **42**, 148–56.

Solomons, N.W. and Keusch, G.T. (1981), Nutritional implications of parasitic infections, *Nutrition Reviews*, **39**, 149–61.

Sprinz, H. (1962), Morphological responses of intestinal mucosa to enteric bacteria and its implication for sprue and Asiatic cholera, *Federation Proceedings*, **21**, 57–64.

Stephenson, L.S., Pond, W.G., Nesheim, M.C., Krock, L.P. and Crompton, D.W.T. (1980), *Ascaris suum*: Nutrient absorption, growth and intestinal pathology in young pigs experimentally infected with 15-day old larvae, *Experimental Parasitology*, **49**, 15–25.

Stephenson, L.S. (1987), *Impact of Helminth Infections on Human Nutrition*, London: Taylor and Francis, 233p.

Stirewalt, M.A. (1963), Chemical biology of secretions of larval helminths, *Annals of New York Academy of Science*, **113**, 36–53.

Symons, L.E.A. (1957), Pathology of infestation of the rat with *Nippostrongylus muris* (Yokogawa). I. Changes in the water content, dry weight and tissue of the small intestine, *Australian Journal of Biological Sciences*, **10**, 374–83.

Symons, L.E.A. (1960a), Pathology of infestation of the rat with *Nippostrongylus muris* (Yokogawa) V. Protein digestion, *Australian Journal of Biological Sciences*, **13**, 578–83.

Symons, L.E.A. (1960b), Pathology of infestation of the rat with *Nippostrongylus muris* (Yokogawa) IV. The absorption of glucose and histidine, *Australian Journal* of Biological Sciences, **13**, 180–7.

Symons, L.E.A. (1961), Pathology of infestation of the rat with *Nippostrongylus muris* (Yokogawa) VI. Absorption *in vivo* from the distal ileum, *Australian Journal of Biological Sciences*, **14**, 165–71.

Symons, L.E.A. (1965), Kinetics of the epithelial cells and morphology of villi and crypts in the jejunum of the rat infected by the nematode *Nippostrongylus brasiliensis*, *Gastroenterology*, **49**, 158–68.

Symons, L.E.A. (1976a), Scanning electron microscopy of the jejunum of the rat infected by the nematode *Nippostrongylus brasiliensis*, *International Journal of Parasitology*, **6**, 107–111.

Symons, L.E.A. (1976b), Malabsorption, in Soulsby, E.J.L. (Ed.), *Pathophysiology of Parasitic Infection*, pp.11–21, New York: Academic Press.

Symons, L.E.A. (1989), *Pathophysiology of Endoparasitic Infection Compared with Ectoparasitic Infestation and Microbial Infection*, New York: Academic Press, 331p.

Symons, L.E.A. and Fairbairn, D. (1962), Pathology, absorption, transport and activity of digestive enzymes in rat jejunum parasitized by the nematode *Nippostrongylus brasiliensis*, *Federation Proceedings*, **21**, 913–8.

Symons, L.E.A. and Jones, W.O. (1970), *Nematospiroides dubius*, *Nippostrongylus brasiliensis* and *Trichostrongylus columbriformis*: protein digestion in infected mammals, *Experimental Parasitology*, **27**, 496–506.

Symons, L.E.A., Gibbins, J.R. and Jones, W.O. (1971), Jejunal malabsorption in the rat infected by the nematode *Nippostrongylus brasiliensis*, *International Journal of Parasitology*, **1**, 179–87.

Takeuchi, A. and Sprinz, H. (1967), Electron-microscope studies of experimental salmonella infection in the preconditioned guinea pig. II. Response of the intestinal mucosa to the invasion of *Salmonella typhimurium*, *American Journal of Pathology*, **51**, 137–43.

von Bonsdorff, B. (1956), Parasitological review: *Diphyllobothrium latum* as a cause of pernicious anemia, *Experimental Parasitology*, **5**, 207–30.

Wakelin, D. (1984), *Immunity to Parasites*, London: Edward Arnold, p. 165.

Walker, W.A., Isselbacher, K.J. and Block, K.J. (1974), Immunologic control of soluble protein absorption from the small intestine: A gut-surface phenomenon, *American Journal of Clinical Nutrition*, **27**, 1434–40.

Wells, P.D. (1962), Mast cell, eosinophil and histamine levels on *Nippostrongylus brasiliensis* infected rats, *Experimental Parasitology*, **12**, 82–101.

Whalen, G.E., Strickland, G.T., Cross, J.H., Uylangco, C., Rosenberg, E.B., Gutman, R.A., Watten, R.H. and Dizon, J.J. (1969), *Intestinal Capillariasis A New Disease in Man*, *The Lancet*, **1**, 13–16.

Whitmire, W.M., Kyle, J.E., Speer, C.A. and Burgess, D.E. (1988), Inhibition of penetration of cultured cells by *Eimeria bovis* sporozoites by monoclonal immunoglobulin G antibodies against the parasite surface protein P20, *Infection and Immunity*, **56**, 2538–43.

Williams, D.M., Sawyer, S. and Remington, J.S. (1976), Role of activated macrophages in resistance of mice to infection with *Trypanosoma cruzi*, *Journal of Infectious Disease*, **134**, 610–4.

Wilson, T.H. (1962), *Intestinal Absorption*, Philadelphia: W.B. Saunders Co., 263 p.

Wing, E.J. and Remington, J.S. (1978), Role of activated macrophages in resistance against *Trichinella spiralis*, *Infection and Immunity*, **21**, 398–404.

World Health Organization (1980), Parasite-related diarrhoeas, *Bulletin of the World Health Organization*, **58**, 819–30.

12. Acquired immunity and epidemiology

R.J. Quinnell and A.E. Keymer

12.1. THE EPIDEMIOLOGY OF PARASITIC INFECTIONS

Introduction

Epidemiology is the study of the population dynamics of parasites. A parasite can be defined as an organism that lives in a close, obligatory association with a host of a different species for at least part of its life; the parasite depends on the host for one or more essential nutrients, and the fitness of the host is thereby decreased. Genetically based defense mechanisms are thus a likely consequence of parasitism. Some defenses are fixed or innate, others are facultative and dependent upon exposure to infection. The presence of facultative defenses gives rise to important differences between the population dynamics of parasites and free-living organisms.

Acquired immunity is the most common form of facultative defense to result in a decrease in parasite fitness as a function of exposure to infection. Many of the epidemiological patterns which are attributed to acquired immunity could, however, be equally well explained by other host-defense mechanisms or by parasite-induced host pathology. Assessment of the actual (rather than the possible) role of acquired immunity will remain a problem with the interpretation of epidemiological data until greater coordination between immunology and epidemiology is achieved. Whilst the molecular and cellular biology of acquired immunity to many parasitic infections is relatively well understood, the epidemiological consequences of the responses are not. In contrast, several well-characterized epidemiological patterns seem likely to have immunological explanations, but the mechanisms responsible have yet to be determined.

Parasites can be broadly categorized into two classes, micro-parasites and macroparasites (Anderson and May, 1979). *Microparasites* (viruses, bacteria, protozoa) are characterized by short generation times and high rates of direct reproduction within the host. *Macroparasites* (helminths and arthropods) are much larger than microparasites, have longer generation times and do not usually multiply directly within the host. The two classes also differ in the dynamics of the immune response they elicit. Immunity to microparasites, particularly viruses and bacteria, can be complete and life-long and is usually maximal after only a single exposure to the parasite. In contrast, immunity to macroparasites is only rarely complete, is usually maximal only after repeated infection and declines in the absence of further exposure. Protozoa, although classed as microparasites, tend to elicit an immune response of the latter type.

The characteristics of host immunity are responsible for many of the differences in population dynamics which exist between the two parasite classes. Microparasitic infections, such as measles, are characterized by rapid changes in the prevalence of infection. Long-lasting, complete immunity in combination with a relatively low rate of host population growth results in a tendency for epidemic infection, with the possibility of stochastic extinction of the parasite if the host population size is small (Anderson and May, 1979). In contrast, incomplete, short-lasting immunity tends to result in the endemic patterns of infection typical of macroparasites (May and Anderson, 1979).

Both field and laboratory studies contribute to the assessment of the role of acquired immunity in parasite epidemiology. Immunization experiments, in which hosts are immunized, often by a drug-abbreviated infection, and challenged, allow direct assessment of the effects of induced immunity and have been favoured by those interested in the mechanisms of the response. Unfortunately, in many cases the intensity of the experimental single-point infection has exceeded natural levels by several orders of magnitude, so that the epidemiological relevance of the conclusions is unclear. Trickle infection experiments, in which hosts are repeatedly infected with small doses of parasites, aim to mimic natural infection. A combination of experimental procedures and appropriate mathematical models is the only method by which the relationship between parasite population processes and observed epidemiological patterns can be ascertained. The extent to which these processes are of relevance for the epidemiology of natural parasite infections can be determined directly if predictions made on the basis of experiments are tested using the limited manipulations appropriate to field studies. Unfortunately, the link between experimental and field epidemiology remains to be made in many cases.

This chapter is organized as follows. First, we discuss the possible ways in which acquired immunity may influence the population dynamics of macro-parasite infections. We then evaluate the relevant experimental evidence in laboratory and agricultural animals and discuss the possible consequences of immunity for human helminth infections and their control. The effects of immunity on microparasite epidemiology have been best characterized for bacterial and viral infections which are beyond the scope of this book. The role of immunity in protozoan epidemiology is discussed very briefly at the end of the chapter, together with the consequences of acquired immunity for mixed-species parasitic infections.

Epidemiology of helminth infection

Anderson and May (1985a) and Anderson (1986) have reviewed the features characteristic of the epidemiology of helminth parasites. They may be summarized as follows: *Stability*—helminth populations are in general remarkably stable as indicated by their response to perturbations of human or climatic origin (Figure 12.1a); *density-dependent constraints*—density-dependent constraints likely to be responsible for the observed population stability have been identified with respect to parasite fecundity, death rate and establishment (Figure 12.1b); *age-intensity profiles*—the average intensity of infection rises in the juvenile host age classes and then either remains at a

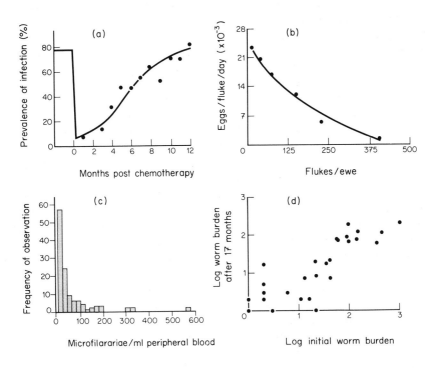

Figure 12.1. (a) Reinfection with *Ascaris lumbricoides* after chemotherapy in the human host (from Thein Hlaing, 1985); (b) Density-dependent fecundity of *Fasciola hepatica* in the sheep host (from Smith, 1984); (c) Frequency distribution of *Wuchereria bancrofti* in humans (from Croll *et al.*, 1982); (d) Predisposition to *Trichuris trichiura* infection in humans, as assessed by the comparison of initial worm burden and worm burden after 17 months of reinfection following chemotherapy (from Bundy and Cooper, 1988).

plateau or declines in the mature age classes, depending on the species of parasite and its transmission rate (See Figure 12.10); *overdispersion*—frequency distributions of the number of parasites per host are almost invariably highly aggregated (overdispersed) rather than random. Most individuals harbour a few or no parasites and only a small proportion are very heavily infected (Figure 12.1c). Often as few as 10% of the hosts carry as many as 70% of the parasites; *predisposition*—data from a number of studies suggest that individual hosts are predisposed to infection with either large or small numbers of worms (Figure 12.1d).

The possible consequences of acquired immunity for helminth epidemiology take two distinct (though interacting) forms, one of relevance to the average intensity of infection within the host population and the other to its variance. The effect of an immune response is to decrease the establishment, rate of development, survival and/or fecundity of the parasite in its host as the latter's exposure to infection increases. Immunity thus has the potential to act in a density-dependent manner to constrain parasite population growth and to promote parasite population stability. In contrast, variation in immunocompetence is likely to influence the distribution of parasites within the host population and may thereby affect stability by modifying the net outcome of density dependence. In Sections 12.2 and 12.3 we review the evidence for the density-dependent nature of acquired immunity and its impact on parasite aggregation.

12.2. IMMUNITY AND DENSITY-DEPENDENCE

Introduction

A schematic macroparasite life cycle is represented in Figure 12.2. Acquired immunity has the potential to regulate any of the population processes occurring within the host and might also affect the survival of the offspring forming the next parasite generation. Its overall epidemiological role is thus a combination of its effects on parasite establishment, development, survival and fecundity.

Establishment can be strictly defined as the proportion of the infecting parasites that successfully invade the host, regardless of whether or not they develop further. However, as immature parasitic stages are frequently very difficult to find, establishment is often assessed on the basis of the proportion of the infecting parasites surviving to maturity and thus encompasses some of the effects of immunity on larval survival. The effect of acquired immunity on establishment is clearly seen in *Haemonchus contortus* infections of sheep, where establishment requires infective third-stage larvae to exsheath in the rumen and then penetrate the mucosa. Penetration is reduced by over 90% in immunized sheep (Miller *et al.*, 1983).

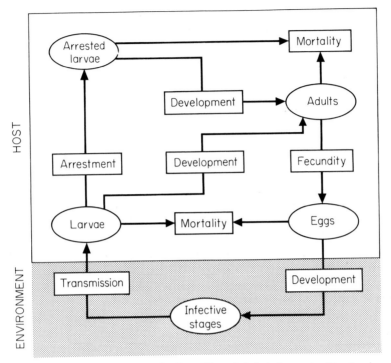

Figure 12.2. Schematic representation of a direct life cycle helminth parasite. Density-dependent constraints could operate on any of the population processes shown (square boxes).

Acquired immunity can act to increase the time taken for larval parasites to reach maturity, either by reducing the rate of development or by the induction of an arrested state. The mucosal stage of *Heligmosomoides polygyrus* (Chapter 8, Figure 8.7), for example, is prolonged in previously exposed mice (Bartlett and Ball, 1974). Arrested development has been demonstrated for many species of trichostrongyle nematode, and although the probability of arrestment is often determined by both larval external environment and genotype, the immunological status of the host also plays an important role (reviewed by Gibbs, 1986).

Acquired immunity has important effects on parasite survival. For example, the mortality rate of adult *H. polygyrus* in immunized NIH mice is an order of magnitude higher than in naive mice, 0.14 as opposed to 0.016/parasite/day (Figure 12.3; Behnke and Robinson, 1985; Robinson *et al.*, 1989).

An effect of immunity on parasite fecundity could come about via changes in the physiology or behaviour of adult worms. The effect is particularly severe in filarial infections, where hosts may become completely amicrofilaraemic as the result of acquired immunity; despite continued adult survival, females cease to liberate larvae (Partono, 1987).

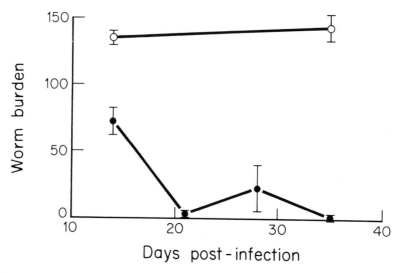

Figure 12.3. Survival of *Heligmosomoides polygyrus* in naive (○) and immunized (●) mice. Immunized mice were given an anthelminthic-abbreviated infection of 250L₃ 28 days before the challenge infection (data from Behnke and Robinson, 1985).

One or more of the above effects of immunity have been demonstrated in most species of parasite studied in laboratory and farm animals. Their relative importance in explaining patterns of parasite population dynamics, however, varies from species to species.

The regulatory role of acquired immunity

Factors other than immunity may well be involved in the density-dependent regulation of helminth population growth, including, for example: 1) Competition between parasites for resources, such as nutrients or space; 2) Inhibitory effects of parasites on each other ('crowding factors'); and 3) Parasite-induced pathology leading to a decrease in the suitability of the environment for parasite survival and reproduction.

Resource competition is probably the most widespread and significant of these factors (Keymer, 1982), although parasite-induced pathology may be of importance in trematode infections (Smith, 1984). Crowding factors have only been isolated from some cestode species (Zavras and Roberts, 1985).

Although these mechanisms are undoubtedly important, evidence for the regulatory role of acquired immunity is widespread. The *H. polygyrus*-laboratory mouse system is a particularly appropriate laboratory model in this context, as there is almost no effect of the immune response during primary infections because of the immunosuppression induced by adult worms (Behnke *et al.*, 1983). There appear to be no density-dependent constraints on the establishment, survival or fecundity of worms in primary infections in many strains of mice, and since immunosuppression seems to correlate with adult worm burden, inverse density-dependence may occur in, for instance, parasite survival (Robinson *et al.*, 1989). Repeated infection, however, induces an immune response which affects worm establishment, survival and fecundity and imposes a very effective density-dependent constraint on parasite population growth (Keymer and Hiorns, 1986). In contrast, *H. polygyrus* infection in malnourished mice, which lack a functional immune response, seems to be regulated only by parasite-mediated host mortality (Slater, 1988).

Acquired immunity can only regulate the parasite population if it acts in a density-dependent manner (for a review of density-dependent regulation see Sinclair, in press). The magnitude of an immune response is generally assumed to be dependent on the level of infection to which the host has been exposed (Figure 12.4), but the precise form of the relationship between the two is not clear. A sigmoid relationship seems likely; many studies suggest the existence of a threshold exposure, below which the effects of immunity are negligible, and the expression of the immune response is expected to reach a maximum above a certain level of antigenic stimulation (as will obviously be the case if immunity ever reaches 100%). Immunity can therefore act in a density-dependent manner, at least over a limited range of exposures. In some cases this range will be wide, although in others maximal immunity may develop after exposure to a very small infecting dose. Some strains of laboratory mouse, for example, become completely immune to reinfection with *Trichuris muris* after exposure to only ten infective eggs (Wakelin, 1973).

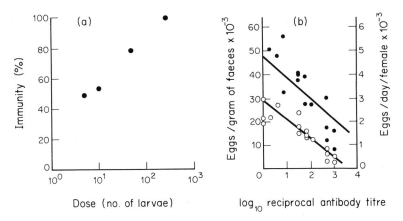

Figure 12.4. (a) The relationship between immunizing dose and immunity of female NIH mice to reinfection with *Heligmosomoides polygyrus* (data from Behnke and Robinson, 1985); (b) The relationship between anti-*H. polygyrus* antibody titre in immune mouse serum passively transferred to recipient Quakenbush mice and the fecundity of *H. polygyrus* on day 21 post-infection. ○ eggs/g faeces; ● eggs/female/day (from Dobson, 1982).

Epidemiological consequences of immunological memory

Acquired immunity differs from most other density-dependent processes in that it is not dependent on *present* parasite density but on the cumulative exposure of the host to previous infection. The presence of this anamnestic (memory) component has important effects on epidemiology which can be illustrated by the dynamics of trickle infection.

The simplest description of parasite population dynamics in a trickle infection is an immigration-death model, such that:

$$dM/dt = \Lambda - \mu M$$

where M is the number of parasites/host, Λ is the rate of infection and μ the rate of parasite mortality. In trickle infection, the value of Λ is experimentally determined. If μ is assumed constant, the number of parasites per host rises monotonically with time to a plateau (M^*) determined by the relative rates of infection and mortality ($M^* = \Lambda/\mu$; Figure 12.5a).

The effect of density-dependence in the rate of parasite mortality (caused, for example, by competition for nutrients or space) can be modelled by considering a simple linear relationship between mortality and worm burden, such that $\mu = a + bM$, where a and b are constants (Anderson and May, 1978). Under this assumption, M^* is reduced in proportion to the rate of infection, but the curves are still monotonic (Figure 12.5b).

An adaptation of this model illustrates the contrasting effect of exposure-dependent immunity. If immunity is assumed to be linearly related to exposure, E, where

$$E = \int_O^t \Lambda(s)ds,$$

and to act by increasing the rate of parasite mortality, then $\mu = c + dE$, where c and d are constants. This model generates convex age-intensity curves at high rates of infection (Figure 12.5c).

Trickle infections with a variety of helminth species have produced convex age-intensity curves (Barger *et al.*, 1985; Crombie and Anderson, 1985; Keymer and Hiorns, 1986; Slater and Keymer, 1986a). These demonstrations of convexity in experiments where the rate of infection is known to be constant thus imply the involvement of some process acting in a manner analogous to that of acquired immunity. Similarly, the convex age-intensity patterns observed in human helminth infection (where infection is variable) can be mimicked by models which incorporate *either* acquired immunity *or* age-dependent changes in the rate of infection.

Immunological memory to macroparasite infections appears to decline with time in the absence of reinfection (Figure 12.6). Unlike trickle experiments, in which the duration of the infection period tends to be short, the long-term dynamics of natural infection are likely to be affected by this decay in immunological memory, whereby the magnitude of the immune response at

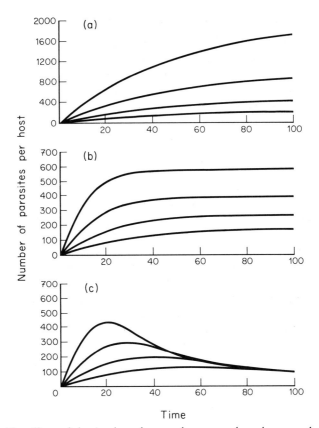

Figure 12.5. The effects of density-dependence and exposure-dependence on the relationship between infection intensity and duration of trickle infection. In each case the curves from bottom to top represent increasing rates of infection in the model described in the text. Other parameter values are arbitrary and constant throughout; (a) constant parasite mortality; (b) density-dependent parasite mortality; (c) exposure-dependent parasite mortality.

any one point in time is not a simple cumulative function of past exposure, but rather a cumulative function of past exposure weighted according to how long ago each exposure occurred. Though of direct significance for quantitative epidemiological studies, this characteristic of immunity plays little part in the qualitative conclusions drawn with respect to its regulatory role.

Mathematical models of immunity

A number of mathematical models of helminth epidemiology incorporating acquired immunity have been formulated (Anderson and May, 1985b; Crombie and Anderson 1985; Berding *et al.*, 1986, 1987; Roberts *et al.*, 1986, 1987; Grenfell *et al.*, 1987a, 1987b; Smith, 1988). Most of these models simplify the effects of immunity by constraining its action to a single developmental

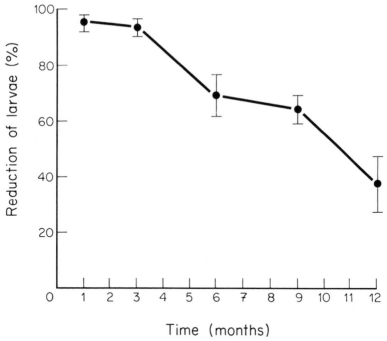

Figure 12.6. Percentage reduction in larvae of *Taenia hydatigena* in sheep at specified periods after immunization. Immune and control sheep were reinfected by grazing on egg-contaminated pasture for eight weeks (from Gemmell and Johnstone, 1981).

stage of the parasite, either a decrease in larval establishment or an increase in adult mortality. The models differ in their assumptions about the relationship between immunity and exposure (linear or sigmoid) and the decay of immunity with time (exponential or step function). They all produce, however, qualitatively similar results, of which the generation of convex age-intensity curves is a characteristic feature.

The models of Grenfell *et al.* (1987a) and Smith (1988) of *Ostertagia ostertagi* and *H. contortus* infections respectively (Chapter 8, Figure 8.7) are more detailed than most in that they consider the dynamics of each developmental stage in the host. The immunological processes regulating parasite population growth in these two parasites are similar but show important quantitative differences. *Ostertagia ostertagi* populations are regulated both by an increase in adult mortality and a gradual decrease in establishment, with a continual turnover of adult worms. In contrast, the establishment of *H. contortus* exhibits a rapid decrease, almost to zero. There is effectively no worm turnover and the surviving adult worms are slowly expelled in a dose-dependent manner (Figure 12.7). *Ostertagia ostertagi* populations are also regulated by severe density-dependent constraints on fecundity, such that egg output is largely independent of population size.

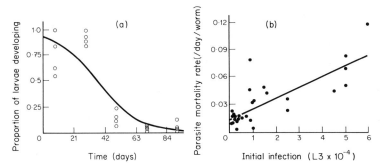

Figure 12.7. Effects of acquired immunity on the population dynamics of *Haemonchus contortus* (from Smith, 1988); (a) change in the proportion of larvae avoiding immune exclusion during a trickle infection; (b) rate of mortality of worms during a primary infection in relation to the initial level of infection.

12.3. IMMUNITY AND HETEROGENEITY IN INFECTION

Immunological causes of heterogeneity

Many factors can contribute to the generation of observed parasite frequency distributions. Immune responses that are uniform between hosts will tend to lead to random distributions. Any differences in immunocompetence, however, will increase the potential for aggregation (Anderson and Gordon, 1982). Heterogeneity in immunocompetence will result if some hosts are genetically unable to mount effective responses (Wakelin, 1986). Hosts may also lose their capacity to mount an immune response for a number of reasons. Loss of immunocompetence can be associated with malnutrition (especially protein malnutrition) or specific nutrient deficiencies (Bundy and Golden, 1987); other infections (see Section 12.7), stress, immunosuppressive drugs or reproductive status may also be involved (Section 13.2).

Through its tendency to create overdispersion, heterogeneity in immunocompetence will increase the severity of density-dependent constraints by raising the average intensity of infection experienced by each of the parasites (Figure 12.8). Parasites which are aggregated in a few hosts will necessarily be subject to, on average, more severe density-dependent constraints on their survival or fecundity than parasites which are randomly distributed. Though the mean parasite burden may be the same in each case, the mean number of parasites encountered sharing the same host *by each of the parasites* will be markedly different.

By creating overdispersion, therefore, immunological variation may increase the stability of parasite populations. All else being equal, however, acquired immunity cannot be responsible both for overdispersion *and also* for the density-dependent regulation of parasite fecundity (Keymer and Slater, 1987). If heterogeneity in immunocompetence is the cause of overdispersion, the

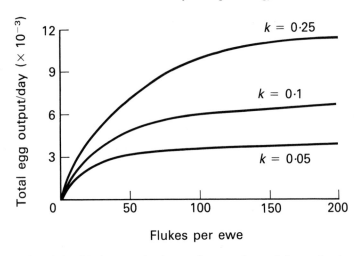

Figure 12.8. The relationship between the degree of aggregation and the predicted total daily egg production in a population of *Fasciola hepatica* (from Smith, 1984). The figure shows the total egg output per host in relation to the intensity of infection for three populations which differ in their level of aggregation. The negative binomial parameter, *k*, gives an inverse measure of aggregation. The differences result from variation in the *net* outcome of density-dependent constraints on fecundity, though the relationship between worm burden and *per capita* fecundity is the same in each case.

heavily infected hosts are by implication immunocompromized. Density-dependent reductions in parasite fecundity caused by immunity thus cannot occur in these hosts. The most obvious way round this paradox is to hypothesize that immunity against establishment or survival (creating overdispersion) and immunity against fecundity (creating density-dependence) are at least partially independent of each other within individual hosts. There is evidence for this from *Trichinella spiralis* (Chapter 6, Figure 6.2) infections in rodents (Grencis and Wakelin, 1983; Wassom *et al.*, 1984).

Genetic control of immunocompetence

Genetic heterogeneity among hosts in their immune response to parasitic infection has been found in all host-parasite systems so far examined (Wakelin and Blackwell, 1988; Grencis, this volume). The widespread presence of genetic heterogeneity implies that there are balanced polymorphisms of resistance genes in the wild (e.g. Wassom *et al.*, 1986). Anderson (1988) has reviewed the mechanisms which could be responsible for the maintenance of such polymorphism. The most important of these are heterozygous advantage and frequency- and density-dependent selection; these both depend on there being some selective disadvantage for resistant genotypes in the absence of infection.

This cost of resistance might arise in a number of ways, depending upon the mechanism of resistance and the way in which the response is controlled (Section 1.2). Unfortunately, apart from the well-known example of sickle-cell anaemia and resistance to *Plasmodium falciparum* in man (see Anderson, 1988), relevant data are extremely limited. The only direct evidence from helminth studies is the negative association found between the resistance of *Biomphalaria glabrata* to *Schistosoma mansoni* and the fitness of snails in the absence of infection (Minchella and Loverde, 1983). In contrast, sheep which have been selectively bred for nematode resistance do not seem to have a lowered productivity (reviewed by Albers and Gray, 1986). Unfortunately, most studies to date have not examined the relative performance of homo- and heterozygotes, though heterozygous advantage has been demonstrated in laboratory host-parasite interactions (Wakelin and Blackwell, 1988; Robinson *et al.*, 1989).

Although the evidence for genetic variation in immunocompetence is substantial, there have been few demonstrations that this variation is associated with overdispersion. The best characterized natural system is the host-parasite interaction between *Hymenolepis citelli* and *Peromyscus maniculatus* in which resistance is controlled by a single dominant gene (Wassom *et al.*, 1974). The frequency of the resistant genotype in natural *Peromyscus* populations is estimated to be 75%, and experimental infections consequently result in some degree of overdispersion (Figure 12.9). Much higher levels of overdispersion, however, are observed in wild populations, and it is clear that although host genetics could play an important role, other factors (such as spatial and behavioural heterogeneity) are also involved in parasite aggregation (Wassom *et al.*, 1986). Likewise, experiments on the transmission of *Heligmosomoides polygyrus* between laboratory mice have shown that even in protein malnourished (and thus uniformly immunocompromized) populations, the distribution of parasites is significantly overdispersed (Slater and Keymer, 1986b).

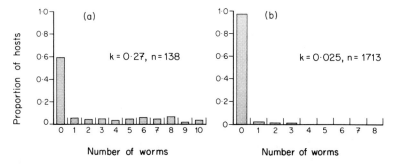

Figure 12.9. Frequency distributions of *Hymenolepis citelli* in *Peromyscus maniculatus* (data from Wassom *et al.*, 1986); (a) laboratory-infected mice; (b) naturally infected mice.

Secondary loss of immunocompetence

Survival within the low responder individuals in a population may be a common strategy for parasites which are rapidly expelled from immunocompetent hosts (Wakelin, 1984a). *Trichuris muris*, for example, is expelled from a majority of laboratory mice before patency, leaving an almost sterile immunity after very low levels of exposure (Wakelin, 1967, 1973). A survey of the intensity of *T. muris* in wild populations indicates that large worm burdens are present only in a few females, probably those with impaired immunocompetence as a result of pregnancy or lactation (Behnke and Wakelin, 1973).

The periparturient rise in faecal egg production noted in most agricultural animals can be regarded as a special case of the effect of immunological heterogeneity on nematode epidemiology, where the heterogeneity between individuals is synchronous through time. Pregnancy and lactation are known to inhibit the immune response to many helminths (Connan, 1970; O'Sullivan and Donald, 1973), resulting in the increased establishment, fecundity and survival of adult worms, as well as the activation and development of arrested larvae. This phenomenon is of particular significance for sheep production. Overwintered ewes have a large proportion of arrested larvae, their arrestment resulting from both environmental changes and the increasing immunity acquired by the sheep over the summer. In spring, the larvae resume development, due both to the genetically programmed termination of arrestment and also to the decay of host immunity in the absence of reinfection. In lactating ewes, this resumption of development is enhanced, as is the survival of adult worms, resulting in a periparturient rise in egg production. In addition, ewes also pick up overwintered larvae from the pasture. The result is an acute rise in pasture larval contamination, sufficient to cause significant mortality of newborn lambs.

12.4. HUMAN HELMINTH INFECTIONS

Introduction

Although the epidemiology of the common human helminth infections has been extensively investigated, the epidemiological significance of acquired immunity remains uncertain (Anderson and May, 1985a; Anderson, 1986; Behnke, 1987a; Bundy, 1988). This is largely the result of two procedural difficulties: the ethical preclusion of experimental intervention and the problems of estimating exposure in the field. An additional difficulty associated particularly with filarial infections is the absence of reliable methods for counting adult worms.

The observed features of human helminth epidemiology (Section 12.1) are all consistent with explanations based on acquired immunity (Anderson and May, 1985a; Anderson, 1985). Immunity has been shown to generate similar patterns in experimental animal infections (Section 12.2). The evidence for acquired immunity to human gastrointestinal helminth infections, however, is largely circumstantial, indirect support being provided by:

1. Comparison of age-intensity profiles between areas with differing transmission rates
2. Comparison of reinfection rates between age classes
3. Comparison of parasite frequency distributions between age classes
4. Immunological correlates of infection
5. Experimental infections.

All of these can provide only equivocal support unless the effect of variation in exposure rate can be controlled; this has so far proved possible only for schistosomiasis.

Age-intensity profiles

Age-intensity profiles in human helminth infection are often convex (see also Chapter 13, Figure 13.1), implying either that exposure is reduced in the older age classes, or that these age classes have a partial immunity. Parasite-induced host mortality is not, in most cases, a viable explanation. If acquired immunity is involved, the convexity of age-intensity profiles should be most extreme in areas with high transmission rates (Anderson and May, 1985b; Berding *et al.*, 1986). A comparison of age-intensity profiles for schistosome and hookworm infection in different communities suggests that this may indeed be the case (Anderson and May, 1985b; Anderson, 1986; Figure 12.10). As the rate of transmission increases (as assessed by the magnitude of the peak intensity), the profiles change from monotonic to convex in form. The alternative explanation, that an age-related decrease in transmission intensity in adults becomes more marked as overall transmission rates increase, seems much less likely.

Rates of reinfection

If acquired immunity is operative, rates of reinfection (relative to precontrol infection intensity) should be reduced in the older age classes (Anderson, 1986). This is clearly the case for schistosomes; the rate of reinfection with *Schistosoma haematobium* after chemotherapy in The Gambia was found to be 1000-fold greater in 5–8 year old children than in adults (Wilkins *et al.*, 1984). Results from studies of geohelminths, however, are much less conclusive. In some studies the rate of reinfection with *Ascaris lumbricoides* has been found to be lower in adults than children (for example Thein Hlaing *et al.*, 1987). However, in other studies reinfection has been found to be independent of age (Elkins *et al.*, 1988).

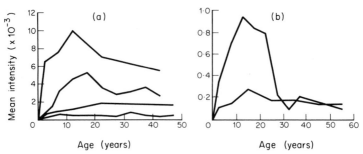

Figure 12.10. Age-related changes in the average intensity of infection (eggs/g faeces) in areas of high and low transmission (from Anderson and May, 1985b), (a) hookworm; (b) schistosomes.

Parasite frequency distributions

Examination of the parasite distribution patterns in different age classes suggests that there is typically an age-dependent decrease in aggregation, with a stable value reached in the adult age classes (Anderson, 1982; Bundy *et al.*, 1988). This pattern can be simulated by a model based on homogeneous immune response between individuals, in which overdispersion is generated by differences in exposure between hosts (Anderson and May, 1985b); as time progresses, heavily infected individuals become immune and thus progressively less heavily infected. This process could also account for the fact that predisposition is more easily demonstrated in younger age classes. However, if genetic heterogeneity in immunocompetence does exist, an increase in overdispersion with age might be expected, as illustrated by the results of trickle infections in laboratory animals (Keymer and Hiorns, 1986; Berding *et al.*, 1987).

Immunological correlates of infection

If acquired immunity is important in generating overdispersion and predisposition, it should be possible to correlate infection intensity with immunological, genetic or nutritional parameters. Direct immunological correlates of reinfection have so far been demonstrated only for schistosome infections (reviewed by Hagan, 1987). Comparison of human HLA antigen frequencies and intensity of infection with *Trichuris trichiura* and *A. lumbricoides* in a Caribbean population has, however, revealed associations between certain Class I and II alleles and infection (Bundy, 1988). Associations between HLA haplotype and response to *Schistosoma japonicum* antigens have also been demonstrated (Kojima *et al.*, 1984). There is also thought to be an association between low levels of plasma zinc (essential for cell-mediated immunity) and high *T. trichiura* burdens (Bundy and Golden, 1987).

Immunological correlates of infection in filariasis are likely to be complex, since immunity can act independently against microfilariae and adult worms.

There are known to be differences between individuals in their immune response to microfilariae of *Brugia malayi* and *Wuchereria bancrofti*, although whether these have a genetic basis is not known (reviewed by Maizels *et al.*, 1987; Piessens *et al.*, 1987). The variations are correlated with the presence or absence of microfilaraemia, and the results from animal models suggest that amicrofilaraemic individuals may be resistant to reinfection (Denham and Fletcher, 1987). Ironically, these individuals may also suffer the most severe pathology (Section 1.8).

Experimental infections

Experimental infections have been for the most part limited to hookworms. A number of studies on volunteer infections with *Necator americanus* have provided little evidence for acquired immunity despite antibody and cellular responses to worm antigen (Ball and Bartlett, 1969; Ogilvie *et al.*, 1978). However, the experiments have been uncontrolled, hookworm doses low and sample sizes very small, giving little scope for extrapolation to the field (Miller, 1979).

Immunology or ecology?

The problem with all the studies discussed above is the distinction between the effects of variability in exposure (between areas, age classes and individuals) and the putative effects of immunity (Warren, 1973). This problem has been to some extent overcome in long-term field studies of *Schistosoma mansoni* in Kenya and *S. haematobium* in The Gambia (reviewed by Butterworth and Hagan, 1987; Hagan, 1987). As schistosomiasis can only be acquired via water contact, behavioural observation allows some degree of exposure quantification (Wilkins, 1987).

The best evidence for acquired immunity to schistosomiasis comes from reinfection studies (Section 12.4). These have highlighted the importance of water contact, and revealed a positive relationship within age groups between the degree of exposure and the intensity of reinfection. At each exposure level, however, a negative relationship between age and reinfection has been found; heavily exposed children in The Gambia were reinfected 100 times faster than comparably exposed adults (Wilkins *et al.*, 1987). The Kenyan study obtained similar results and demonstrated the existence of subpopulations of resistant children with high exposure but minimal reinfection (Butterworth *et al.*, 1984, 1985). There is thus convincing epidemiological evidence for an age-dependent development of resistance to reinfection with schistosomes. This resistance is likely to be due to acquired immunity.

There is an obvious need for similar field studies involving some estimate of exposure for different individuals or age groups on other human helminths. Exposure to soil-transmitted helminths is notoriously difficult to quantify, but two methods may produce useful results. Ingestion of soil (geophagia) is thought to be an important method of transmission of geohelminths in children. A method for quantification of geophagia has recently been developed, and it may be possible to look for correlations between geophagia and reinfection (Wong *et al.*, 1988). Exposure estimates have also been obtained for hookworms in West Bengal. Most exposure seems to occur focally around defaecation sites in the wet season (Schad *et al.*, 1983) and can be measured using a damp pad technique to quantify infective larvae in the soil (Hominick *et al.*, 1987). Preliminary evidence suggests that some individuals are predisposed to infection by behavioural factors; further longitudinal studies using this technique would be of interest.

12.5.   IMMUNITY, EPIDEMIOLOGY AND PARASITE CONTROL

The accumulating evidence in favour of a role for immunity in the epidemiology of helminth infection has prompted some consideration of the implications of acquired immunity for the design of parasite control programs for both people and livestock. For example, Anderson and May (1985b) used a model incorporating acquired immunity (Section 12.2) to examine the effect of mass and selective chemotherapy on helminth populations in human communities. The results suggest that repeated mass chemotherapy may act to reduce the level of immunity within a population, such that the average worm burdens in the older age classes of the population are raised above their precontrol levels (Figure 12.11). This effect is likely to be most marked when immunity is relatively strong and long lasting.

Some confirmation of these predictions has come from the results of a control program for echinococcosis and cysticercosis in New Zealand. The aim of this program was to break the transmission link from intermediate to final host by means of a reduction in the frequency with which dogs were allowed to feed on contaminated sheep products. The program was successful but led to an increase in worm burdens of *Taenia hydatigena* in older sheep as a result of their declining immunity (Gemmell *et al.*, 1986; Roberts *et al.*, 1987). A concomitant and similar increase in infection with *T. ovis* was observed, presumably as the result of cross-immunity between the two species (Gemmell *et al.*, 1987; Section 12.7).

Age-dependent increases in infection intensity could turn out to be a major disadvantage of mass chemotherapy programmes, since the coverage necessary for parasite eradication is unlikely to be achieved. The use of selective chemotherapy (whereby those heavily infected are singled out for treatment) avoids this problem while retaining the potential to generate a significant reduction in the overall average worm burden (Anderson and May, 1985b; Anderson and Medley, 1985).

Vaccination against human helminth infections, although not yet possible, seems to be a likely future development. However, a potential problem for

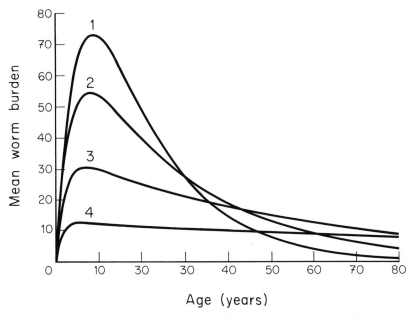

Figure 12.11. Predicted impact of chemotherapy on the intensity of infection in a homogeneous population (with respect to immunological responsiveness and exposure to infection) in which drug treatment is applied randomly (independent of parasite load) and repeatedly. The four curves denote the age-intensity profile for the precontrol situation (curve 1) and three increasing levels of drug application (curves 2 to 4) (from Anderson and May 1985b).

parasite control based on vaccination is that of protecting the 'low responders' in a population (Wakelin, 1984a, 1986; Anderson and May 1985b). If there is heterogeneity in immune responsiveness between human hosts, then the successful deployment of vaccines will depend critically upon overcoming the lack of naturally-acquired protective immunity in low responder hosts (Section 6.6). There is little relevant evidence from animal models, but the observation that even low responder strains of mice can be immunized against *H. polygyrus* is encouraging (Behnke and Robinson, 1985). Vaccination of malnourished hosts may prove to be even more of a problem (Wakelin, 1986; Slater and Keymer, 1988).

12.6.   PROTOZOAN INFECTIONS

The incorporation of acquired immunity into models of the epidemiology of microparasitic infection has a long history (see Bailey, 1975). Specific development of these models with respect to protozoan infection has, however, been more recent (see Aron and May, 1982). Most models have been constructed on the basis that immunity is complete, fixed in duration and independent of continued exposure (Dietz *et al.*, 1974; Rogers, 1988); only recently have models been refined to include the dynamics of immunity boosted by reinfection (Aron, 1983, 1988). Analysis of these models has emphasized the fact that the epidemiology of acquired immunity can complicate disease control. As with helminth parasites (Section 12.5), partially effective control has the potential to be harmful, as a reduction in transmission may lead to a decrease in naturally acquired immunity and thus to an increased prevalence of disease (Figure 12.12). As stressed by Aron and May (1982), the epidemiological understanding of acquired immunity (including population genetics as well as population dynamics) is just as important as molecular and cellular study with respect to the prospects for disease control.

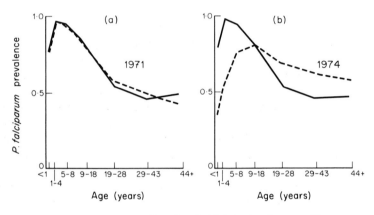

Figure 12.12. (a) Age-prevalence profile of malarial infection in two village groups in 1971 showing no difference between the two groups (solid and broken line). In 1972 and 1973 there was massive anti-malarial intervention in one of the groups, subsequently halted for adults in 1974; (b) age-prevalence profile of infection in 1974. The adults in the previously protected group (broken line) show significantly higher prevalences than those in the unprotected villages (solid line; from Aron and May, 1982).

12.7. INTERACTIONS BETWEEN PARASITE SPECIES

The previous discussion has concentrated on infections consisting of a single parasite species. Many hosts in natural populations, however, are infected with a number of different parasites, and interspecific interactions may be important. Host immunity may be involved in both negative and positive interactions.

Cross-immunity between parasite species has been frequently demonstrated in the laboratory (Dobson, 1985) and can be generated in a number of ways. For example, there may be specific cross-immunity between two parasite species which share protective antigens. This kind of interaction is likely to be most common between closely related, probably congeneric, species. The immunity may be reciprocal, as it is, for example, against *Taenia hydatigena* and *T. ovis* in the sheep (Gemmell, 1969), or it may operate in one direction only. Thus *Oesophagostomum columbianum* can immunize sheep against infection with *O. venulosum*, but the reverse is not true (Dash, 1981). In other cases, cross-immunity may be mediated by a non-specific effector response and thus require the presence of the immunizing species. The response elicited by *Trichinella spiralis* in mice, for example, produces a marked inflammation which will expel concurrent infections of other intestinal parasites (Wakelin, 1984b). Cross resistance that does not depend on an immune response may occur if the pathology induced by one species causes resistance to another.

The effects of cross-immunity on epidemiology can be readily seen when the population of one species is reduced by control. For example, *Taenia hydatigena* has a greater intrinsic rate of reproduction than *T. ovis* and before control measures were introduced in New Zealand was much the commoner of the two. Sheep developed immunity to this species at an early age (Gemmell *et al.*, 1986; Roberts *et al.*, 1987). After the introduction of control measures, the population of *T. hydatigena* decreased, reducing the immunological constraints on *T. ovis* and resulting in a marked increase in infection levels (Roberts *et al.*, 1987). Similarly, control of *Oesophagostomum columbianum* in New South Wales created an increase in the number of hosts susceptible to infection with *O. venulosum*, which subsequently increased in prevalence (Dash, 1981).

The effects of cross-immunity on the distribution of parasites within host populations are not well known; it is often difficult to extrapolate from laboratory experiments based on large discrete infections to the field where hosts are subject to continual reinfection with small numbers of several different parasites. If cross-immunity operates within a uniformly immunocompetent host population, a negative correlation between parasite species within individual hosts would be expected. However, if immunity is under genetic control and if genetic predisposition operates similarly for both parasite species, positive associations may result. Positive correlations between different helminth species have been recorded in sheep (Barger, 1984) and man (Haswell-Elkins *et al.*, 1987), but the causative mechanisms for these patterns are not known.

Positive interspecific interactions can also result if, instead of inducing cross-immunity, parasites of one species suppress the host immune response to heterologous antigens. This would be expected to occur as a side effect of non-specific immunosuppression induced by a parasite to aid its own survival. Examples of both protozoan and helminth infections that cause immunosuppression are known (Cox, 1986; Behnke, 1987b; Richie, 1988). Concurrent *H. polygyrus* infection, for example, prolongs the survival of at least five other helminth species in the laboratory (Section 13.2) although interactions in the field remain equivocal (Kisielewska, 1970).

ACKNOWLEDGEMENTS

We would like to thank M.G. Shorthose for the modelling in Figure 12.5 and M. Amphlett for preparing the figures. B.T. Grenfell, A.F. Read, J.R. Clarke, A.F.G. Slater and R.D. Gregory all made helpful comments on earlier drafts.

REFERENCES

Albers, G.A.A. and Gray, G.D. (1986), Breeding for worm resistance: a perspective. in Howell, M.J. (Ed.), *Parasitology – Quo Vadit?* pp.559–66. Canberra: Australian Academy of Science.

Anderson, R.M. (1982), The population dynamics and control of hookworm and roundworm infections, in Anderson, R.M. (Ed.), *Population Dynamics of Infectious Diseases* pp.67–106, London: Chapman and Hall.

Anderson, R.M. (1985), Mathematical models in the study of the epidemiology and control of ascariasis in man, in Crompton, D.W.T., Nesheim, M.C. and Pawlowski, Z.S. (Eds.), *Ascariasis and Its Public Health Significance*, pp.39–67, London: Taylor & Francis.

Anderson, R.M. (1986), The population dynamics and epidemiology of intestinal nematode infections, *Transactions of the Royal Society of Tropical Medicine and Hygiene*, **80**, 686–96.

Anderson, R.M. (1988), The population biology and genetics of resistance to infection. in Wakelin, D. and Blackwell, J. (Eds.), *The Genetics of Resistance to Bacterial and Parasitic Infection*, pp.233–64. London: Taylor & Francis.

Anderson, R.M. and May, R.M. (1978), Regulation and stability of host-parasite population interactions. I. Regulatory processes, *Journal of Animal Ecology*, **47**, 219–47.

Anderson, R.M. and May, R.M. (1979), Population biology of infectious diseases: Part I., *Nature*, **280**, 361–7.

Anderson, R.M. and Gordon, D.M. (1982), Processes influencing the distribution of parasite numbers within host populations with special emphasis on parasite-induced host mortalities, *Parasitology*, **85**, 373–98.

Anderson, R.M. and May, R.M. (1985a), Helminth infections of humans: mathematical models, population dynamics and control, *Advances in Parasitology*, **24**, 1–101.

Anderson, R.M. and May, R.M. (1985b), Herd immunity to helminth infection and implications for parasite control, *Nature*, **315**, 493–6.

Anderson, R.M. and Medley, G.F. (1985), Community control of helminth infections of man by mass and selective chemotherapy, *Parasitology*, **90**, 629–60.

Aron, J.L. (1983), Dynamics of immunity boosted by exposure to infection, *Mathematical Biosciences*, **64**, 249–59.

Aron, J.L. (1988), Acquired immunity dependent upon exposure in an SIRS epidemic model, *Mathematical Biosciences*, **88**, 37–47.

Aron, J.L. and May, R.M. (1982), The population dynamics of malaria, in Anderson, R.M. (Ed.), *Population Dynamics of Infectious Diseases*, pp.139–79, London: Chapman and Hall.

Bailey, N.T.J. (1975), *The Mathematical Theory of Infectious Disease and Its Applications*, London: Griffin.

Ball, P.A.J. and Bartlett, A. (1969), Serological reactions to infection with *Necator americanus*, *Transactions of the Royal Society of Tropical Medicine and Hygiene*, **63**, 362–9.

Barger, I.A. (1984), Correlations between numbers of enteric nematode parasites in grazing lambs, *International Journal for Parasitology*, **14**, 587–9.

Barger, I.A., Le Jambre, L.F., Georgi, J.R. and Davies, H.I. (1985), Regulation of *Haemonchus contortus* populations in sheep exposed to continuous infection, *International Journal for Parasitology*, **15**, 529–33.

Bartlett, A. and Ball, P.A.J. (1974), The immune response of the mouse to larvae and adults of *Nematospiroides dubius*, *International Journal for Parasitology*, **4**, 463–70.

Behnke, J.M. (1987a). Do hookworms elicit protective immunity in man? *Parasitology Today*, **3**, 200–6.

Behnke, J.M. (1987b), Evasion of immunity by nematode parasites causing chronic infections, *Advances in Parasitology*, **26**, 1–17.

Behnke, J.M. and Wakelin, D. (1973), The survival of *Trichuris muris* in wild populations of its natural hosts, *Parasitology*, **67**, 157–64.

Behnke, J.M. and Robinson, M. (1985), Genetic control of immunity to *Nematospiroides dubius*: a 9-day anthelminthic abbreviated immunising regime which separates weak and strong responder strains of mice, *Parasite Immunology*, **7**, 235–53.

Behnke, J.M., Hannah, J. and Pritchard, D.I. (1983), Evidence that adult *Nematospiroides dubius* impair the immune response to a challenge infection, *Parasite Immunology*, **5**, 397–408.

Berding, C., Keymer, A.E., Murray, J.D. and Slater, A.F.G. (1986), The population dynamics of acquired immunity to helminth infection, *Journal of Theoretical Biology*, **122**, 459–71.

Berding, C., Keymer, A.E., Murray, J.D. and Slater, A.F.G. (1987), The population dynamics of acquired immunity to helminth infection: experimental and natural transmission, *Journal of Theoretical Biology*, **126**, 167–82.

Bundy, D.A.P. (1988), Population ecology of intestinal helminth infections in human communities, *Philosophical Transactions of the Royal Society of London*, **B 321**, 405–20.

Bundy, D.A.P. and Golden, M.H.N. (1987), The impact of host nutrition on gastrointestinal helminth populations, *Parasitology*, **95**, 623–35.

Bundy, D.A.P. and Cooper, E.S. (1988), The evidence for predisposition to trichuriasis in humans: comparison of institutional and community studies, *Annals of Tropical Medicine and Parasitology*, **82**, 251–6.

Bundy, D.A.P., Kan, S.P. and Rose, R. (1988), Age-related prevalence, intensity and frequency distribution of gastrointestinal helminth infection in urban slum children from Kuala Lumpur, Malaysia, *Transactions of the Royal Society of Tropical Medicine and Hygiene*, **82**, 289–94.

Butterworth, A.E. and Hagan, P. (1987), Immunity in human schistosomiasis, *Parasitology Today*, **3**, 11–16.

Butterworth A.E., Dalton, P.R., Dunne, D.W., Mugambi, M., Ouma, J.H., Richardson, B.A., Arap Siongok, T.K. and Sturrock, R.F. (1984), Immunity after treatment of human schistosomiasis mansoni. I. Study design, pretreatment observations and results of treatment, *Transactions of the Royal Society of Tropical Medicine and Hygiene*, **78**, 108–23.

Butterworth, A.E., Capron, M., Cordingley, J.S., Dalton, P.R., Dunne, D.W., Kariuki, H.C., Kimani, G., Koech, D., Mugambi, M., Ouma, J.H., Prentice, M.A., Richardson, B,A,. Arap Siongok, T.K., Sturrock, R.F. and Taylor, D.W. (1985), Immunity after treatment of human schistosomiasis mansoni. II. Identification of resistant individuals, and analysis of their immune responses, *Transactions of the Royal Society of Tropical Medicine and Hygiene*, **79**, 393–408.

Connan, R.M. (1970), The effect of host lactation on the self cure of *Nippostrongylus brasiliensis* in rats, *Parasitology*, **61**, 27–33.

Cox, F.E.G. (1986), Interactions in protozoan infections, in Howell, M.J. (Ed.), *Parasitology – Quo Vadit?*, pp. 569–75, Canberra: Australian Academy of Science.

Croll, N.A., Ghadirian, E. and Sukul, N.C. (1982), Exposure and susceptibility in human helminthiasis, *Tropical Doctor*, **12**, 136–8.

Crombie, J.A. and Anderson, R.M. (1985), Population dynamics of *Schistosoma mansoni* in mice repeatedly exposed to infection, *Nature*, **315**, 491–3.

Dash, K.M. (1981), Interaction between *Oesophagostomum columbianum* and *Oesophagostomum venulosum* in sheep, *International Journal for Parasitology*, **11**, 210–17.

Denham, D.A. and Fletcher, C. (1987), The cat infected with *Brugia pahangi* as a model of human filariasis, in Evered, D. and Clark S. (Eds.), *Filariasis (Ciba Foundation Symposium 127)*, pp. 225–30, Chichester: John Wiley & Sons.

Dietz, K., Molineaux, L. and Thomas, A. (1974), A malaria model tested in the African savannah, *Bulletin of the World Health Organisation*, **50**, 347–57.

Dobson, C. (1982), Passive transfer of immunity with serum in mice infected with *Nematospiroides dubius*: influence of quality and quantity of immune serum, *International Journal for Parasitology*, **12**, 207–13.

Dobson, A.P. (1985), The population dynamics of competition between parasites, *Parasitology*, **91**, 317–47.

Elkins, D.B., Haswell-Elkins, M. and Anderson, R.M. (1988), The importance of host age and sex to patterns of reinfection with *Ascaris lumbricoides* following mass anthelminthic treatment in a South Indian fishing community, *Parasitology*, **96**, 171–84.

Gemmell, M.A. (1969), Hydatidosis and cysticercosis, 1. Acquired resistance to the larval phase, *Australian Veterinary Journal*, **45**, 521–4.

Gemmell, M.A. and Johnstone, P.D. (1981), Factors regulating tapeworm populations: estimation of the duration of acquired immunity by sheep to *Taenia hydatigena*, *Research in Veterinary Science*, **30**, 53–6.

Gemmell, M.A., Lawson, J.R. and Roberts, M.G. (1987), Population dynamics in echinococcosis and cysticercosis: evaluation of the biological parameters of *Taenia hydatigena* and *T. ovis* and comparison with those of *Echinococcus granulosus*, *Parasitology*, **94**, 161–80.

Gemmell, M.A., Lawson, J.R., Roberts, M.G., Kerin, B.R. and Mason, C.J. (1986), Population dynamics in echinococcosis and cysticercosis: comparison of the response of *Echinococcus granulosus*, *Taenia hydatigena* and *T. ovis* to control, *Parasitology*, **93**, 357–69.

Gibbs, H.C. (1986), Hypobiosis in parasitic nematodes – an update, *Advances in Parasitology*, **25**, 129–74.

Grencis, R.K. and Wakelin, D. (1983), Immunity to *Trichinella spiralis* in mice. Factors involved in direct anti-worm effects, *Wiadomosci Parazytologiczne*, **29**, 387.

Grenfell, B.T., Smith, G. and Anderson, R.M. (1987a), The regulation of *Ostertagia ostertagi* populations in calves: the effect of past and current experience of infection on proportional establishment and parasite survival, *Parasitology*, **95**, 363–72.

Grenfell, B.T., Smith, G. and Anderson, R.M. (1987b), A mathematical model of the population biology of *Ostertagia ostertagi* in calves and yearlings, *Parasitology*, **95**, 389–406.

Hagan, P. (1987), Human immune response. in Rollinson, D. and Simpson, A.J.G. (Eds.), *The Biology of Schistosomes*, pp.295–320, London: Academic Press.

Haswell-Elkins, M.R., Elkins, D.B. and Anderson, R.M. (1987), Evidence for predisposition in humans to infection with *Ascaris*, hookworms, *Enterobius* and *Trichuris* in a South Indian Fishing Community, *Parasitology*, **95**, 323–37.

Hominick, W.M., Dean, C.G. and Schad, G.A. (1987), Population biology of hookworms in West Bengal: analysis of numbers of infective larvae recovered from damp pads applied to the soil surface at defaecation sites, *Transactions of the Royal Society of Tropical Medicine and Hygiene*, **81**, 978–87.

Keymer, A.E. (1982), Density-dependent mechanisms in the regulation of intestinal helminth populations, *Parasitology*, **84**, 573–87.

Keymer, A.E. and Hiorns, R.W. (1986), *Heligmosomoides polygyrus* (Nematoda): the dynamics of primary and repeated infection in outbred mice, *Proceedings of the Royal Society of London*, B **229**, 47–67.

Keymer, A.E. and Slater, A.F.G. (1987), Helminth fecundity: density dependence or statistical illusion? *Parasitology Today*, **3**, 56–8.

Kisielewska, K. (1970), Ecological organisation of intestinal helminth groupings in *Clethrionomys glareolus*, *Parasitologia Polonica*, **18**, 121–208.

Kojima, S., Yano, A., Sasazuki, T. and Ohta, N. (1984), Associations between HLA and immune responses in individuals with chronic schistosomiasis japonica, *Transactions of the Royal Society of Tropical Medicine and Hygiene*, **78**, 325–9.

Maizels, R.M., Selkirk, M.E., Sutanto, I. and Partono, F. (1987), Antibody responses to human lymphatic filarial parasites, in Evered, D. and Clark, S. (Eds.), *Filariasis (Ciba Foundation Symposium 127)*, pp. 189–99, Chichester: John Wiley & Sons.

May, R.M. and Anderson, R.M. (1979), Population biology of infectious diseases: Part II, *Nature*, **280**, 455–61.

Miller, H.R.P., Jackson, F., Newlands, G. and Appleyard, W.T. (1983), Immune exclusion, a mechanism of protection against the ovine nematode *Haemonchus contortus*, *Research in Veterinary Science*, **35**, 357–63.

Miller, T.A. (1979), Hookworm infection in man, *Advances in Parasitology*, **17**, 315–84.

Minchella, D.J. and Loverde, P.T. (1983), Laboratory comparison of the relative success of *Biomphalaria glabrata* stocks which are susceptible and insusceptible to *Schistosoma mansoni*, *Parasitology*, **86**, 335–44.

Ogilvie, B.M., Bartlett, A., Godfrey, R.C., Turton, J.A., Worms, M.J. and Yeates, R.A. (1978), Antibody responses in self-infections with *Necator americanus*, *Transactions of the Royal Society of Tropical Medicine and Hygiene*, **72**, 66–71.

O'Sullivan, B.M. and Donald, A.D. (1973), Responses to infection with *Haemonchus contortus* and *Trichostrongylus colubriformis* in ewes of different reproductive status, *International Journal for Parasitology*, **3**, 521–30.

Partono, F. (1987), The spectrum of disease in lymphatic filariasis, in Evered, D. and Clark, S. (Eds.), *Filariasis (Ciba Foundation Symposium 127)* pp. 15–27, Chichester: John Wiley & Sons,.

Piessens, W.F., Wadee, A.A. and Kurniawan, L. (1987), Regulation of immune responses in lymphatic filariasis in Evered, D. and Clark, S. (Eds.), *Filariasis (Ciba Foundation Symposium 127)*, pp. 164–72, Chichester: John Wiley & Sons.

Richie, T.L. (1988), Interactions between malaria parasites infecting the same vertebrate host, *Parasitology*, **96**, 607–41.

Roberts, M.G., Lawson, J.R. and Gemmell, M.A. (1986), Population dynamics in echinococcosis and cysticercosis: mathematical model of the life-cycle of *Echinococcus granulosus*, *Parasitology*, **92**, 621–41.

Roberts, M.G., Lawson, J.R. and Gemmell, M.A. (1987), Population dynamics in echinococcosis and cysticercosis: mathematical models of the life-cycles of *Taenia hydatigena* and *T. ovis*, *Parasitology*, **94**, 181–97.

Robinson, M., Wahid, F.N., Behnke, J.M. and Gilbert, F.S. (1989), Immunological relationships during primary infection with *Heligmosomoides polygyrus (Nematospiroides dubius)*: dose-dependent expulsion of adult worms, *Parasitology*, **98**, 115–24.

Rogers, D.J. (1988), A general model for the African trypanosomiases, *Parasitology*, **97**, 193–212.

Schad, G.A., Nawalinski, T.A. and Kochar, V.K. (1983), Human ecology and the distribution and abundance of hookworm populations, in Croll, N.A. and Cross, J. (Eds.), *Human Ecology and Infectious Disease*, pp. 187–223, New York: Academic Press.

Sinclair, A.R.E., in press, Population regulation, in Cherret, J.M. (Ed.), *Ecological Concepts: The Contribution of Ecology to an Understanding of the Natural World*, Oxford: Blackwell Scientific Publications.

Slater, A.F.G. (1988), The influence of dietary protein on the experimental epidemiology of *Heligmosomoides polygyrus* (Nematoda) in the laboratory mouse, *Proceedings of the Royal Society of London*, **B 234**, 239–54.

Slater, A.F.G. and Keymer, A.E. (1986a), *Heligmosomoides polygyrus* (Nematoda): The influence of dietary protein on the dynamics of repeated infection, *Proceedings of the Royal Society of London*, **B 229**, 69–83.

Slater, A.F.G. and Keymer, A.E. (1986b), Epidemiology of *Heligmosomoides polygyrus* in mice: experiments on natural transmission, *Parasitology*, **93**, 177–87.

Slater, A.F.G. and Keymer, A.E. (1988), The influence of protein deficiency on immunity to *Heligmosomoides polygyrus* (Nematoda) in mice, *Parasite Immunology*, **10**, 507–22.

Smith, G. (1984), Density-dependent mechanisms in the regulation of *Fasciola hepatica* populations in sheep, *Parasitology*, **88**, 449–61.

Smith, G. (1988), The population biology of the parasitic stages of *Haemonchus contortus*, *Parasitology*, **96**, 185–95.

Thein Hlaing (1985), *Ascaris lumbricoides* infections in Burma, in Crompton, D.W.T., Nesheim, M.C. and Pawlowski, Z.S. (Eds.), *Ascariasis and its public health significance*, pp. 83–112, London: Taylor & Francis.

Thein Hlaing, Than Saw and Myint Lwin. (1987), Reinfection of people with *Ascaris lumbricoides* following single, 6-month and 12-month interval mass chemotherapy in Okpo village, rural Burma, *Transactions of the Royal Society of Tropical Medicine and Hygiene*, **81**, 140–6.

Wakelin, D. (1967), Acquired immunity to *Trichuris muris* in the albino laboratory mouse, *Parasitology*, **57**, 515–24.

Wakelin, D. (1973), The stimulation of immunity to *Trichuris muris* in mice exposed to low-level infections, *Parasitology*, **66**, 181–9.

Wakelin, D. (1984a), Evasion of the immune response: survival within low responder individuals of the host population, *Parasitology*, **88**, 639–57.

Wakelin, D. (1984b), *Immunity to Parasites*, London: Edward Arnold.

Wakelin, D. (1986), Genetic and other constraints on resistance to infection with gastrointestinal nematodes, *Transactions of the Royal Society of Tropical Medicine and Hygiene*, **80**, 142–8.

Wakelin, D. and Blackwell, J. (1988), *The Genetics of Resistance to Bacterial and Parasitic Infection*, London: Taylor & Francis.

Warren, K.S. (1973), Regulation of the prevalence and intensity of schistosomiasis in man: immunology or ecology? *Journal of Infectious Diseases*, **127**, 595–609.

Wassom, D.L., de Witt, C.W. and Grundmann, A.W. (1974), Immunity to *Hymenolepis citelli* by *Peromyscus maniculatus*: Genetic control and ecological implications, *Journal of Parasitology*, **60**, 47–52.

Wassom, D.L., Wakelin, D., Brooks, B.O., Krco, C.J. and David, C.S. (1984), Genetic control of immunity to *Trichinella spiralis* infections in mice. Hypothesis to explain the role of H-2 genes in primary and challenge infections, *Immunology*, **51**, 625–31.

Wassom, D.L., Dick, T.A., Arnason, N., Strickland, D. and Grundmann, A.W. (1986), Host Genetics: A key factor in regulating the distribution of parasites in natural host populations, *Journal of Parasitology*, **72**, 334–7.

Wilkins, H.A. (1987), The epidemiology of schistosome infections in man, in Rollinson, D. and Simpson, A.J.G. (Eds.), *The Biology of Schistosomes*, pp. 379–97, London: Academic Press.

Wilkins, H.A., Goll, P.H., Marshall, T.F. de C. and Moore, P.J. (1984), Dynamics of *Schistosoma haematobium* infection in a Gambian community. III. Acquisition and loss of infection, *Transactions of the Royal Society of Tropical Medicine and Hygiene*, **78**, 227–32.

Wilkins, H.A., Blumenthal, U.J., Hagan, P., Tulloch, S. and Hayes, R.J. (1987), Resistance to reinfection after treatment for urinary schistosomiasis, *Transactions of the Royal Society of Tropical Medicine and Hygiene*, **81**, 29–35.

Wong, M.S., Bundy, D.A.P. and Golden, M.H.N. (1988), Quantitative assessment of geophagic behaviour as a potential source of exposure to geohelminth infection, *Transactions of the Royal Society of Tropical Medicine and Hygiene*, **82**, 621–5.

Zavras, E.T. and Roberts, L.S. (1985), Developmental physiology of cestodes. XVIII. Characterization of putative crowding factors in *Hymenolepis diminuta*, *Journal of Parasitology*, **71**, 96–105.

13. Evasion of host immunity

Jerzy M. Behnke

13.1. INTRODUCTION

Despite the array of host-defense mechanisms which have been described in the preceding chapters, parasites frequently survive for many years in otherwise fully immunocompetent hosts, giving rise to chronic infections with insiduously progressive pathology. Indeed one of the hallmarks of human parasitic infections is their longevity. It is readily apparent that the host does not have things totally its own way; the immune system is far from infallible in protecting against parasitic invasion. The self-limiting infections popular in laboratory studies (e.g. *Nippostrongylus brasiliensis, Trichinella spiralis* (Chapter 8, Figure 8.7 and Chapter 6, Figure 6.2)) are by no means typical; in fact quite the reverse.

One of the consequences of long-lived parasites is that pathology may take years to develop and the association between disease and exposure to infection may not be immediately obvious to affected individuals. With the exception of acute infections, such as those associated with malaria, *Theileria, Babesia* and trypanosomiasis, death of the host is seldom caused directly by the parasitic organisms. Frequently, when the host dies it is from secondary causes. Susceptibility to other infectious organisms may be enhanced or impairment of health may predispose the individual to accidents; lack of fitness may lead to unemployment, poverty and malnutrition (Chapter 1). In *Onchocerca volvulus* (Chapter 8, Figure 8.9) mortality seldom follows directly from infection, but the disease causes blindness and this leaves the individual prone to accidents during the course of everyday activities (Kirkwood *et al.*, 1983b).

Onchocerciasis is also a good example of a parasite which is extremely long lived. Its lifespan has been estimated at 11–18 years (Roberts *et al.*, 1967),

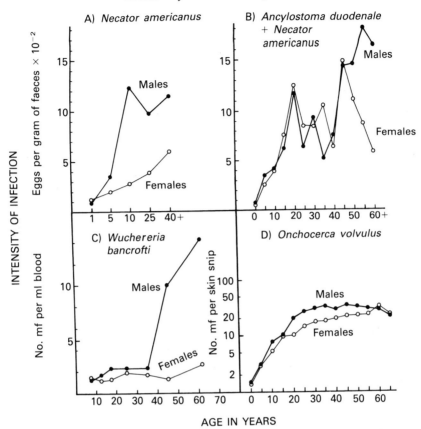

Figure 13.1. Examples of age-intensity profiles showing stable or increasing intensities of infection throughout life, in communities affected by (a) *Necator americanus* (Gambia; Knight and Merrett, 1981), (b) mixed hookworm infection but predominantly *Ancylostoma duodenale* (Taiwan; Hsieh, 1970), (c) *Wuchereria bancrofti* (Philippines; Grove *et al.*, 1978) and (d) *Onchocerca volvulus* (West Africa; Kirkwood *et al.*, 1983a).

and in endemic areas people harbour worms throughout life. At the population level, there is little evidence of a reduction in either the prevalence or intensity of infection with age (Figure 13.1). Worms are acquired in early life and as they die others replace them to be replaced in turn themselves in surviving people. Further examples of parasitic organisms, in which there is no decline in infection intensity with age of the host population as would be expected had strong resistance been elicited by continuous exposure to infection, are also depicted in Figure 13.1 and these contrast with schistosomiasis (Figure 13.2) where, quite clearly, the heaviest worm burdens are carried by teenaged children. In schistosomiasis the subsequent decline in infection may be related to age-associated changes in behaviour leading to less contact with the infective stages (children may have higher water contact rates than adult through play), but there is equally compelling evidence that parasite burdens are controlled by age-dependent acquired immunity resulting from continuous exposure in the field (Butterworth and Hagan, 1987). A state of total resistance, however is not achieved and all age groups, even the oldest, continue to harbour parasites and to pass eggs, playing a vital role in transmission.

The remarkable feature of the age-intensity profiles for infections with hookworms and filarial parasites (Figure 13.1) is that there is no apparent falling off of the mean intensity with age. The age-intensity curves are stable across the age groups, once the plateau has been achieved or continue to rise, probably reflecting the success of the parasite in avoiding host immunity and the failure of the host to control infection effectively. These are intriguing phenomena. Why is the immune system seemingly powerless and how do the parasites achieve the upperhand? Various strategies by which parasites gain an advantage over host immunity have been investigated experimentally and hypotheses linking laboratory results with field observations have been formulated. Some of the concepts have already been introduced in Chapter 12. Here, I will analyse the scope of available strategies under two broad categories: exploitation of permanent or temporary weaknesses in host

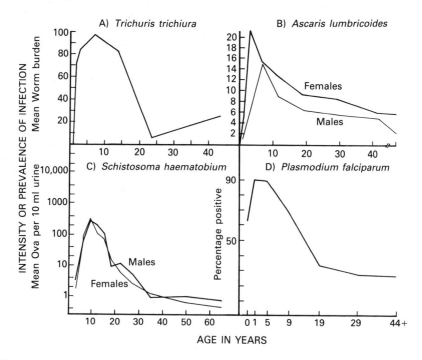

Figure 13.2. Examples of age-intensity and age-prevalence profiles showing decreasing trends in older age groups. Communities affected by (a) *Trichuris trichiura* (Caribbean; Bundy *et al.*, 1987), (b) *Ascaris lumbricoides* (Southern India; Elkins *et al.*, 1986), (c) *Schistosoma haematobium* (Gambia; Wilkins *et al.*, 1984) and (d) prevalence of *Plasmodium falciparum* (Nigeria; Fleming *et al.*, 1979).

defensive strategies, which I have called host-derived mechanisms and secondly, mechanisms evolved by the parasites themselves in response to host resistance, parasite-derived mechanisms.

13.2. HOST-DERIVED MECHANISMS

Genetically determined unresponsiveness

There is good evidence from epidemiological studies that a small proportion of any wild host population shows exceptional susceptibility to infection and may harbour very heavy parasite loads. It has been estimated that for nematode parasites up to 70% of the parasite population may reside in fewer than 15% of the hosts (Anderson and Medley, 1985). Figure 13.3 presents a selection of pertinent examples to illustrate this phenomenon. The causes of non-responsive, 'wormy' individuals are still poorly understood (Chapter 6) but are undoubtedly multifactorial. Experimental studies, however, have provided firm evidence that genetically determined inability to mount an appropriate host-protective response may be one explanation for the field observations.

When mice of the randomly bred Schofield strain were exposed to infection with *Trichuris muris*, about 72% resisted infection and the remaining 28% developed patent worm burdens (Wakelin, 1975). Furthermore the susceptible individuals also failed to resist a secondary infection whereas all the resistant animals showed evidence of a rapid anamnestic response. Genetic control of this phenomenon was implicated through the results of breeding experiments (Table 13.1). Susceptible mice were distinguished from resistant through the presence of eggs in their faeces five weeks post-infection and an initial batch of 20 male and female mice was segregated according to their responder status. Two lines were subsequently established in each of which the mice were brother-sister mated. All the responder individuals encountered in the susceptible line were eliminated and the breeding pairs for the next generation were established from susceptible mice only and vice versa in the resistant line. A representative sample from each generation of both lines was tested for responder status and it was found that after only two generations all the progeny of the responder line were totally resistant. It proved more difficult to select for a non-responder status, and a 100% non-responder strain was not achieved even after nine generations.

Another vivid illustration of this phenomenon comes from the work of Wassom *et al.*, (1973) who found that a small proportion of wild deer mice *Peromyscus maniculatus* carried patent infections with the tapeworm *Hymenolepis citelli* (1–4%). In the laboratory some infected deer mice did not expel worms and even secondary infections were tolerated by such mice (Section 12.3).

These experiments illustrate that, under both laboratory and field conditions, individuals, seemingly incapable of expressing resistance to a particular species

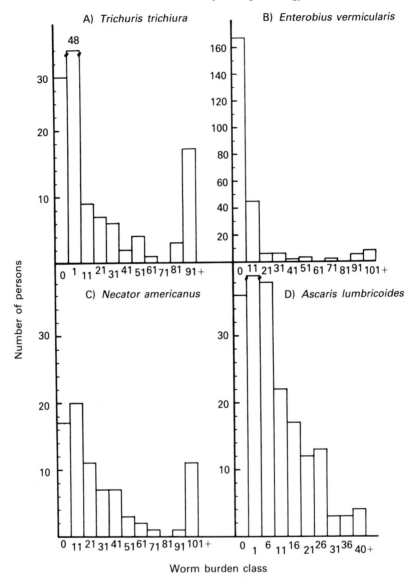

Figure 13.3. Examples of frequency distributions of intestinal nematode parasites affecting man. (a) *Trichuris trichiura* (Caribbean; Bundy, 1986) (b) *Enterobius vermicularis* (Southern India; Haswell-Elkins *et al.*, 1987), (c) *Necator americanus* (Iran; Croll and Ghadirian, 1981) and (d) *Ascaris lumbricoides* (Southern India; Elkins *et al.*, 1986).

of parasite, exist because they do not possess the genetically encoded information which is necessary to generate host-protective immunity. They may be exploited by parasites, and, as Wassom *et al.* (1973) discovered, a proportion of the population may consequently carry heavy infections in nature.

Table 13.1. Selection for immunity to *Trichuris muris* in successive generations of Schofield strain mice. The parental Schofield strain stock mice were tested for patent infections on day 35 to 42 following infection. Mice showing eggs in their faeces were considered non-responsive and were brother-sister mated at each generation. The responder line was selected on the basis of absence of patent infections five to six weeks post-infection (Wakelin, 1975).

Generation	Non-responder line % of non-responders	Responder line % of responders
Parental	28	72
F1	32	95
F2	27	100
F3	75	100
F4	79	100
F5	88	100
F6	75	100

It is not yet possible to evaluate the significance of laboratory findings on a broader front. Non-responder individuals would carry parasites throughout life, and as long as a constant proportion of the population is influenced by non-responder genes, the relative constancy of age-intensity curves may be explained. However, the situation is clearly much more complex, and, as will be seen in subsequent sections of this chapter, there are other explanations for failure of host-protective responses. Nevertheless, genetic constraints undoubtedly exert a significant influence and, as stated earlier, constitute an important target for current epidemiological research programmes.

Neonatal unresponsiveness

It is widely recognized that in many parasitic infections young animals show greater susceptibility to infection than those which have become sexually mature (Gray, 1972; Rajasekariah and Howell, 1977). The onset of increased resistance usually coincides with the development of maturity and is probably attributable to the direct effect of, as well as the interaction between, changing hormonal balance (Solomon, 1969) and the immune system (Alexander and Stimson, 1988). Regrettably, few models have been analysed thoroughly enough to establish conclusively the mechanisms involved.

Figure 13.4a illustrates the results of an experiment in which the timecourse of the infection with *Nippostrongylus brasiliensis* was monitored in seven-day-old and mature rats by regular faecal egg counts. The young rats maintained worms into adulthood and did not eliminate the parasites within the usual three-week period observed in mature animals. This phenomenon was first described by Jarrett *et al.* (1966) and Kassai and Aitken (1967) and was likened to a state of immunological tolerance. However, classical tolerance was not the correct explanation because rats infected during the neonatal

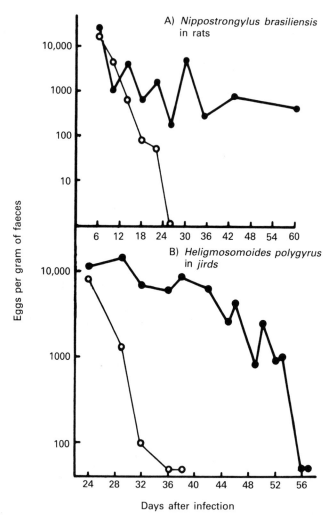

Figure 13.4. Delayed expulsion of (a) *Nippostrongylus brasiliensis* (Dineen and Kelly, 1973) and (b) *Heligmosomoides polygyrus* (Hannah and Behnke, 1983) from rats and jirds, respectively, infected in neonatal life. Mature animals (○); 7-day-old rats or 12-day-old jirds (●).

period responded normally to a challenge infection in adult life (Jarrett and Urquhart, 1969). Furthermore a degree of dose dependence was observed when young rats were infected with more than 250 larvae, a proportion of the worm population being lost, albeit slowly in comparison with mature hosts (Jarrett *et al.*, 1968).

Experiments by Australian parasitologists attempted to pin point the lesion in neonatal rats and Table 13.2 illustrates the details of their experiments. It was established that, despite their relative longevity, the adult worms developing in neonatally infected rats were damaged immunologically. The antibody response of neonatal rats was shown to be normal and when provided with immune lymphocytes, neonatal rats expelled *N. brasiliensis* quite effectively. Thus all the processes, except the lymphocyte-mediated control of the final effector mechanism were unimpaired. However the explanation for the defect in the cellular arm of the response has never been conclusively established. Two possibilities are apparent. It is conceivable that insufficient quantities of reactive cells were available within the mesenteric lymph nodes or that effector cells were down regulated through suppressive regulatory mechanisms in neonatal rodents.

Neonatal unresponsiveness may be of significance in the agricultural industry. Lambs fail to develop resistance to *Haemonchus contortus* but do eventually control worms in adult life (Manton *et al.*, 1962). In experimental trials mature sheep were successfully vaccinated against *H. contortus* using an irradiated larval vaccine, but lambs of less than five months were not protected (Urquhart *et al.*, 1966; Benitez-Usher *et al.*, 1977). Unfortunately lambs constitute an important component of the sheep industry, animals being generally slaughtered at about six to eight months for meat. Here, neonatal unresponsiveness constitutes a major hurdle to the introduction of a vaccine against an economically important parasite.

The mechanisms involved in immunological unresponsiveness in neonatal animals are still poorly understood. Most rodents are born with a very immature immune system, but significantly elevated levels of non-specific suppressor cells (natural suppressor cells) have also been demonstrated in the first weeks after birth (Mosier and Johnson, 1975; Calkins and Stutman,

---

Table 13.2. Evidence that neonatal rats have a defect in the cellular mechanism responsible for rejection of *N. brasiliensis*. Donor rats, whether adult or neonatal, were infected with 500 larvae of *N. brasiliensis* and were killed on day 17 for preparation of mesenteric lymph node cell suspensions. Recipient rats were challenged with 1000 larvae and were killed on day 10 post-infection at a time when normal rejection of primary infections from adult rats has not yet taken place. Reductions in worm burden at this time-point, therefore, reflect accelerated expulsion of worms (Dineen and Kelly, 1973).

Treatment	Effect
Adult rats were given mesenteric lymph node cells from either infected neonatal rats or infected adult rats.	
None	No loss of worms
Cells from adult rats	Accelerated expulsion
Cells from neonatal rats	No loss of worms
Adult or neonatal rats were given mesenteric lymph node cells from infected adult rats.	
Adult rats, none	No loss of worms
Adult rats, cells from adult rats	Accelerated expulsion
Neonatal rats, none	No loss of worms
Neonatal rats, cells from adult rats	Accelerated expulsion

1978; Strober 1984) and defective antigen presentation by neonatal macrophages has been reported (Lu *et al.*, 1979). It is thought that during this period of neonatal unresponsiveness, rodents may acquire tolerance to self antigens, but it is also conceivable that this temporary period of immundepression safeguards the suckling young against becoming sensitized to proteins in the mother's milk. Parasites may capitalize on this 'Achilles' heel' and become established during a critical period in the host's life.

Reduced immunocompetence in mature hosts

Parasites may exploit temporary or prolonged periods when the host's immune system is incapacitated. A variety of factors may be responsible.

Lactation

The effect of lactation in allowing parasites to avoid expulsion from the host is best illustrated again by reference to *N. brasiliensis* (Figure 13.5), but the phenomenon has been observed in a variety of animal models (e.g. *T. colubriformis*, O'Sullivan, 1974; *T. spiralis*, Ngwenya, 1980; *T. muris*, Selby and Wakelin, 1975) and in the field (O'Sullivan and Donald, 1970). Connan (1970) found that lactating rats sustained considerably longer infections than would normally be expected in adult life. The effect was best observed when lactating rats were infected during the first two weeks following parturition and was dependent on suckling, since rats whose young were removed after birth did not harbour extended infections. The immunodepressive influence of lactation was potent enough to interfere with the expression of immunity in rats which had already been sensitized by a previous infection (Connan, 1972).

It was again the Australian workers who analysed the phenomenon by adoptive transfer techniques, concluding that potentially reactive cells were present and that effector cells could express an inflammatory response in lactating rats (Dineen and Kelly, 1972). The antibody response in lactating rats appeared to be unaffected since the characteristic signs of damage to adult worms were evident at about the normal time following infection. However, the activity of potentially reactive cells was substantially inhibited in lactating animals, most probably through the presence of lactogenic hormone. One possibility is that the hormone prolactin (which increases 10–12 times in concentration in the blood during lactation) has an adverse effect on the differentiation of lymphoid cells (Kelly and Dineen, 1973). A principal role for prolactin has still not been conclusively established, but there is little doubt that immunodepression during lactation is primarily endocrinal in origin (Connan, 1976).

Pregnancy

The expulsion of *N. brasiliensis* is unaffected by pregnancy itself (Connan, 1970), but a parasite which exploits the hormonal perturbations associated

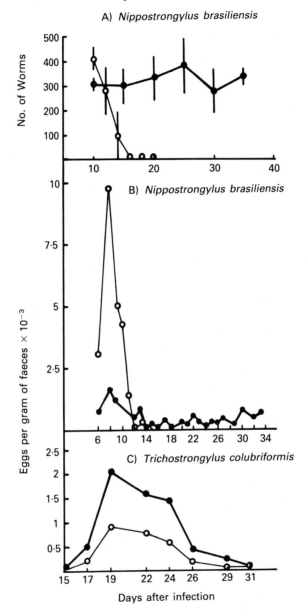

Figure 13.5. Delayed expulsion of (a and b) *Nippostrongylus brasiliensis* (Connan, 1970, 1973) and (c) *Trichostrongylus colubriformis* (O'Sullivan, 1974) from lactating rats (a and b) and guinea pigs (c). Non-lactating animals (○); lactating animals (●).

with pregnancy and their associated effects on the immune system is *Toxocara canis*. Arrested larvae in bitches become activated in pregnancy and infect the young foetus *in utero* (Lloyd *et al.*, 1983). During lactation, the bitch also passes eggs indicating that she is susceptible to infection during this period.

It is still not clear whether this is because some arrested larvae complete development in the bitch's intestine or whether, as seems more likely, she is infected by larvae passing out of the pups' faeces after birth. Nevertheless, this is a striking example of a parasite which has evolved to exploit the reproductive cycle of its host and concomitant temporary periods of reduced immunocompetence, for its own transmission.

Another parasite which may exploit immunological perturbations associated with hormonal changes during pregnancy is *Trypanosma musculi*. Initial infection with this species is followed by a temporary parasitaemia which is eventually cleared and the mice become resistant to further challenge. However, some parasites survive in the vasa recta capillaries of the kidneys and in the placenta of pregnant mice. Furthermore, the parasite reappears in peripheral capillary blood during pregnancy, and an opportunity is created for transmission to the progeny through a suitable flea vector.

The association between pregnancy, lactation and neonatal susceptibility to infection would appear to favour parasite transmission in that infective stages resulting from organisms developing in the lactating mother would find a suitable home in the susceptible and unresponsive neonates. The overall balance would at first sight seem to be heavily biased in favour of the parasite. However, there is an important process which serves to limit transmission, or at least the survival, of the parasites in young offspring. The colostrum and milk of lactating rodents and man contains host-protective antibodies (Haneberg, 1974; Perry 1974) and although these may be of little benefit to the mother, the protective role of passively transferred antibodies is well documented in *T. spiralis* and *N. brasiliensis*. In man antibodies (IgG) may also pass across the placenta and confer resistance against malaria and trypanosomiasis to offspring of resistant mothers (Takayanagi *et al.*, 1978). Passive protection via antibodies transferred in the milk, colostrum or across the placenta would therefore appear to be strategies to offset mechanisms favouring parasite transmission.

Stress and old age

Few experimental studies have specifically addressed the role of stress in affecting host susceptibility to infection, but the link between stress and immunodepression is well recognized (Ghoneum *et al.*, 1987). Various hormones, including the immunodepressive corticosteroids, are released when the host is subjected to stress (De Souza and van Loon, 1985). Mice which have been preimmunized with *T. spiralis* show significantly depressed secondary responses when challenged whilst simultaneously being subjected to stress (Robinson, 1961).

Old age is associated with progressive decline in the functional capacity of both T and B lymphocytes and in the number of cells responding to antigenic signals, despite relatively normal stem and accessory cell function (Menon *et al.*, 1974; Nordin and Makinodan, 1974). Weakened resistance, however, cannot be explained solely in terms of a numerical depletion of functional lymphocytes but is probably also attributable to less tight homeostatic regulation of the immune system (Hess and Knapp, 1978; Wade and Szewczuk, 1984), or even to down regulation through increased suppressor cell activity (Makinodan and Kay, 1980). However, the mucosal immune system appears to be less affected and remains vigorous throughout life in mice (Wade and Szewczuk, 1984), an observation which reinforces the concept that the immune system is compartmentalized. Interestingly in one of the few examples using a parasitic model, Crandall (1975) found that old $C_{57}Bl_6$ mice (22 months) were significantly more susceptible to *T. spiralis* than four-month-old animals and did not develop high IgG antibody titres.

Nutrition

Diet is an important factor in influencing infection with gastrointestinal, as well as other, parasitic organisms. Some dietary constituents may actually be beneficial to the host by causing indirect or direct anti-parasitic effects (e.g. fresh grass has been implicated in self-cure of *H. contortus*, Allonby and Urquhart, 1973). Deprivation of essential proteins in the diet can lead to profound immunodepression, which interacts synergistically with increased exposure rates to infectious organisms causing enhanced disease and pathology. Some parasites may disturb intestinal function to such a degree that the host does not make full use of dietary contents. However, the absorptive capacity of the intestine has enormous reserve and if the diet is adequate, can generally compensate for such disturbances (Castro *et al.*, 1989). If the diet itself is deficient, the resulting malnutrition and associated immunodepression may make resistance an impossible task (Stephenson, 1987).

Although the reasons for immunodepression in malnourished hosts are not fully understood, it has been recognized for some time that vaccination with BCG may be less successful in protein energy malnourished children compared with better nourished individuals living in the same environment (Keusch *et al.*, 1983). In humans and in experimentally malnourished animals, the most severe effects are seen in T-dependent regions of the secondary lymphoid organs (Wing *et al.*, 1988). Malnutrition is associated with a reduction in circulating T cells, reduced T-cell function, defects in T-cell responses to immunoregulatory signals and with an overall increase in suppressor rather than helper activity (Hoffman-Goetz *et al.*, 1986; Wing *et al.*, 1988).

Antibody production is less susceptible to perturbations resulting from malnutrition, and B-cell dependent areas of the secondary lymphoid organs generally remain normal. However, some depression has been reported in the case of the response to protein H antigen and polysaccharide O antigen from *Salmonella typhi*. Antibody responses often require T-cell help in order to function normally, and in the presence of a defective T-cell capability depressed antibody responses may be a secondary phenomenon. Vaccines against measles, smallpox and polio appear to generate normal levels of protection against disease, even in severely malnourished people. However, intestinal disturbances, such as diarrhoea, arising from mucosal infections are common, indicating that the mechanisms protecting mucosal surfaces are weakened in malnourished hosts. The intestinal IgA antibody response in particular is severely compromized (Chandra, 1975; McMurray *et al.*, 1977; Koster and Pierce, 1985; McGee and McMurray, 1988a), and in mice a role for IgA specific suppressor T cells has been suggested (McGee and McMurray, 1988b).

Experiments with *N. brasiliensis* in rats have shown that parasites survive longer in malnourished animals and that the host-protective response which normally terminates the primary infection is depressed and delayed (Figure 13.6). Mice exposed to trickle infections with *Heligmosomoides polygyrus* fail to control the infection, and in contrast to animals maintained on a normal diet, those fed on only a 2% protein diet developed accumulating worm burdens with each successive inoculation (Figure 13.6; Slater, 1988). Thus the survival of gastrointestinal parasitic nematodes may be prolonged in malnourished hosts. In the wild, where animals are subject to environmental stress with greater and lesser food availability in relation to seasonal abundance of suitable crops, situations are likely to arise which parasites may exploit to their own advantage. In social animals, subordinate individuals may have less access to suitable food and in combination with the social stress to which they would be subjected (Esch *et al.*, 1975; Wittenberger, 1981), the outcome may be both malnourishment and increased susceptibility to disease (Davis and Read, 1958). Having made these points, it is necessary to emphasize that many parasitic organisms survive totally adequately in well-nourished hosts, and therefore the contribution of malnutrition to creating opportunities for prolonged survival of parasites must be taken in context. Exploitation of this type of host weakness is unlikely to be the sole cause of chronicity for any single species of parasite.

Concurrent infection

In concurrent infections, parasitic organisms may interact in a variety of ways, some promoting each others survival, others competing. The subject has been reviewed by Holmes (1973) and more recently by Christensen *et al.* (1987) and Richie (1988). Parasites may interfere with each other's presence via the immune system, and such interactions can be grouped under three headings (Table 13.3; 12.7). The present section is concerned only with the last of these categories, namely interactive protection from host immunity. It is clear that certain species can conspire to increase host susceptibility to heterologous infection (Christensen *et al.*, 1987). A particularly interesting example in this context, is *H. polygyrus* which itself causes chronic infections in mice. This nematode has a marked down-regulatory effect on immunological events in the intestinal tract and in concurrent infections with a variety of other species, the heterologous organisms benefit by surviving longer than

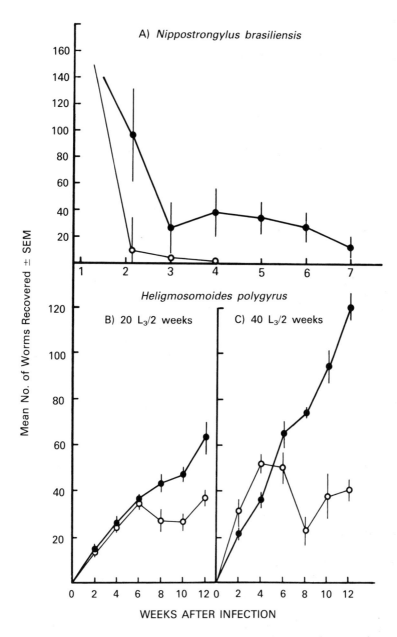

Figure 13.6. The effect of malnutrition on the course of infection with intestinal nematodes. (a) A single-pulse infection with *Nippostrongylus brasiliensis* in rats (Bolin *et al.*, 1977). Repeated administration of 20 (b) or 40 (c) larvae of *Heligmosomoides polygyrus* per two weeks to mice (Slater and Keymer, 1986). Animals fed a normal diet (○); Animals maintained on a protein-deficient diet (●).

Table 13.3. Interactions between parasites residing in the same host via the immune system (Mitchell, 1979).

---

1. Cross-immunity between heterologous species.
   Beneficial to the host. Dependent on the possession of similar or identical antigens by different species of parasites (e.g. *T. spiralis* and *T. muris*: Lee *et al.*, 1982)
2. Interactive immunity.
   Beneficial to the host. One species initiates a host response and the subsequent effector components act non-specifically to the detriment of an unrelated species residing in the same host (e.g. malaria and *Babesia*; Cox, 1978: *T. spiralis* and *H. diminuta* or *H. microstoma*; Behnke *et al.*, 1977; Christie *et al.*, 1979).
3. Interactive protection.
   Detrimental to the host. The presence of one parasite species extends the duration of the infection with others or increases host susceptibilty to infection by heterologous species (Behnke *et al.*, 1978; Behnke, 1987; Christensen *et al.*, 1987).

---

they would do normally (e.g. Figure 13.7, *T. spiralis*, *T. muris*, *H. diminuta*, *N. brasiliensis*).

The practical implications of observations based on experimental infections in laboratory animals are not obvious from the literature. Attempts to demonstrate enhanced susceptibility to infection by a particular species, in humans harbouring a heavy heterologous infection, have met with equivocal success. Croll and Ghadirian (1981) failed to find a significant association between *A. lumbricoides*, *T. trichiura* and *Trichostrongylus* but established a significant positive association between *T. trichiura* and *Trichostrongylus* species. With the exception of the latter combination and the general absence of concurrent heavy infections with more than one species, it was considered that 'wormy' persons did not have an overall increased susceptibility to infections of all kinds and in practical terms parasites in the human gut did not interact to benefit each other's survival. However, contrasting conclusions were arrived at by a recent study in India. Haswell-Elkins *et al.* (1987) found a significant correlation between five of the six possible pair-wise comparisons involving *A. lumbricoides*, hookworms, *Trichuris* and *Enterobius*. The relationship between *Ascaris* and *Trichuris* (via faecal egg counts) showed a particularly strong correlation. There are other examples in the literature. Higher worm burdens with *T. axei* have been described in sheep infected with *H. contortus* and *O. circumcincta* than in animals without the latter species (Turner *et al.*, 1962), and more recently mixed infections with *O. leptospicularis* and *O. ostertagi* were found to result in a synergistic enhancement of the establishment of both species in calves (Al Saqur *et al.*, 1984). There is also some indication that gastrointestinal worms may increase susceptibility to lungworm infection (Kloosterman and Frankena, 1988). The factors involved in these positive interactions are still incompletely understood, but it is conceivable that an explanation for some may lie in one species exploiting the evasive strategies of the other and so extending the duration of its own survival.

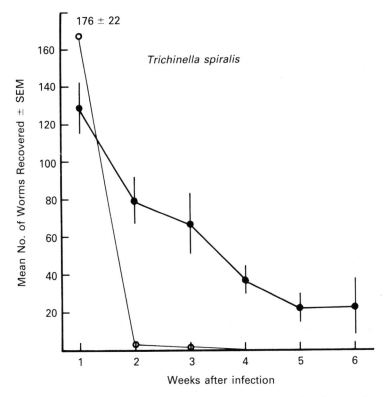

Figure 13.7. The effect of concurrent infection with two different species of nematode parasites. The figure shows delayed expulsion of *Trichinella spiralis* from mice concurrently infected with *Heligmosomoides polygyrus* (Behnke *et al.*, 1978). Mice infected with *T. spiralis* alone (○); mice infected with *T. spiralis* and *H. polygyrus* (●).

An interaction between unrelated infectious organisms which has far-reaching consequences for people living in the tropics is that between malaria and Burkitt's lymphoma. The latter is a sarcoma of the jaw (Burkitt, 1969) commonly encountered in warm, moist, tropical countries such as central Africa, New Guinea and South America. The lymphoma is believed to be associated with the Epstein-Barr (EB) virus (Herpes group), but the virus is not essential to the pathogenesis of the lymphoma. EB virus is also responsible for the ubiquitously distributed but relatively mild infection, glandular fever (Tosato, 1987).

The geographical distribution of Burkitt's lymphoma coincides remarkably with that of malaria, and it is implied that an interaction exists. Five lines of evidence support this conclusion:

1. A high incidence of Burkitt's lymphoma is found only in areas where malaria is endemic. There is a lower incidence at higher altitudes in endemic areas and this coincides with less malaria.

2. Persistent malaria precipitates the onset of Burkitt's lymphoma.
3. There is a seasonal pattern of onset of Burkitt's lymphoma again coinciding with the malaria transmission season.
4. The incidence of Burkitt's lymphoma drops when malaria is controlled.
5. The haemoglobin genotype AS is under represented in patients with Burkitt's lymphoma.

A causal connection between Burkitt's lymphoma and malaria is therefore implied (Morrow *et al.*, 1970). It is suggested that malarious people are more prone to develop Burkitt's lymphoma when infected with the EB virus, and this may be related to the malaria-induced changes in the lymphoid and reticulo-endothelial systems (Figure 13.8).

A comparable situation has been described in mice infected concurrently with *P. berghei yoelii* and with Moloney lymphomagenic virus (MLV). About

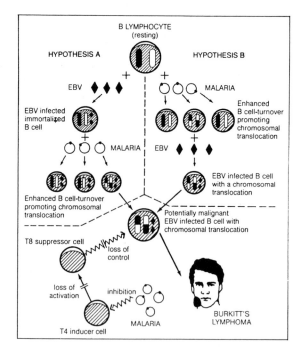

Figure 13.8. Two possible mechanisms for the interaction of malaria and Epstein Barr Virus (EBV) in the pathogenesis of Burkitt's lymphoma. The lymphoma seems to be caused by uncontrolled growth of B cells infected with EBV which show a chromosomal translocation affecting cellular growth and differentiation. The inducer-suppressor T cell is inhibited by malaria, allowing the tumour to develop.

Malaria also acts in the initiation of B-cell transformation. In hypothesis A, the B cells are first immortalized by infection with EBV and their multiplication is then stimulated by repeated malaria infections—thus increasing the likelihood of chromosomal translocation. In hypothesis B, the primary event would be malaria infection promoting B-cell multiplication and chromosomal translocation, followed by EBV infection which immortalizes the transformed cells (Greenwood, 1987).

25% of normal mice develop lymphoma when exposed to MLV but the percentage rises to almost 100% among mice concurrently infected with malaria (Wedderburn, 1974).

These examples serve to illustrate that in nature, man and wild animals harbour a variety of different parasites concurrently (Figure 13.9) and that infectious organisms can interact in a way which may enhance the host's susceptibility to heterologous infection. Such interactions may be mediated through the immune system and in some cases may be beneficial to the organisms involved by extending the period of infection and hence the opportunities for transmission.

13.3. PARASITE-DERIVED MECHANISMS

Manipulation of parasite antigens to reduce overall immunogenicity to the host

Lack of antigenicity in parasite molecules

It has been argued in the past that parasites may fail to elicit protective responses in their hosts because the molecules on their surface or in their excretory products may be non-immunogenic, presumably through some peculiarity of their biochemical structure. There is little evidence to support

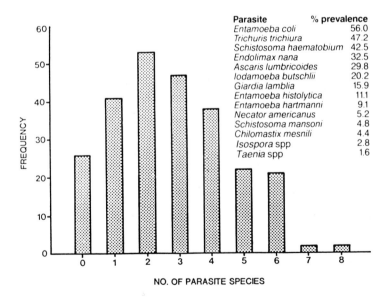

Figure 13.9. The number of parasite species per child in a small primary school in the Valley of a Thousand Hills near Durban in South Africa (Kvalsvig, 1988; by kind permission of Dr. Kvalsvig and Dr. R.M. Cooppan).

such a view. Every species which has been examined, using contemporary techniques, has been shown to be intrinsically capable of stimulating the host's immune system and of eliciting a response. The resultant immune response may not always be protective to the host, thereby enabling the parasite to survive. An insight into the significance of these interactions may be gained through experiments utilizing abnormal host-parasite relationships. Provided the development of the species concerned is not excessively abnormal, the balance of the relationship may be swayed in favour of the host, and a protective response ensues. Parasites are finely tuned to the physiology of their definitive hosts, and even a slight change in this relationship may tip the balance in the host's direction. Three examples will serve to illustrate the point.

Heligmosomoides polygyrus is a long-lived parasite in laboratory mice usually causing a stable chronic infection. In the Mongolian jird, *Meriones unguiculatus*, the parasite establishes, grows to maturity and initiates egg production. However, within five weeks of infection, all the worms are rejected (Figure 13.10d) unless the host is immunocompromized (Jenkins, 1977; Hannah and Behnke, 1982). Therefore, this parasite is immunogenic,

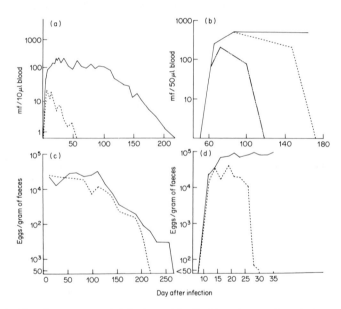

Figure 13.10. The time course of infection with filarial and gastrointestinal nematodes of laboratory animals (reproduced from Behnke, 1987). (a) *Acanthocheilonema (Dipetalonema) viteae* microfilaraemia in BALB/b (—) and C57BL/10 (__) mice, following transplantation of five female worms (Storey *et al.*, 1985). (b) *A. viteae* microfilaraemia in LSH and CB (—), LCH and MHA (__) and LVG and PD4 (— -- —) hamsters following infection with third-stage larvae (Neilson, 1978). (c) *Heligmosomoides polygyrus* in C57BL/10 (—) and NIH (__) mice following infection and 200 larvae (Behnke and Robinson, 1985). (d) *H. polygyrus* in CFLP mice (—) and jirds (__) following infection with 150 larvae (Hannah and Behnke, 1982).

and in order to evoke a response by the jird it must release appropriate antigens in the host. Presumably the same antigens are also released during infections in mice, but in this latter environment, the normal definitive host, a comparable immune response is not initiated. It is conceivable, as argued earlier, that the mouse strains used may have been susceptible strains without the genes required to develop a host-protective response to *H. polygyrus*, but this is clearly not so because the same mouse strains can be made responsive by repeated infection. The failure of mice to resist a primary infection is believed to be associated with the immunomodulatory activity of adult worms. Whilst jirds provide the nutritional environment for normal development, they remain insusceptible to this evasion strategy which is finely tuned to the mouse system and consequently successfully eliminate the parasites within a few weeks of infection (Behnke, 1987).

Another example is given by *Acanthocheilonema (Dipetalonema) vitae*, a filarial parasite of wild jirds. In the normal definitive host the adult worms live for a long time (Johnson *et al.*, 1974) and microfilarial counts rise during infection to a plateau which is maintained for many months (Beaver *et al.*, 1974; Weiss, 1970). In hamsters the parasite develops normally and microfilariae appear in the circulation, but within three months of infection an antibody response is generated and microfilaraemia is cleared (Haque *et al.*, 1978; Neilson, 1978; Weiss, 1978).

A final example is represented by *Hymenolepis diminuta* which survives for years in rats, its normal host, but not in mice. A protective immune response is initiated in mice, and the worms are cleared from the intestine before patency (see p. 382).

There are many comparable examples in the literature. One notable host-parasite system which has been investigated quite thoroughly over the years is represented by infections with *Schistosoma mansoni* in a variety of different hosts including mice, rats, hamsters, guinea pigs and even man. In each host the response shows a distinct pattern. In hamsters, for example, innate resistance at the skin is low, whereas in rats it is extremely high. A discussion of this topic is beyond the scope of the present chapter, and the reader is referred to the literature on this subject (Clegg and Smithers, 1968; Capron and Capron, 1986).

Quantitative effects—the threshold phenomenon

It has been known for a long time that in parasitic nematode infection of the intestinal tract, a certain minimum threshold of intensity in worm numbers is required to elicit a host response. In *N. brasiliensis* this may correspond to about 50 worms per rat although the threshold clearly varies between laboratory rat strains (Ovington, 1986).

The threshold phenomenon is well documented in mice infected with *T. muris* where again the exact threshold varies from one mouse strain to another (Behnke *et al.*, 1984). In BALB/c mice given 20 eggs, some larvae developed into patent adults and were not expelled, but when the dose was increased to 100 eggs no patent adults were found at autopsy on day 35. The number of eggs required to elicit an immune response is variable between strains. In some (e.g. $C_{57}Bl_{10}$) it may be higher, whereas in others, (e.g. NIH), lower. Furthermore, the threshold may be altered by concurrent infection, particularly when the second species is *H. polygyrus*; the threshold for eliciting a response is significantly elevated (Behnke *et al.*, 1984).

Low-level infections may be an effective strategy in evading host resistance. Epidemiological data confirm that in man and wild animals low intensity infections are usually encountered with only a few hosts harbouring heavy worm burdens. Observations on a wild house mouse population affected by *T. muris* confirm that low-intensity infections, one to five worms, were normal in nature. Infection intensities of this order would not be rejected by most strains of laboratory mice (Behnke and Wakelin, 1973).

The evolution of host-like antigens on parasites—molecular mimicry

The concept of parasitic organisms evolving so as to minimize their immunogenicity to the host by gradual reduction of the antigenic disparity between themselves and the tissues of their hosts originated in the 1960s through publication by Sprent (1962) and Dineen (1963) and was refined by Damian (1964) who reviewed the evidence available for related molecules being expressed by both host and parasite.

One of the best known and most often quoted examples was the discovery of A2 macroglobulin on the tegument of *S. mansoni* (Damian, 1967). This molecule which is seen by antisera raised in rhesus monkeys againt pure mouse A2 macroglobulin, does not cross react with primate A2 macroglobulin and is observed on worms grown both in mice and baboons. The fact that worms developing in baboons express this antigen makes it unlikely that it is an antigen of host origin and it must be concluded that this particular molecule is synthesized by the parasites themselves. However, worms have survived unaffected in primates vaccinated against this antigen, and therefore specific antibody responses do not mediate protection against infection. Despite the lack of serological cross reactivity between antisera specific for murine and primate A2 macroglobulin, the molecules are likely to be similar in some respects and in this context the hypothesis put forward by Grossman *et al.*, (1986) would seem to have particular appeal in explaining their biological significance.

Grossman *et al.*, (1986) have argued that antigens which resemble host antigens, when available locally in sites of parasite residence, at optimal densities and for prolonged periods, favour the generation of lymphocyte clones which proliferate without maturation into effector cells. These gain a selective advantage because lymphocytes which react proliferatively with certain classes of abundant self antigens (i.e. the host-reference antigens which the parasites attempt to mimick) are selected and by a process of feedback regulation dynamically suppress the proliferation and activities of other clones.

The lymphocytes positively selected by self recognition have relatively low affinities for the driving antigens (self antigens or parasite host-like antigens). Suppression of host-protective immunity ensues when such clones have numerical dominance over other clones. Since these clones have proliferated with little or no maturation, the outcome is non-responsiveness and active suppression of the other potentially effector clones. Furthermore, it is envisaged that this mechanism operates locally in the immediate vicinity of the parasite and thus the hypothesis predicts concomitant immunity. The infective stages of schistosomes penetrating at a distant skin site would be open to attack through unaffected high-affinity clones facilitating effector responses against invasion. The parasites thus create in the host locally restricted sites where they prosper, but superinfection of the host is minimized by the resistance to further infection. This hypothesis is amenable to experimental testing and its predictions will undoubtedly be examined. It is particularly interesting in the present context because it predicts concomitant immunity, local aggregations of parasites, polyclonal activation and also auto immunity, all of which are readily apparent manifestations of many parasitic infections.

Manipulation of parasite antigens to enable evasion of the host-effector mechanisms

Antigens on the surface of parasites constitute the direct interface between host and parasite and as such represent the most apparent and easily accessible target for a host response. Effector responses, involving antibody (complement-mediated lysis or antibody-dependent cellular cytotoxicity, ADCC) targeted at the parasite surface, constitute the most obvious line of attack (Chapter 5). Parasites, however, have evolved strategies for minimizing such attacks by rapid turnover or regular change in the antigens exposed on the surface (Turner, 1984).

Antigenic variation

Perhaps the best example of antigenic variation is given by Salivarian trypanosomes. The historic report by Ross and Tomson (1910) was the first to draw attention to the regular fluctuations in parasite numbers which characterize trypanosome infections in man (Figure 13.11). Each peak of parasitaemia is now recognized as constituting essentially one antigenic variant of the parasite, the dominant form at any one time being referred to as the homotype. Declining parasitaemia reflects a host response specific to the homotype. The elimination of this variant by an antibody response is followed soon by the appearance of another variant which is unaffected by the antibodies responsible for clearing the previous homotype. The replacement variant arises from the several heterotypes (suppressed variants which are present in the host in small numbers at any one time, and once a dominant

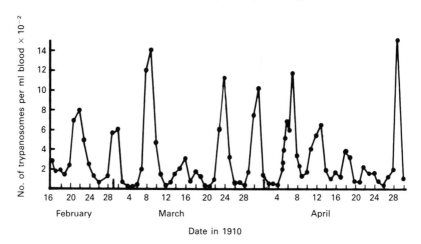

Figure 13.11. Regular fluctuations in the concentration of trypanosomes in the blood of a patient who was exposed to *Trypanosoma gambiense* in Africa and then monitored closely during treatment in the Southern Hospital in Liverpool (Ross and Tomson, 1910).

form has been established, the other heterotypes are kept suppressed until this second homotype is itself eliminated by a specific antibody response (Figure 13.12).

The structure of these antigens has already been described in Chapter 2. Each trypanosome has encoded in its genome the information required to express any of the hundreds of variants (variable surface glycoproteins—VSG) which characterized a particular clone. Only one is expressed at any one time point, but this may be shed by the activity of an enzyme which cleaves the molecule from the plasma membrane where the conserved tail region of each VSG penetrates the lipid bilayers. A replacement is then synthesized, expressed in the position of its predecessor and, temporarily at least, the parasite is free from the host response (Donelson and Turner, 1985).

The repertoire of antigenic variants seems to be inexhaustible. It is thought that upwards of 1000 distinct VSGs may be encoded in the genome of each isolate with additional variants arising through gene conversions creating hybrid genes (Steinart and Pays, 1986; Langley and Roth, 1987). Not only do different species of trypanosomes express their own characteristic variants, but strains of each species representing distinct geographical isolates can also be distinguished on the basis of their VSG repertoire. Thus the problem for immunological intervention is immense (Nantulya, 1986; Section 14.5). Originally hopes for introducing an effective vaccine were based on the observation that on passage through the tsetse fly, irrespective of which VSG was taken up, the parasites reverted to a basic form which was also expressed during the first peak of parasitaemia after transmission to the mammalian host. Further, original studies by Gray (1965) had suggested that the sequence of the

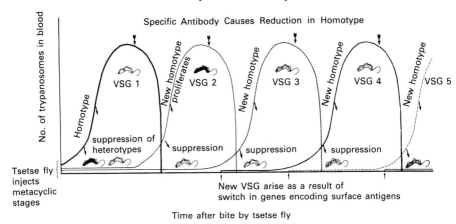

Figure 13.12. Schematic diagram showing a hypothetical sequence of events following the exposure of an individual to the bite of a tsetse fly infected with trypanosomes. During the blood meal a variety of metacyclic forms are injected. Some of these will be of one antigenic type expressing variable surface glycoprotein 1 (VSG1) and this variant, referred to as the homotype, will cause the first peak of parasitaemia. In the meantime other antigenic variants, the heterotypes, are kept suppressed until the homotype is recognized by the immune system and specific antibody begins to reduce the intensity of parasitaemia. One of the heterotypes now begins to proliferate and replaces the original homotype as the dominant variant in the circulation. During a prolonged infection, new variants arise (VSG4 and VSG5 in the figure) which were not expressed in the metacyclic population. These will also persist in very low intensity until an opportunity arises for them to proliferate more rapidly and replace an existing homotype.

appearance of VATs was totally predictable for a given clone. It was thus argued that a vaccine which generated protection in the host against the basic form and the VATs expected in the first peaks of parasitaemia would be protective. However, the pattern of appearance of variants was later shown not to be totally predictable, although there was a high statistical probability of a particular VSG being expressed in the relapse population of a specific clone (Miller and Turner, 1981). Attempts to vaccinate against the basic metacyclic form and the most likely relapse VATs were unsuccessful because the parasites simply reverted to a VAT which had not been catered for in the vaccine.

The metacyclic forms in the tsetse fly are not uniform, and whilst the homotype accounts for the majority, the salivary gland population is also contaminated by representatives of other VATs usually expressed in subsequent peaks of parasitaemia (Barry et al., 1979). However, the total pool of VATs which may be expressed in the metacyclic population is not unlimited (Barry, 1986). Metacyclic forms represent a subset of the entire antigenic repertoire for a given isolate (Barry et al., 1983). Thus the possibility still exists of devising a cocktail vaccine which covers the entire spectrum of forms likely to be encountered at transmission. Such a vaccine may prevent the parasites from establishing a foothold in the host by eliminating the organisms before they have an opportunity to proliferate (Steinart and Pays, 1986).

Antigenic variation has been reported in other protozoan infections, notably malaria and *leishmania*. In malaria, antigenically disparate forms arise when relapse populations appear weeks or months after a primary episode of infection. Studies with *P. cynomolgi* in monkeys have demonstrated that such populations are antigenically distinct when compared to the original inoculum (Voller and Rossan, 1969). However, the pattern of antigenic variation in malaria is quite different from that observed in trypanosomiasis. The frequency of appearance of antigenic variants is considerably lower and each gives rise to a bout of malaria which may last for days or even weeks but certainly does not cause as intense an infection as the original inoculum. Thus there appears to be a degree of immunity which transcends antigenic variation, and this is believed to reside in the T-lymphocyte population which initiates a secondary rather than a primary response on encountering relapse variants in sufficient quantity (Brown, 1974; Howard, 1984; Hommel, 1985).

Antigenic variation has also been described in helminth parasites although the situation here is not directly comparable to that observed in the protozoa. The first species to be studied in this context was *N. brasiliensis*. Worms developing in hosts resistant to infection adapt to the immune environment and can be distinguished readily from primary infection worms with no experience of host immunity. Adapted worms are stunted and do not show the severe internal disruption which is evident in primary infection worms when they are eventually subjected to the host response. Table 13.4 summarizes the differences between normal, damaged and adapted worms.

Perhaps the most interesting finding is that the pattern of secretion of acetylcholinesterase (ACH) is dramatically altered. ACH has been identified in many intestinal nematode species (Ogilvie *et al.*, 1973) and some have been

Table 13.4. Comparison of the properties and characteristics of normal, damaged and adapted *N. brasiliensis* (Jenkins, 1972; Ogilvie, 1974; Jenkins *et al.*, 1976).

Feature compared	Worm population Normal	Damaged	Adapted
Size	large	large	small
Fecundity	high	very low	intermediate
Distribution in the intestine	mid-gut	anterior	anterior
Survival in naive recipient rats	7–10 days	2–6 days	more than 2 weeks
Immunogenicity	high	?	low
Acetycholinesterase secretion a) Qualitative			
Isoenzymes secreted	A,B,C	B,C	A,B,C
b) Quantitative			
Relative amount secreted	1	3.6	1.9

found to secrete enormous quantities of this enzyme (e.g. *Nematodirus battus*; Lee and Martin, 1976; Douch *et al.*, 1988). The metabolic effort required to synthesize the quantities involved implies that the role of this enzyme must be quite important to the parasites. Nevertheless, ACH is not universally secreted and even closely related organisms differ, e.g. *N. americanus* secretes large amounts of ACH; *A. tubaeformae* and *A. ceylanicum* do not, and yet all three occupy quite comparable niches in the intestine. The role of ACH in intestinal parasites is still a controversial issue. Originally it was considered that this enzyme acted as a biological holdfast for the parasites, slowing down intestinal peristalsis and creating a more amenable environment for the worms. More recently Philipp (1984) has pointed out that ACH has potent inhibitory effects against mucus secretion in the intestine, and since mucus secretion is thought to play a vital role in resistance, especially of secondary infections (Miller, 1987), it makes sense for parasites to suppress mucus secretion as a counter strategy. It is possible that the different isoenzymes of ACH have varying efficacy in inhibiting mucus secretion or affecting peristaltic activity and that modulation of the ratio of isoenzymes secreted is a direct response to the state of immunity of the host in an effort to prolong worm survival.

Adult *N. brasiliensis* developing in immune rats are less immunogenic than normal worms during a primary infection. This can be seen in two ways. Firstly, when comparable numbers of normal and immune adapted worms are transplanted into naive rats, the former generate more intense secondary responses to subsequent challenge with larvae (Jenkins, 1972). Secondly, when normal and immune-adapted worms are transplanted into naive rats, the former population is expelled in the normal way whilst the latter persists significantly longer, maintaining egg production after it has totally ceased in rats given normal worms (Figure 13.13).

So far as is known, these observations appear to be unique to *N. brasiliensis*. Comparable experiments with *T. spiralis* or *S. ratti* have not given similar results. The latter species may become stunted in secondary infections but show no enzymatic variation nor antigenic changes of the type described for *N. brasiliensis*.

Shedding of parasite antigens

Although the outer surface of nematodes is seemingly composed of a tough, protective cuticle, there is good evidence it is metabolically active, both absorbing extrinsic molecules and secreting others into the parasite's environment, e.g. during *in vitro* incubation the surface cuticular antigens may be released (Philipp and Rumjaneck, 1984). *Trichinella spiralis* was the first to be studied in this context, but other species include *N. brasiliensis*, *H. polygyrus* and *N. americanus*. I^{125}-labelled worms of these species release surface antigens which can be detected in culture medium as radioactivity.

One immediately apparent purpose of this strategy may be to foil the host's attempts to encapsulate the parasite through ADCC reaction or through

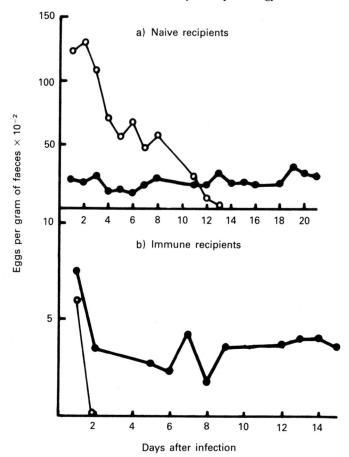

Figure 13.13. Prolonged survival of immune-adapted *Nippostrongylus brasiliensis* on transplantation to a) naive or b) immune rats (Jenkins and Phillipson, 1972). Both graphs show faecal egg counts taken at intervals from rats infected with either normal worms (○) or immune-adapted worms after laparotomy (●).

complement-mediated cytotoxicity. Specific responses, mediated via antibody would be incapacitated if the target antigens were to be only weakly or transiently attached to the parasite surface. Nevertheless, antibody dependent cytoadherence (ADCA) can be readily demonstrated with most species using *in vitro* techniques. Infective larvae of *T. spiralis* and *N. brasiliensis* become enmeshed in adherent effector cells when incubated in their presence together with complement. However, *in vitro* conditions can never mimick totally the *in vivo* environment, and it is as well to remember that it is the *in vivo* conditions which a parasite has to face which have dictated its evolution and are responsible for the adaptations which it exhibits. *In vivo*, where the parasite may be better tuned to the events taking place in the tissues, antigen shedding may serve to delay entrapment and in the case of migrating larvae, which move rapidly through the host, may provide an evasive strategy of sufficient potency to allow significant numbers of worms to survive unimpaired.

Some of the most interesting observations on this subject were made by Vetter and Klaver-Wesseling (1978) who showed that antibody to surface determinants of *A. caninum* would only bind to frozen worms and not to live parasites. Binding to live worms could only be demonstrated when their metabolic processes were inhibited by azide.

Schistosomes have a complex outer membrane overlying the tegument which comprizes a double lipid bilayer (Hockley and McLaren 1973; McLaren and Hockley, 1977; and Hockley *et al.*, 1975). The synthesis of this membrane has received a lot of attention because of its unusual structure and of the host antigens which are attached to it. The outer bilayer was also shown to be continuously turning over with new membrane replacing shed material. Initial *in vitro* estimates of the rate of turnover gave half lives of less than 6 h (Wilson and Barnes, 1977; Roberts *et al.*, 1983) which were not compatible with studies *in vivo*. Worms grown in mice (and therefore with acquired mouse erythrocyte antigens) required three to seven days residence after transplantation to naive monkeys to replace sufficient mouse antigen by monkey erythrocyte antigen to enable survival on subsequent retransplantation to monkeys immunized against mouse erythrocyte antigens (Smithers *et al.*, 1969). Worms transplanted directly from mice to monkeys immunized against mouse antigens were rapidly destroyed (Smithers and Terry 1969, 1976; Smithers *et al.*, 1969). However, the development of more sensitive *in vitro* techniques for measuring host antigens on parasite surfaces gave estimates which were in close agreement with *in vivo* studies. Experiments in which worms were implanted into hamsters and subsequently recovered for measurement of residual mouse erythrocyte antigens estimated a half-life of 5.4 days (Figure 13.14; Saunders *et al.*, 1987). It is thus unlikely that the relatively slow rate of turnover would serve to frustrate the adherence of antibodies or host-effector cells to schistosomes, and it was concluded that the process was probably more important in repairing damage to the parasite surface than actively evading immunity.

A particularly fascinating and probably unique adaptation has been described in pentastomid parasites, which exploit specialized glands (frontal, hook and subparietal) to secrete a lamellate secretion onto the cuticle. The continuous discharge of this material, especially in those areas which have intimate contact with the host, and its special properties are believed to protect the parasite from host-immune effectors, probably by frustrating the attempts of leucocytes to adhere to the parasite surface (Riley *et al.*, 1979; Riley, 1986).

Stage-specific antigens

The development of surface radio-labelling techniques using I^{125} and their successful application to investigations of the antigenic nature of the nematode cuticle surface was followed by the discovery that in some species the

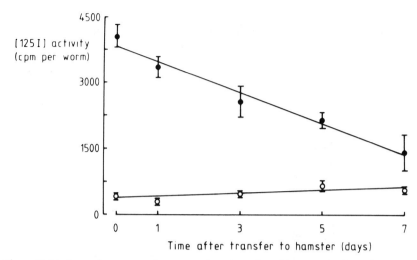

Figure 13.14. Loss of mouse erythrocyte antigen from the schistosome surface after transfer of worms from mice to hamsters. Mouse erythrocyte antigens were measured by an immunoradiometric assay utilizing rabbit anti-mouse erythrocyte serum. (○) Mean control values (worms assayed with normal rabbit serum); (●) mean of experimental values (worms assayed with rabbit anti-mouse erythrocyte serum). Bars denote associated standard errors (Saunders *et al.*, 1987).

dominant antigenic molecules varied from one developmental stage to the next (Figure 13.15; Maizels *et al.*, 1983). Thus in *T. spiralis* a repertoire of antigenic molecules varying in size were recognized as distinctive of specific developmental stages (Parkhouse and Ortega-Pierres, 1984). Each time the parasite moulted the host had to initiate the process of recognition and generation of effector response anew (Philipp and Rumjaneck, 1984). Such a mechanism would seem to be appropriate for *T. spiralis* where the host is likely to encounter numerous infective larvae at one meal, e.g. by the consumption of infected meat. The larvae would subsequently develop in unison, much as in a single pulse laboratory infection, and because the antigens available to the host change with each moult, the larval stages would be protected until no more moults were possible. In *T. spiralis* the adult worms are generally short lived and are eliminated effectively, but the muscle larvae persist in the safety of their intramuscular sites of encystment. This mechanism can be envisaged as delaying the host-protective response by a few days perhaps but certainly not in the long term because nematodes only undergo four moults and once the adult stage has developed, no more changes are possible. In sheep and cattle continually exposed to small numbers of trichostrongyle larvae during grazing, a mechanism of this sort would have little overall protective value as the parasite burden would not be synchronized in development.

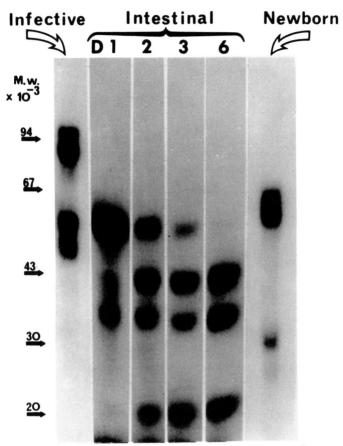

Figure 13.15. Stage-specific surface antigens of *Trichinella spiralis*. The figure shows an autoradiograph of an SDS-PAGE gel of [125]Iodine-labelled surface proteins of the three stages of *T. spiralis*: infective larvae, intestinal worms (removed from the small intestine at days 1, 2, 3 and 6 after oral infection) and newborn larvae (Philipp *et al.*, 1980); reprinted by permission from *Nature* Vol. **287**, p. 539, copyright © 1980, Macmillan Magazines Ltd.

Manipulation of host molecules by parasites as an antigenic disguise: masquerading as self

One possible solution to the problem of being recognized by the host as a foreign tissue is to confuse the immune system by camouflaging the parasite surface with host molecules. Such a mechanism is recognized as offering protection to schisotosomes from host immunity, and over the last two decades this has received considerable attention from many laboratories because of the scope for elaborate and elegant experimentation as well as the intellectual challenge involved in unravelling the component processes (McLaren, 1984).

The pioneering work of Smithers and Terry with *S. mansoni* established the concept of concomitant immunity in which adult worms were shown to survive in rhesus monkeys while eliciting resistance to challenge infection with cercariae. Thus, whilst the host expressed fully protective responses to cercarial challenge, adult worms could not be rejected (Smithers and Terry, 1976).

This was followed by an exciting period during which the source of antigenic stimulation was identified as emanating principally from adult worms, and eventually the concept of disguise was tested by the classic experiments in which worms were transplanted from mice into anti-mouse monkeys (monkeys immunized against mouse antigens) and were shown to be killed in these recipients whilst surviving in non-immunized monkeys (Table 13.5). Monkey-derived worms transplanted into anti-mouse monkeys were also unaffected by the immunization. The source of the host antigens was identified as the major blood group antigens (Clegg *et al.*, 1971), and the location on the parasite was shown to be the outer lipid bilayer of the tegument (Smithers *et al.*, 1969).

Host antigens are believed to be acquired within four days of the passage of cercaria through the host skin, and their appearance on the tegument coincides with growing resistance to immune effector mechanisms. Thus four hour schistosomula are susceptible to antibody-mediated cellular cytotoxicity and do not express host antigens, but their surface is rich in parasite antigens. In contrast four-day-old schistosomula express large quantities of host erythrocyte antigen but are only weakly identified by sera with specificity for parasite antigens. Although glycolipid major blood group antigens A,B, H and Lewis[b+] are the principal molecules utilized by schistosomes in man,

Table 13.5. Evidence that schistosomes utilize host blood group antigens to evade the host-immune response. Schistosomes were grown either in A) donor monkeys or donor mice and subsequently transplanted into naive recipient monkeys or monkeys vaccinated against mouse antigens (Smithers *et al.*, 1969; see also Smithers and Terry, 1969); or B) *in vitro* in human blood (group A or B) and were implanted into naive recipient monkeys or monkeys vaccinated against human blood group A (Goldring *et al.*, 1976). In each case the final column of the table shows the percentage recovery of worms from individual monkeys at autopsy four to five weeks later.

Donor species	Recipient monkey	Percentage of worms recovered
A) Monkey	Normal	60,
Mouse	Normal	66, 89
Monkey	Anti-Mouse	83, 100, 95
Mouse	Anti-Mouse	0, 0, 0
Blood group used for culture		
B) Group A	Anti-A	1, 3, 5, 0
Group A	Normal	81, 79, 48, 69,
Group B	Anti-A	80, 79

other host molecules are also incorporated, including MHC gene products (Sher *et al*, 1978). Host antigens are believed to mask schistosome surface antigens, interdigitating between parasite molecules in the tegumental outer membrane and hence make the worms 'invisible' to the immune system, but the disguise is probably not entirely complete because adult worms bind anti-schistosome antibody, albeit weakly (McLaren, 1984).

The search for host molecules on parasites has been extended to other species, but no other parasite has been discovered using a strategy directly comparable to that employed by schistosomes. Attempts to repeat the Smithers and Terry experiments with *B. pahangi* were unsuccessful (McGreevy *et al.*, 1975) but human blood group substances A and B have been reported on the microfilariae of *Wuchereria bancrofti* and *Loa loa* (Ridley and Hedge, 1977). Other studies have revealed host albumin on the cuticle of microfilariae, the functional significance of which is not clear. *Wuchereria bancrofti* and *B. malayi* have very similar host-parasite relationships and life cycles and yet the Mf of *W. bancrofti* express human albumin in association with the cuticle and those of *B. malayi* do not (Maizels *et al.*, 1984). Host albumin has also been recognized on the Mf of *O. gibsoni* (Forsyth *et al.*, 1984). This molecule may mask all other antigenic epitopes on the cuticle of the Mf, allowing them free movement through host skin, unhindered by host resistance.

Manipulation of the host immune system

In the course of the continuous contest between host and parasite, mechanisms have evolved through which parasites manipulate the host's immune system in a manner advantageous to themselves.

Misdirection of the immune response

LOW AFFINITY RESPONSES
One of the outstanding successes of the last decade has been the development of recombinant DNA technology, and the application of these techniques to the study of malaria antigens has led to the discovery that many of the surface and released antigens have common epitopes and an unusual repetitive molecular structure (Chapter 2). A characteristic feature of the malaria antigens is the regular occurrence of repeated segments, comprizing hydrophilic and hydrophobic amino acids alternating in tandem repeats at set intervals throughout the molecule (Anders *et al.*, 1986). Anders (1986) has suggested that these antigens may be involved in protecting the parasite by misdirecting the host response in a direction leading to the synthesis of low-affinity antibodies. It is proposed that when presented with an array of immunodominant, interrelated antigenic epitopes with varying degrees of cross reactivity, the host generates a spectrum of antibodies mostly with low affinity for the antigens in question. Thus instead of selecting B-cell clones with high affinity for parasite antigens, with resultant host-protective immunity, clones producing

low-affinity antibodies, which would otherwise be deleted, are allowed to persist (Manser *et al.*, 1985). The net outcome is that a low-key response is generated and the failure of the host to detect specific targets and to concentrate effector mechanisms against these, benefits parasite survival.

As far as is known, such a strategy appears to be unique to malaria, but it is likely that other species have evolved comparable mechanisms for minimizing their immunogenicity. Recently a 29 kDa protein with extensive regions of a repeated 5-amino acid motif has been cloned from *Brugia* (Selkirk *et al.*, 1988). Besides proteins, carbohydrates and lipids are also located on the surface of a variety of different parasitic species, including nematodes. These molecules consist of polymers of sugar units or fatty acids and may also appear to the host as a spectrum of similar epitopes on the parasite surface, with conceivably the same consequence as in malaria.

BLOCKING ANTIBODIES

A related strategy has been recognized in schistosome infections in man. For many years blocking antibodies have been suspected and indeed experimental studies have alluded to their existence (Grzych *et al.*, 1984). Butterworth and Hagan (1987) have put forward a hypothesis based on blocking antibodies to explain the epidemiological patterns of infection as reflected in age-intensity profiles of affected communities (Figure 13.2c). It is postulated that in early life, as children acquire worms through continuous exposure to small quantities of cercariae, the major antigenic stimulus is provided by the eggs. The fecundity of adult worms is immense (female worms produce 300 or more eggs per day), and hence eggs accumulate in tissue sites, constituting in time a greater mass than that represented by the adult worms themselves. The antigens released by the eggs include the major egg polysaccharide, particularly an antigen known as K3. These elicit antibody responses which are cross reactive with antigens on the schistosomular surface (see p. 33, 38 kDa), but because IgM antibodies are elicited preferentially, instead of mediating protective immunity, the cellular cytotoxic mechanisms are blocked. Children concurrently mount antibody responses, including IgG and IgE isotypes, against the same antigens, and although these could mediate protective effector mechanisms, they are prevented from doing so by IgM antibodies to the same antigenic determinants. In early childhood, this blocking mechanism appears to be dominant, but by the teenage years, the balance shifts in favour of the host and protective responses override the blocking mechanism, preventing further infection. Consequently, the characteristic downward trend in the average age-intensity curve is generated. Blocking antibodies and serum factors have also been recognized in other parasitic diseases, notably in filariasis (Ottesen, 1984; Behnke, 1987).

Immunodepression

Immunodepression is perhaps the most difficult overall strategy to outline, because to date no clear picture has emerged and there are few generalizations which can be made. It would be true to say that most parasites can subject their hosts to a period of temporary immunodepression, and this can be demonstrated in a variety of experimental and diagnostic ways. Assays involving the response of host lymphocytes to mitogens show a non-specific depression of responsiveness. Other experimental approaches include injection of unrelated antigens into infected animals (e.g. sheep erythrocytes SRBC) and subsequent measurement of the ensuing response to the heterologous antigen. Direct measurement of lymphocyte or effector cell numbers in the host or at the site of infection are also ways of evaluating the host potential for responding. Depressed host immunocompetence can be readily demonstrated using these techniques in malaria, trypanosomiasis, filariasis and in a variety of other systems (Askonas, 1984). In some instances it may stem from the pathology induced by infection; in others from an exhaustion of the immune system, following continuous stimulation by an array of changing antigens (e.g. trypanosomiasis). The membrane molecules of trypanosomes possess mitogenic properties, the result of which is that lymphocytes with specificities unrelated to parasite antigens are also induced to proliferate in the host and in due course the system becomes overloaded and the reactivity to a given set of experimentally applied antigens may be diminished (Vickerman and Barry, 1982).

There is little long-term benefit to the parasite in compromising the host's immune system non-specifically and thus laying the host open to attack by other pathogens, as in the case of AIDS. Host survival would be drastically curtailed, and with the death of the host the parasite would also perish. There are exceptions to this, as for example in the case of parasites which are transmitted through predator-prey food chains which may benefit from weakening the host in a way which impairs the escape response to predation by definitive hosts. It may also be that parasite reproduction and transmission is satisfied in the intervening period from infection to total incapacitation, and here the death of the host may be of little consequence to the parasite. However, for chronic species, which depend on prolonged survival in the definitive host to provide them with the opportunity for the production of the quantities of eggs which are necessary to overcome the hazards involved in transmission, non-specific immunodepression of the host would be a self-defeating strategy.

Of more relevance are those studies which suggest that overall host immune reactivity is preserved but the responses to parasite antigens are diminished. In this context filariasis would appear to provide the best examples although it must be emphasized that this is still a controversial area. Specific unresponsiveness of lymphocytes to parasite antigens has been demonstrated in hosts with microfilaraemia (Ottesen *et al.*, 1977, 1984; Piessens *et al.*, 1980a, 1980b; Haque *et al.*, 1983), whilst overall reactivity to other non-parasite-related antigens is unaffected (Figure 13.16; Narayanan *et al.*, 1986). Microfilaraemic hosts do not show antibodies to microfilarial antigens as long as microfilariae persist in the circulation, but when microfilaraemia is cleared, antimicrofilarial antibodies can be detected and lymphocyte responsiveness to

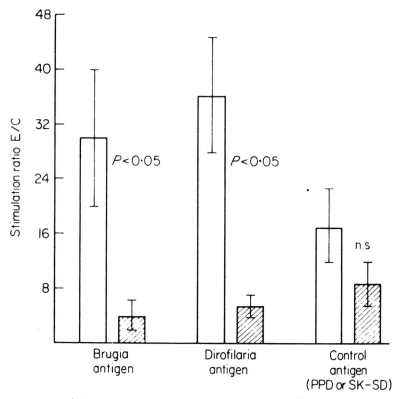

Figure 13.16. Lymphocyte responsiveness of children living on a Pacific Island (Mauke, Cook Islands) where bancroftian filariasis is endemic to filarial and non-parasite antigens. The responses of four infected, microfilaraemic children (hatched columns) are compared to those of six children living in the same environment but having no clinical or parasitological evidence of infection (open columns). Columns define the arithmetic means of the stimulation ratios in each group and the error bars record 1 standard error on each side of the mean. Lymphocytes from all individuals were challenged with filarial antigens, tuberculin protein (PPD) and streptokinase-streptodornase (SK-SD) (Ottesen *et al.*, 1977).

Mf antigens is restored (Ponnuduria *et al.*, 1974; Subrahmanyam *et al.*, 1978). A mechanism along these lines would clearly benefit the parasite through providing an opportunity for transmission without exposing the host to the onslaught of other pathogens.

As has already been emphasized, local immunomodulatory activity by parasites may be another avenue through which survival is prolonged in otherwise immunocompetent hosts. Parasitic worms, whether living in the gut or in the tissues, often accumulate in sites in the host, e.g. *Litomosoides carinii* in pleural cavity, *O. volvulus* in nodules, intestinal worms aggregate at points along the length of the intestine. In addition to exploiting their preferred sites for nutritional benefits, such aggregations may pay dividends if the organisms prolong their survival in the host by invoking immunomodulatory strategies. There is no conclusive evidence in support of this hypothesis, but circumstantial evidence is available. Immunoregulatory molecules have been identified in *Onchocerca gibsoni* and *Oesophagostomum radiatum* (10 and 25–35 kDa). It is likely that parasite factors interfering with host immunoregulation are small and relatively labile (Behnke, 1987), and this may be why there has been little success in isolating and characterizing candidate molecules. Nevertheless, the hypothesis that parasites aggregate in order to produce areas of relative safety from the host's immune system provides an explanation for the observed accumulation of parasites in specific organ sites, particularly when under immunological stress, and is one which is readily testable experimentally.

To date only two systems have been described in which it has been shown unequivocally that the consequences of the presence of adult worms leads to an increase in host susceptibility to infection. Klei *et al.*, (1980, 1981) demonstrated that jirds carrying patent infections with *B. pahangi* develop more worms on challenge than controls, an observation which may be explicable by parasite-induced structural changes in the lymphatic system in which the adult worms reside or by diminished specific responsiveness to parasite antigens brought about by the activities of the established worms.

Another example is given by the demonstration that mice infected with adult *H. polygyrus* could not be immunized by an irradiated larval vaccine which was shown to be more than 90% protective in uninfected mice (Behnke *et al.*, 1983). It was suggested that the immunomodulatory activity of adult worms was responsible, down regulating local intestinal responses to the parasite. *Heligmosomoides polygyrus* is known to have the capacity to interfere with a variety of responses associated with the gut. There appears to be little mast cell infiltration into the lamina propria during primary infection, and in mice concurrently infected with *T. spiralis*, a species which normally itself generates a vigorous mastocytosis, mast cell infiltration is severely depressed and delayed, and moreover, *T. spiralis* persists considerably longer than normal (Figure 13.7; Dehlawi *et al.*, 1987). The most likely target for the putative immunomodulatory factors produced by *H. polygyrus* are T lymphocytes mediating local inflammatory responses, but the existence of the molecules has yet to be confirmed and their mode of action remains speculative (Dehlawi and Wakelin, 1988). There is little argument, however, about the consequences of infection with this species. The effects are non-specific in so far as other parasites residing in the intestine are also affected and their survival is equally prolonged in concurrent infections.

The role of immunodepression in facilitating parasites to survive in the face of host resistance is a subject which has received a lot of attention over the last decade and still unresolved questions abound. Potentially it is a minefield for all-encompassing hypotheses because of the complexities involved in unravelling the components of homeostatic regulation of the immune system and because of the variety of different parasitic organisms with their individual requirements for survival and transmission. For each system it is important to establish whether immunodepression is an inevitable consequence of perturbed immunoregulation and temporary imbalance during the generation of a parasite-specific response or whether it is a strategy primarily initiated by the parasite to benefit its own survival. In other words, as Barriga (1984) emphasized, immunodepression may be an 'epiphenomenon' rather than an evasive strategy. This aspect of parasite immunology presents many exciting challenges or future experimental research. Foremost among the objectives must be to isolate, identify and characterize parasite molecules, with immunomodulatory properties and to vaccinate hosts in order to protect them against the immunomodulatory effects of such molecules. This may be achieved through manipulation of the molecules by steric changes which maintain immunogenicity but reduce the functional immunodepressive properties. In the absence of suppression, host resistance may then develop naturally following exposure to infection.

Resistance to host-effector mechanisms

Biochemical processes

Although the mechanisms which result in parasite attrition are generally invoked specifically through the immune system, the effector components are non-specific and involve in most cases chemical events which are either directly damaging to the parasite or create an environment in which parasite survival is untenable. An opportunity is therefore available for parasites to interfere with the biochemical pathways associated with effector mechanisms. *Leishmania* invade and develop in host macrophages. Multiplication occurs within lysosomes which contain a battery of toxic enzymes whose primary purpose is to destroy the phagosomal contents. However, the toxic mechanism itself is not blocked because organisms, such as *Listeria monocytogenes*, are killed when present concurrently within *Leishmania* infected phagosomes. Thus *Leishmania* resist the effector mechanism at the individual cellular level. Resistance is specific for the macrophages of the natural host and does not endow protection against macrophages of abnormal hosts (Mauel, 1984; Chapters 4,7,8).

ANTI-OXIDANT ENZYMES

Recent attention has been focused on the oxidative bursts employed by macrophages and other cells to generate oxygen radicals, the latter being lethal to cellular membranes and cytoplasmic processes (Hughes, 1988). Oxygen radicals are normally rapidly inactivated by a variety of enzymes of which the most important are superoxide dismutase, catalase and glutathione peroxidase. Otherwise the oxygen species would be equally damaging to host tissues (Chapter 7).

Resistance to oxygen radicals has been investigated in a number of helminths, but most attention has focused on *Trichinella spiralis*. Newborn larvae have less superoxide dismutase and glutathione peroxidase than either adult worms or muscle larvae and show correspondingly enhanced susceptibility to killing by leucocytes and in cell-free systems where oxidants are generated by chemical reactions. All stages of *T. spiralis* are low in catalase levels (Callahan *et al.*, 1988).

Interestingly, it has been demonstrated that *H. polygyrus*, which causes chronic intestinal infections, has significantly higher concentrations of these enzymes than *N. brasiliensis*, a parasite which generally succumbs to intestinal responses in its rat host. *Heligmosomoides polygyrus* has three times the level of catalase, two times the level of superoxide dismutase and four times the level of glutathione reductase normally present in *N. brasiliensis*. The relative importance of oxygen radicals as components of the inflammatory response in the gut is difficult to evaluate, but it is conceivable that such molecules contribute to the damage sustained by *N. brasiliensis* prior to expulsion from the rat, and that *H. polygyrus* survives because it is better able to resist the damage that could result from the release of free radicals during intestinal inflammation (Smith and Bryant, 1986). However, the weight of evidence in my view is on the side of active interference by this species with host-immune processes rather than on resistance to effector mechanisms as the main strategy enabling chronic survival.

Resistance to oxidant killing has also been described in *S. mansoni*. Adult worms are relatively resistant, possess higher levels of oxidant scavengers and have a greater capacity for removing hydrogen peroxide than schistosomula which, by comparison, are susceptible to the effects of the oxygen radicals. Unlike *H. polygyrus*, *S. mansoni* does not have the enzyme catalase but instead resistance is thought to be mediated through exceptionally high levels of cytochrome C peroxidase and glutathione (Mkoji *et al.*, 1988).

ANTIBODY NEUTRALIZING PROCESSES
Protection from the host-effector mechanisms may be achieved by releasing enzymes which cleave antibodies attached to the parasite surface. Two different systems have been described. In *T. cruzi* and *Fasciola hepatica* antibodies bound to parasite antigens by Fab portions of the molecule are digested by proteases, leaving the antigenic sites saturated by Fab, but no Fc portions through which the antibody may exert its biological effect (Chapman and Mitchell, 1983). Schistosomes are believed to be at least partially protected by a similar but distinct mechanism during their development in the lungs and residence in the liver. In this case, Fc portions attached to Fc receptors on the tegument are left behind after enzymatic cleavage and, furthermore, the free Fab fragments down regulate phagocytic and cytotoxic activity of effector cells in the parasites' vicinity (Capron *et al.*, 1980; Auriault *et al.*, 1981; Cohen, 1982). The capacity to secrete these 'Fabulating' enzymes appears to be transient, older worms relying more on the antigenic disguise discussed earlier.

For both micro- and macroparasites inhabiting mucosal surfaces, IgA antibodies present a particular threat. Certain species of bacteria have evolved enzymes which can cleave this molecule specifically. Thus human IgA1 can be digested by *Streptococcus sanguisa* and *Neisseria gonorrhoeae*, but IgA2 is not susceptible to cleavage because the structure of this latter isotype does not have the amino acid sequences which are the targets of the bacterial enzymes (Plaut, 1983). There is no evidence that macro-parasites employ similar strategies for inactivating IgA but it is known that several species secrete protease enzymes (McKerrow and Doenhoff, 1988), and it is conceivable that some of these have specific activity for IgA (Leid *et al.*, 1987; Lightowlers and Rickard, 1988).

ANTI-COMPLEMENTARY FACTORS

Larval cestodes need to survive for a long time in their intermediate hosts, since the route to the definitive host is via the food chain. In *Taenia taeniaeformis* the developing parasites show anti-complementary activity. A highly sulphated polyanionic proteoglycan is secreted, and this consumes complement via the alternative pathway. Leid *et al.*, (1987) have recently purified an enzyme which has been called 'Taeniaestatin', and this, in addition to inactivating complement, is also inhibitory to trypsin, chymotrypsin and inhibits lymphocyte proliferation. Complement-fixing activity has been shown in the hydatid cyst fluid of *Echinococcus granulosus*, and it may be that the factor responsible also has more wide-ranging immunodepressive properties than were originally envisaged (Hammerberg *et al.*, 1977; Perricone *et al.*, 1980).

OTHER MECHANISMS

McLaren and Smithers (1982) found that whilst five day through to three-week-old lung schistosomula were recognized by antisera against host erythrocyte antigens, such worms could not be killed by eosinophils. Under identical culture conditions, antibodies and eosinophils collaborated to destroy adult worms within a 24 h period of incubation *in vitro*. Eosinophils were shown to attach to five-day lungworms in moderate quantities and even to discharge the contents of their cytoplasmic vacuoles onto the worm surface, but the viability of lung parasites was not affected. It was concluded that lungworms must possess a second defense mechanism which enables them to resist the toxic mediators liberated by eosinophils. The mechanism may be related to some aspect of the changing surface configuration of the parasite as it develops in the lungs, but the details are still not understood (McLaren and Terry, 1982).

Rapid repair

Platyhelminth parasites, in particular, are surrounded by a live cytoplasmic tegument which is continually being replaced. Studies on *Hymenolepis diminuta* in mice have shown that immediately before expulsion, lesions appear over the strobilar surface, which under the electron microscope can be seen to involve disruption of cytoplasmic organization, reduction of microtriches, and disorganization of ribosomal and mitochondrial structure. However, these lesions are rapidly repaired, and on transfer to a suitable medium, tapeworms recover within one hour or so (Befus and Threadgold, 1975). In the mouse, a host from which parasites are rejected, the immune response gains an upper hand and the worms sustain damage faster than can be repaired. In the rat, which tolerates a chronic infection with this species and to which presumably the parasite is better tuned biochemically, the rate at which damage is sustained is less than the rate of repair and the balance in this system is in favour of the parasite (McCaigue *et al.*, 1986). *Hymenolepis diminuta* can actually outlive its host, as was shown by experiments in which worms were transplanted from one generation of rats to the next (Read, 1967).

CONTINUOUS RELOCATION IN THE HOST

Although most parasites have clearly defined preferred sites in their hosts, many embark on regular movements within the boundaries of their environment. Hookworms, for example, are thought to stay attached to a feeding site in the gut for four to six hours before moving on to another location (Kalkofen, 1970). If the host response is a local one, continuous relocation may leave the worms one step ahead of the local immune activity, until an opportunity is presented for entrapment or until the whole of the intestine responds at the organ level, leaving no hiding place for parasites (Behnke, 1987). Some species of filarial worms continue to migrate around the host during the patent period. Whilst *W. bancrofti* are generally confined to the lymphatics and *O. volvulus* in nodules, *A. viteae*, which live subcutaneously, migrate around the tissues of the host quite actively. After transplantation to an abdominal site on a naive host, adult worms may be encountered in the head region and even in the pleural cavity (Court *et al.*, 1988; Storey and Behnke unpublished observations). Another parasite which is reputed to regularly shift its location is *Loa loa* (Orihel and Eberhard, 1985). In these species destruction of the worms is preceded by encapsulation and the development of a fibrotic abscess around the parasite. As long as the worms remain active and move through host tissues, surface adherent cells may be sloughed off and ADCC reactions may be prevented.

13.4. CONCLUDING REMARKS

In this chapter, I have discussed some of the strategies which parasites use to prolong their existence in the host. This has not been a detailed treatise on all the available strategies nor have I considered the more speculative possibilites for which there is little experimental evidence. It is important to realize that at the end of the day the rate at which parasites die must equal the rate at which they are replaced. The latter component involves both the rate of reproduction and transmission efficiency. Thus the need to survive for a particular period in the host is dependent on losses sustained at other points in the life cycle. Ultimately each reproductive organism must replace itself in the next generation if overall stability of parasites is to be maintained in the host population.

In order to keep pace with the host population, parasites have had to respond to the defensive strategies which evolved to protect the hosts, and they have accomplished this in a variety of different ways. It is almost inevitable that research work has focused on those species which are economically or medically important or show fascinating or unusual solutions. Schistosomes, it is now realized, employ a battery of evasion strategies to counter the comparable array of defensive effector mechanisms available to the host. Each infected host represents a unique host-parasite combination and, environmental influences excluded, the balance of the relationship will be determined by the genetically encoded information available to each participant. In similar conditions other animals of the same species may behave differently, and in each case the outcome will reflect the particular circumstances of the relationship.

There are undoubtedly strategies still awaiting to be discovered, and others which will evolve in the future as the arms race continues. Just as our appreciation of the complexity of the immunoregulatory pathways which control the immune system matures, so also new opportunities will be presented for elucidating how the network can be manipulated by parasites to secure, in evolutionary terms, a temporary advantage over the host.

REFERENCES

Alexander, J. and Stimson, W.H. (1988), Sex hormones and the course of parasitic infection, *Parasitology Today*, **4**, 189–93.

Allonby, E.W. and Urquhart, G.M. (1973), Self-cure of *Haemonchus contortus* infections under field conditions, *Parasitology*, **66**, 43–53.

Al Saqur, I., Armour, J., Bairden, K., Dunn, A.M., Jennings, F.W. and Murray, M. (1984), Experimental studies on the interaction between infections of *Ostertagia leptospicularis* and other bovine *Ostertagia* species, *Zeitschrift fur Parasitenkunde*, **70**, 809–17.

Anders, R.F. (1986), Multiple cross-reactivities amongst antigens of *Plasmodium falciparum* impair the development of protective immunity against malaria, *Parasite Immunology*, **8**, 529–39.

Anders, R.F., Shi, P-T., Scanlon, D.B., Leach, S.J., Coppel, R.L., Brown, G.V., Stahl, H-D. and Kemp, D.J. (1986), Antigenic repeat structures in proteins of *Plasmodium falciparum*, in Wheelan, J. (Ed.), *Synthetic Peptides as Antigens, Ciba Foundation Symposium*, **No. 119**, 164–75.

Anderson, R.M. and Medley, G.F. (1985), Community control of helminth infection of man by mass and selective chemotherapy, *Parasitology*, **60**, 629–60.

Askonas, B.A. (1984), Interference in general immune function by parasite infections; African trypanosomiasis as a model system, *Parasitology*, **88**, 633–8.

Auriault, C., Quaissi, M.A., Torpier, G., Eisen, H. and Capron, A. (1981),

Proteolytic cleavage of IgG bound to the Fc receptor of *Schistosoma mansoni* schistosomula, *Parasite Immunology*, **3**, 33–44.

Barriga, O.O. (1984), Immunomodulation by nematodes: a review, *Veterinary Parasitology*, **14**, 299–320.

Barry, J.D. (1986), Surface antigens of African trypanosomes in the Tsetse fly, *Parasitology Today*, **2**, 143–5.

Barry, J.D., Crowe, J.S. and Vickerman, K. (1983), Instability of the *Trypanosoma brucei rhodesiense* metacyclic variable antigen repertoire, *Nature*, **306**, 699–701.

Barry, J.D., Hajduk, S.L., Vickerman, K. and Le Ray, D. (1979), Detection of multiple antigen types in metacyclic populations of *Trypanosoma brucei*, *Transactions of the Royal Society of Tropical Medicine and Hygiene*, **73**, 205–8.

Beaver, P.V., Orihel, T.C. and Johnson, M.H. (1974), *Dipetalonema viteae* in the experimentally infected jird *Meriones unguiculatus*. III Microfilaraemia in relation to worm burden, *Journal of Parasitology*, **60**, 310–5.

Befus, A.D. and Threadgold, L.T. (1975), Possible immunological damage to the tegument of *Hymenolepis diminuta* in mice and rats, *Parasitology*, **71**, 525–34.

Behnke, J.M. (1987), Evasion of immunity by nematode parasites causing chronic infections, *Advances in Parasitology*, **26**, 1–71.

Behnke, J.M. and Wakelin, D. (1973), The survival of *Trichuris muris* in wild populations of its natural host, *Parasitology*, **67**, 157–64.

Behnke, J.M. and Robinson, M. (1985), Genetic control of immunity to *Nematospiroides dubius*; a 9 day anthelmintic abbreviated immunizing regime which separates weak and strong responder strains of mice, *Parasite Immunology*, **7**, 235–53.

Behnke, J.M., Bland, P.W. and Wakelin, D. (1977), Effect of the expulsion phase of *Trichinella spiralis* on *Hymenolepis diminuta* infection in mice, *Parasitology*, **75**, 79–88.

Behnke, J.M., Wakelin, D. and Wilson, M.M. (1978), *Trichinella spiralis*: delayed rejection in mice concurrently infected with *Nematospiroides dubius*, *Experimental Parasitology*, **46**, 121–30.

Behnke, J.M., Hannah, J. and Pritchard, D.I. (1983), *Nematospiroides dubius* in the mouse: evidence that adult worms depress the expression of homologous immunity, *Parasite Immunology*, **5**, 397–408.

Behnke, J.M., Ali, N.M.H. and Jenkins, S.N. (1984), Survival to patency of low level infections with *Trichuris muris* in mice concurrently infected with *Nematospiroides dubius*, *Annals of Tropical Medicine and Parasitology*, **78**, 509–17.

Benitez-Usher, C., Armour, J., Duncan, J.L., Urquhart, G.M. and Gettenby, G. (1977), A study of some factors influencing immunization of sheep against *Haemonchus contortus* using attenuated larvae, *Veterinary Parasitology*, **3**, 327–42.

Bolin, T.D., Davis, A.E., Cummins, A.G., Duncombe, V.M. and Kelly, J.D. (1977), Effect of iron and protein deficiency on the expulsion of *Nippostrongylus brasiliensis* from the small intestine of the rat, *Gut*, **18**, 182–6.

Brown, K.N. (1974), Antigenic variation and immunity to malaria, in *Parasites in the Immunized Host: Mechanisms of Survival, Ciba Foundation Symposium*, **25**, 35–51.

Bruce, R.G. and Wakelin, D. (1977), Immunological interaction between *Trichinella spiralis* and *Trichuris muris* in the intestine of the mouse, *Parasitology*, **74**, 163–73.

Bundy, D.A.P. (1986), Epidemiological aspects of *Trichuris* and trichuriasis in Caribbean communities, *Transactions of the Royal Society of Tropical Medicine and Hygiene*, **80**, 706–18.

Bundy, D.A.P., Cooper, E.S., Thompson, D.E., Anderson, R.M. and Didier-Blanchard, J.M. (1987), Age-related prevalence and intensity of *Trichuris trichiura* in St. Lucia, *Transactions of the Royal Society of Tropical Medicine and Hygiene*, **81**, 85–94.

Burkitt, D.P. (1969), Etiology of Burkitt's lymphoma—an alternative to a vectored virus, *Journal of the National Cancer Institute*, **42**, 19–28.

Butterworth, A.E. and Hagan, P. (1987), Immunity in human schistosomiasis, *Parasitology Today*, **3**, 11–6.

Calkins, C.E. and Stutman, O. (1978), Changes in suppressor mechanisms during postnatal development in mice, *Journal of Experimental Medicine*, **147**, 87–97.

Callahan, H.L., Crouch, R.K. and James, E.R. (1988), Helminth anti-oxidant enzymes: a protective mechanism against host oxidants, *Parasitology Today*, **4**, 218–25.

Capron, M. and Capron, A. (1986), Rats, mice and men—models for immune effector mechanisms against schistosomiasis, *Parasitology Today*, **2**, 69–75.

Capron, A., Auriault, C., Mazingue, C., Capron, M. and Torpier G. (1980), Schistosome mechanisms of evasion, *Jansens Research Foundation Series*, **2**, 217–25.

Castro, G.A., Behnke, J.M. and Weisbrodt, N.W. (1990), Hookworm infection and malabsorption: where do we stand, in Schad, G.A. and Warren, K.S. (Eds.), *Hookworm Disease: Current Status and New Directions*, London: Taylor & Francis.

Chandra, R.K. (1975), Reduced secretory antibody response to live attenuated measles and polio virus vaccines in malnourished children, *British Medical Journal*, **2**, 583–5.

Chapman, C.B. and Mitchell, G.F. (1983), Proteolytic cleavage of immunoglobulin by enzymes released by *Fasciola hepatica*, *Veterinary Parasitology*, **11**, 165–78.

Christensen, N.O., Nansen, P., Fagbemi, B.O. and Monrad, J. (1987), Heterologous antagonistic and synergistic interactions between helminths and between helminths and protozoans in concurrent experimental infection of mammalian hosts, *Parasitology Research*, **73**, 387–410.

Christie, P.R., Wakelin, D. and Wilson, M.M. (1979), The effect of the expulsion phase of *Trichinella spiralis* on *Hymenolepis diminuta* in rats, *Parasitology*, **78**, 323–30.

Clegg, J.A. and Smithers, S.R. (1968), Death of schistosome cercariae during penetration of mammalian skin by *Schistosoma mansoni*, *Parasitology* **58**, 111–28.

Clegg, J.A., Smithers, S.R. and Terry, R.J. (1971), Acquisition of human antigens by *Schistosoma mansoni* during cultivation *in vitro*, *Nature*, **232**, 653–4.

Cohen, S. (1982), Survival of parasites in the immunocompetent host. in Cohen, S. and Warren, K.S. (Eds.), *Immunology of Parasitic Infections*, 2nd edition, pp. 138–61, Oxford: Blackwell Scientific Publications.

Connan, R.M. (1970), The effect of host lactation on the self-cure of *Nippostrongylus brasiliensis* in rats, *Parasitology*, **61**, 27–33.

Connan, R.M. (1972), The effect of host lactation on a second infection of *Nippostrongylus brasiliensis* in rats, *Parasitology*, **64**, 229–33.

Connan, R.M. (1973), The immune response of the lactating rat to *Nippostrongylus brasiliensis*, *Immunology*, **25**, 261–7.

Connan, R.M. (1976), Effect of lactation on the immune response to gastrointestinal nematodes, *Veterinary Record*, **99**, 476–7.

Court, J.P., Stables, J.N., Lees, G.M., Martin-Short, M.R. and Rankin, R. (1988), *Dipetalonema viteae* and *Brugia pahangi* transplant infections in gerbils for use in antifilarial screening, *Journal of Helminthology*, **62**, 1–9.

Cox, F.E.G. (1978), Heterologous immunity between piroplasms and malaria parasites: the simultaneous elimination of *Plasmodium vinckei* and *Babesia microti* from the blood of doubly infected mice, *Parasitology*, **76**, 55–60.

Crandall, R.B. (1975), Decreased resistance to *Trichinella spiralis* in aged mice, *Journal of Parasitology*, **61**, 566–7.

Croll, N.A. and Ghadirian, E. (1981), Wormy persons: contributions to the nature and patterns of overdispersion with *Ascaris lumbricoides*, *Ancylostoma duodenale*,

Necator americanus and *Trichuris trichiura*, *Tropical and Geographical Medicine*, **33**, 241–8.

Damian, R.T. (1964), Molecular mimicry: antigen sharing by parasite and host and its consequences, *The American Naturalist*, **98**, 129–49.

Damian, R.T. (1967), Common antigens between *Schistosoma mansoni* and the laboratory mouse, *Journal of Parasitology*, **53**, 60–4.

Damian, R.T., Greene, N.D. and Hubbard, W.J. (1973), Occurrence of mouse α2 macroglobulin antigenic determinants on *Schistosoma mansoni* adults with evidence on their nature, *Journal of Parasitology*, **59**, 64–73.

Davis, D.E. and Read, C.P. (1958), Effect of behaviour on the development of resistance in trichinosis, *Proceedings of the Society of Experimental Biology and Medicine*, **99**, 269–72.

Dehlawi, M.S. and Wakelin, D. (1988), Suppression of mucosal mastocytosis by *Nematospiroides dubius* results from an adult worm mediated effect upon host lymphocytes, *Parasite Immunology*, **10**, 85–95.

Dehlawi, M.S., Wakelin, D. and Behnke, J.M. (1987), Suppression of mucosal mastocytosis by infection with the intestinal nematode *Nematospiroides dubius*, *Parasite Immunology*, **9**, 187–94.

de Souza, E.B. and van Loon, G.R. (1985), Differential plasma B-endorphin, B-lipotropin and adrenocorticotropin respones to stress in rats, *Endocrinology*, **116**, 1577–86.

Dineen, J.K. (1963), Antigenic relationships between host and parasite, *Nature*, (*London*), **197**, 471–2.

Dineen, J.K. and Kelly, J.D. (1972), The suppression of rejection of *Nippostrongylus brasiliensis* in lactating rats: the nature of the immunological defect, *Immunology*, **22**, 1–12.

Dineen, J.K. and Kelly, J.D. (1973), Immunological unresponsiveness of neonatal rats to infection with *Nippostrongylus brasiliensis*. The competence of neonatal lymphoid cells in worm expulsion, *Immunology*, **25**, 141–50.

Donelson, J.E. and Turner, M.J. (1985), How the trypanosome changes its coat, *Scientific American*, **252**, February, 32–9.

Douch, P.G.C., Harrison, G.B.L., Buchanan, L.L. and Greer, K.S. (1988), Relationship of nematode cholinesterase activity and nematode burdens to the development of resistance to Trichostrongyle infections in sheep, *Veterinary Parasitology*, **27**, 291–308.

Elkins, D.B., Haswell-Elkins, M. and Anderson, R.M. (1986), The epidemiology and control of intestinal helminths in the Pulicat Lake region of Southern India. 1. Study design and pre- and post-treatment observations on *Ascaris lumbricoides* infection, *Transactions of the Royal Society of Tropical Medicine and Hygiene*, **80**, 774–92.

Esch, G.W., Gibbons, J.W. and Bourque, J.E. (1975), An analysis of the relationship between stress and parasitism, *American Midland Naturalist*, **93**, 339–53.

Fleming, A.F., Storey, J., Molineaux, L., Iroko, E.A. and Attai, E.D.E. (1979), Abnormal haemoglobin in the Sudan Savanna of Nigeria. 1. Prevalence of haemoglobins and relationships between sickle cell trait, malaria and survival, *Annals of Tropical Medicine and Parasitology*, **73**, 161–72.

Forsyth, K.P., Copeman, D.B. and Mitchell, G.F. (1984), Differences in the surface radio iodinated proteins of skin and uterine microfilariae of *Onchocerca gibsoni*, *Molecular and Biochemical Parasitology*, **10**, 217–29.

Ghoneum, M., Gill, G., Assanah, P. and Stevens, W. (1987), Susceptibility of natural killer cell activity of old rats to stress, *Immunology*, **60**, 461–5.

Goldring, O.I., Clegg, J.A., Smithers, S.T. and Terry, R.J. (1976), Acquisition of human blood group antigens by *Schistosoma mansoni*, *Clinical and Experimental Immunology*, **26**, 181–7.

Gray, A.R. (1965), Antigenic variation in clones of *Trypanosome brucei*. I. Immunological relationships of the clones, *Annals of Tropical Medicine and Parasitology*, **59**, 27–36.

Gray, J.S. (1972), The effect of host age on the course of infection of *Raillietina cesticillus* (Molin, 1858) and the fowl, *Parasitology*, **65**, 235–41.

Greenwood, B.M. (1987), Asymptomatic malaria infections—do they matter? *Parasitology Today*, **3**, 206–14.

Grossman, Z., Greenblatt, C.L. and Cohen, I.R. (1986), Parasite immunology and lymphocyte population dynamics, *Journal of Theoretical Biology*, **121**, 129–39.

Grove, D.I., Valeza, F.S. and Cabrera, B.D. (1978), Bancroftian filariasis in a Philippine village: clinical, parasitological, immunological and social aspects, *Bulletin of the World Health Organization*, **56**, 975–84.

Grzych, J.M., Capron, M., Dissous, C. and Capron, A. (1984), Blocking activity of rat monoclonal antibodies in experimental schistosomiasis, *Journal of Immunology*, **133**, 998–1004.

Hammerberg, B., Musoke, A.J. and Williams, J.F. (1977), Activation of complement by hydatid cyst fluid of *Echinococcus granulosus*, *Journal of Parasitology*, **63**, 327–31.

Haneberg, B. (1974), Human milk immunoglobulins and agglutinins to rabbit erythrocytes, *International Archives of Allergy and Applied Immunology*, **47**, 716–29.

Hannah, J. and Behnke, J.M. (1982), *Nematospiroides dubius* in the jird, *Meriones unguiculatus*: factors affecting the course of a primary infection, *Journal of Helminthology*, **56**, 329–38.

Haque, A., Lefebvre, M.N., Ogilvie, B.M. and Capron, A. (1978), *Dipetalonema viteae* in hamsters: effect of antiserum or immunization with parasite extracts on production of microfilariae, *Parasitology*, **76**, 61–75.

Haque, A., Capron, A., Quaissi, A., Kouemeni, L., Lejeune, J.P., Bonnel, B. and Pierce, R. (1983), Immune unresponsiveness and its possible relation to filarial disease, *Contributions to Microbiology and Immunology*, **7**, 9–21.

Haswell-Elkins, M.R., Elkins, D.B. and Anderson, R.M. (1987), Evidence for predisposition of humans to infection with *Ascaris*, hookworm, *Enterobius* and *Trichuris* in a South Indian fishing community, *Parasitology*, **95**, 323–37.

Haswell-Elkins, M.R., Elkins, D.B., Manjula, K., Michael, E. and Anderson, R.M. (1987), The distribution and abundance of *Enterobius vermicularis* in a South Indian fishing community, *Parasitology*, **95**, 339–54.

Hess, E.V. and Knapp, D. (1978), The immune system and ageing: a case of the cart before the horse, *Journal of Chronic Diseases*, **31**, 647–9.

Hockley, D.J. and McLaren, D.J. (1973), *Schistosoma mansoni*: changes in the outer membrane of the tegument during development from cercaria to adult worms, *International Journal for Parasitology*, **3**, 13–25.

Hockley, D.J., McLaren, D.J., Ward, B.J. and Nermut, M.V. (1975), A freeze-fracture study of the tegumental membranes of *Schistosoma mansoni* (Platyhelminthes: Trematoda), *Tissue and Cell*, **7**, 485–96.

Hoffman-Goetz, L., Keir, R. and Young, C. (1986), Modulation of cellular immunity in malnutrition: effect of interleukin 1 on suppressor T cell activity. *Clinical and Experimental Immunology*, **65**, 381–6.

Holmes, J.C. (1973), Site selection by parasitic helminths: interspecific interactions, site segregation and their importance to the development of the helminth communities, *Canadian Journal of Zoology*, **51**, 333–47.

Hommel, M. (1985), Antigenic variation in malaria parasites, *Immunology Today*, **6**, 28–33.

Howard, R.J. (1984), Antigenic variation of bloodstage malaria parasites. *Philosophical Transactions of the Royal Society of London*, **B307**, 141–58.

Hsieh, H.C. (1970), Studies on endemic hookworm. 1. Survey and longitudinal observations in Taiwan, *Japanese Journal of Parasitology*, **19**, 508–22.

Hughes, H.P.A. (1988), Oxidative killing of intracellular parasites mediated by macrophages, *Parasitology Today*, **4**, 340–7.

Jarrett, E.E.E. and Urquhart, G.M. (1969), Immunological unresponsiveness to helminth parasites. III. Challenge of rats previously infected at an early age with *Nippostrongylus brasiliensis*, *Experimental Parasitology*, **25**, 245–57.

Jarrett, E.E.E., Jarrett, W.F.H. and Urquhart, G.M. (1966), Immunological unresponsiveness in adult rats to the nematode *Nippostrongylus brasiliensis* induced by infection in early life, *Nature (London)*, **211**, 1310–1.

Jarrett, E.E.E., Jarrett, W.F.H. and Urquhart, G.M. (1968), Immunological unresponsiveness to helminth parasites. I. The pattern of *Nippostrongylus brasiliensis* infection in young rats, *Experimental Parasitology*, **23**, 151–60.

Jenkins, D.C. (1972), *Nippostrongylus brasiliensis*: observations on the comparative immunogenicity of adult worms from primary and immune adapted infections, *Parasitology*, **65**, 547–50.

Jenkins, D.C. (1977), *Nematospiroides dubius*: the course of primary and challenge infections in the jird, *Meriones unguiculatus*, *Experimental Parasitology*, **41**, 335–40.

Jenkins, D.C. and Phillipson, R.F. (1972), Evidence that the nematode *Nippostrongylus brasiliensis* can adapt to and overcome the effects of host immunity, *International Journal for Parasitology*, **2**, 353–9.

Jenkins, D.C., Ogilvie, B.M. and McLaren, D.J. (1976), The effects of immunity and the mode of infection on the development of *Nippostrongylus brasiliensis* in rats, in van den Bossche, H. (Ed.), *Biochemistry of Parasites and Host–Parasite Relationships*, pp. 299–306, Amsterdam: Elsevier North Holland.

Johnson, M.H., Orihel, T.C. and Beaver, P.C. (1974), *Dipetalonema viteae* in the experimentally infected jird, *Meriones unguiculatus*. 1. Insemination, development from egg to microfilaria, reinsemination and longevity of mated and unmated worms, *Journal of Parasitology*, **60**, 302–9.

Kalkofen, U.P. (1970), Attachment and feeding behaviour of *Ancylostoma caninum*, *Zeitschrift für Parasitenkunde*, **33**, 339–54.

Kassai, T. and Aitken, I.D. (1967), Induction of immunological tolerance in rats to *Nippostrongylus brasiliensis* infection, *Parasitology*, **57**, 403–18.

Kelly, J.D. and Dineen, J.K. (1973), The suppression of rejection of *Nippostrongylus brasiliensis* in Lewis strain rats treated with ovine prolactin. The site of the immunological defect, *Immunology*, **24**, 551–8.

Kemp, D.J., Coppel, R.L. and Anders, R.F. (1987), Repetitive proteins and genes of malaria, *Annual Reviews of Microbiology*, **41**, 181–208.

Kemp, W.M., Damian, R.T., Greene, N.D. and Lushbaugh, W.B. (1976), Immunocytochemical localization of mouse alpha 2- macroglobulin like antigenic determinants on *Schistosoma mansoni* adults, *Journal of Parasitology*, **62**, 413–9.

Keusch, G.T. (1985), Nutrition and immune function, in Warren, K.S. and Mahmoud, A.A.F. (Eds.), *Tropical and Geographical Medicine*, pp. 212–8, New York: McGraw-Hill Book Co.

Keusch, G.T., Wilson, C.S. and Waksal, S.D. (1983), Nutrition, host defences and the lymphoid system, in Sallin, J.I. and Fauci, A.S. (Eds.), *Advances in Host Defense Mechanisms*, Vol. 2, New York: Plenum Press.

Kirkwood, B., Smith, P., Marshall, T. and Prost, A. (1983a), Variations in the prevalence and intensity of microfilarial infections by age, sex, place and time in the area of the Onchocerciasis Control Programme, *Transactions of the Royal Society of Tropical Medicine and Hygiene*, **77**, 857–61.

Kirkwood, B., Smith, P., Marshall, T. and Prost, A. (1983b), Relationships between mortality, visual acuity and microfilarial load in the area of the Onchocerciasis

Control Programme, *Transactions of the Royal Society of Tropical Medicine and Hygiene*, 77, 862–3.

Klei, T.R., McCall, J.W. and Malone, J.B. (1980), Evidence for increased susceptibility of *Brugia pahangi*-infected jirds (*Meriones unguiculatus*) to subsequent homologous infections, *Journal of Helminthology*, 54, 161–5.

Klei, T.R., Enright, E.M., Blanchard, D.P. and Uhl, S.A. (1981), Specific hyporesponsive granulomatous tissue reactions in *Brugia pahangi*-infected jirds, *Acta Tropica*, 38, 267–76.

Kloosterman, A. and Frankena, K. (1988), Interactions between lungworms and gastrointestinal worms in calves, *Veterinary Parasitology*, 26, 305–20.

Knight, R. and Merrett, T.G. (1981), Hookworm infection in rural Gambia. Seasonal changes, morbidity and total IgE levels, *Annals of Tropical Medicine and Parasitology*, 75, 299–314.

Koster, F. and Pierce, N.F. (1985), Effect of protein deprivation on immuno-regulatory cells in the rat mucosal immune response, *Clinical and Experimental Immunology*, 60, 217–24.

Kvalsvig, J.D. (1988), The effects of parasitic infection on cognitive performance, *Parasitology Today*, 4, 206–8.

Langley, G. and Roth, C. (1987), Antigenic variation in parasitic protozoa, *Microbiological Sciences*, 4, 280–5.

Lee, D.L. and Martin, J. (1976), Changes in *Nematodirus battus* associated with the development of immunity to this nematode in lambs, in van den Bossche, H. (Ed.), *Biochemistry of Parasites and Host–Parasite Relationships*, pp. 311–8, Amsterdam: Elsevier North Holland.

Lee, T.D.G., Grencis, R.K. and Wakelin, D. (1982), Specific cross-immunity between *Trichinella spiralis* and *Trichuris muris*: immunization with heterologous infections and antigens and transfer of immunity with heterologous immune mesenteric lymph node cells, *Parasitology*, 84, 381–9.

Leid, S.W., Suquet, C.M. and Tanigoshi, L. (1987), Parasite defense mechanisms for evasion of host attack: a review *Veterinary Parasitology*, 25, 147–62.

Lightowlers, M.W. and Rickard, M.D. (1988), Excretory-secretory products of helminth parasites: effect on host immune responses, *Parasitology*, 96, Supplement, 123–66.

Lloyd, S., Amerasinghe, P.H. and Soulsby, E.J.L. (1983), Periparturient immuno-suppression in the bitch and its influence on infection with *Toxocara canis*. *Journal of Small Animal Practice*, 24, 237–47.

Lu, C.Y., Calamai, E.G. and Unanue, E.R. (1979), A defect in the antigen presenting function of macrophages from neonatal mice, *Nature*, 282, 327–9.

Maizels, R.M., Meghji, M. and Ogilvie, B.M. (1983), Restricted sets of parasite antigens from the surface of different stages and sexes of the nematode *Nippostrongylus brasiliensis*, *Immunology*, 48, 107–21.

Maizels, R.M., Philipp, M., Dasgupta, A. and Partono, F. (1984), Human serum albumin is a major component on the surface of microfilariae of *Wuchereria bancrofti*, *Parasite Immunology*, 6, 185–90.

Makinodan, T. and Kay, M.M.B. (1980), Age influence on the immune system, *Advances in Immunology*, 29, 287–330.

Manser, T., Wysocki, L.J., Gridley, T., Near, R.I. and Gefter, M.L. (1985), The molecular evolution of the immune response, *Immunology Today*, 6, 94–101.

Manton, V.J.A., Peacock, R., Poynter, D., Silverman, P.H. and Terry, R.J. (1962), The influence of age on naturally acquired resistance to *Haemonchus contortus* in lambs, *Research in Veterinary Science*, 3, 308–14.

Mauel, J. (1984), Mechanisms of survival of protozoan parasites in mononuclear phagocytes, *Parasitology*, 88, 579–92.

McCaigue, M.D., Halton, D.W. and Hopkins, C.A. (1986), *Hymenolepis diminuta*:

ultrastructural abnormalities in worms from C57 mice, *Experimental Parasitology*, **62**, 51–60.

McGee, D.W. and McMurray, D.N. (1988a), The effect of protein malnutrition on the IgA immune response in mice, *Immunology*, **63**, 25–9.

McGee, D.W. and McMurray, D.N. (1988b), Protein malnutrition reduces the IgA immune response to oral antigen by altering B-cell and suppressor T-cell function, *Immunology*, **64**, in press.

McGreevy, P.B., Ismail, M.M., Phillips, T.M. and Denham, D.A. (1975), Studies with *Brugia pahangi*, 10. An attempt to demonstrate the sharing of antigenic determinants between the worm and its hosts, *Journal of Helminthology*, **49**, 107–13.

McKerrow, J.H. and Doenhoff, M.J. (1988), Schistosome proteases, *Parasitology Today*, **4**, 334–9.

McLaren, D.J. (1984), Disguise as an evasive stratagem of parasitic organisms, *Parasitology*, **88**, 597–611.

McLaren, D.J. and Hockley, D.J. (1977), Blood flukes have a double outer membrane. *Nature*, **269**, 147–9.

McLaren, D.J. and Terry, R.J. (1982), The protective role of acquired host antigens during schistosome maturation, *Parasite Immunology*, **4**, 129–48.

McLaren, D.J., Clegg, J.A. and Smithers S.R. (1975), Acquisition of host antigens by young *Schistosoma mansoni* in mice: correlation with failure to bind antibody *in vitro*, *Parasitology*, **70**, 67–75.

McMurray, D.N., Rey, H., Casazza, L.J. and Watson, R.R. (1977), Effect of moderate malnutrition on concentrations of immunoglobulins and enzymes in tears and saliva of young Columbian children, *American Journal of Clinical Nutrition*, **30**, 1944–8.

Menon, M., Jaroslow, B.N. and Koesterer, R. (1974), The decline of cell-mediated immunity in aging mice, *Journal of Gerontology*, **29**, 499–505.

Miller, H.R.P. (1987), Gastrointestinal mucus, a medium for survival and for elimination of parasitic nematodes and protozoa, *Parasitology*, **94**, Supplement, 77–100.

Miller, E.N. and Turner, M.J. (1981), Analysis of antigenic types appearing in first relapse populations of clones of *Trypanosoma brucei*, *Parasitology*, **82**, 63–80.

Mitchell, G.F. (1979), Effector cells, molecules and mechanisms in host-protective immunity to parasites, *Immunology*, **38**, 209–23.

Mkoji, G.M., Smith, J.M. and Prichard, R.K. (1988), Antioxidant systems in *Schistosoma mansoni*: evidence for their role in protection of adult worms against oxidant killing, *International Journal for Parasitology*, **18**, 667–73.

Morrow, R.H., Kisuulo, A., Pike, M.C. and Smith, P.G. (1970), Burkitt's lymphoma in the Mengo districts of Uganda: epidemiologic features and their relationship to malaria, *Journal of the National Cancer Institute*, **56**, 479–83.

Mosier, D.E. and Johnson, B.M. (1975), Ontogeny of mouse lymphocyte function. II. Development of the ability to produce antibody is modulated by T-lymphocytes, *Journal of Experimental Medicine*, **141**, 216–26.

Nantulya, V.M. (1986), Immunological approaches to the control of animal trypanosomiasis, *Parasitology Today*, **2**, 168–73.

Narayanan, P.R., Vanamala, C.R., Alamelu, R., Kumaraswamy, V., Tripathy, S.P. and Prabhakar, R. (1986), Reduced lymphocyte responses to mitogens in patients with Bancroftian filariasis, *Transactions of the Royal Society of Tropical Medicine and Hygiene*, **80**, 78–84.

Neilson, J.T.M. (1978), Primary infections of *Dipetalonema viteae* in an outbred and five inbred strains of golden hamsters, *Journal of Parasitology*, **64**, 378–80.

Ngwenya, B.Z. (1980), Altered lysophospholipase B responsiveness in lactating mice infected with intestinal nematode parasites, *Parasitology*, **81**, 17–26.

Nordin, A.A. and Makinodan, T. (1974), Humoral immunity in aging, *Federation Proceedings*, **33**, 2033–5.

Ogilvie, B.M. (1974), Antigenic variation in the nematode *Nippostrongylus brasiliensis*, in Porter, R. and Knight, J. (Eds.), *Parasites in the Immunized Host: Mechanisms of Survival*, pp. 81–100, *Ciba Foundation Symposium 25*, Amsterdam: Elsevier North Holland.

Ogilvie, B.M., Rothwell, T.L.W., Bremner, K.C., Schnitzerling, H.J., Nolan, J. and Keith, R.K. (1973), Acetylcholinesterase secretion by parasitic nematodes. I. Evidence for secretion of the enzyme by a number of species, *International Journal for Parasitology*, **3**, 589–97.

Orihel, H.C. and Eberhard, M.L. (1985), *Loa loa*; development and course of patency in experimentally-infected primates, *Tropical Medicine and Parasitology*, **36**, 215–24.

O'Sullivan, B.M. (1974), Effects of lactation on *Trichostrongylus colubriformis* infection in the guinea-pig, *International Journal for Parasitology*, **4**, 177–81.

O'Sullivan, B.M. and Donald, A.D. (1970), A field study of nematode parasite populations in the lactating ewe, *Parasitology*, **61**, 301–15.

Ottesen, E.A. (1984), Immunological aspects of lymphatic filariasis and onchocerciasis in man, *Transactions of the Royal Society of Tropical Medicine and Hygiene*, **78**, Supplement, 9–18.

Ottesen, E.A., Weller, P.F. and Heck, L. (1977), Specific cellular immune unresponsiveness in human filariasis, *Immunology*, **33**, 413–21.

Ovington, K.S. (1986), Trickle infections of *Nippostrongylus brasiliensis* in rats, *Zeitschrift für Parasitenkunde*, **72**, 851–3.

Parkhouse, R.M.E. and Ortega-Pierres, G. (1984), Stage-specific antigens of *Trichinella spiralis*, *Parasitology*, **88**, 623–30.

Perricone, R.L., Fontana, C., de Carolis, and Ottaviani, P. (1980), Activation of alternative complement pathway by fluid from hydatid cysts, *New England Journal of Medicine*, **302**, 808–9.

Perry, R.H. (1974), Transfer of immunity to *Trichinella spiralis* from mother to offspring, *Journal of Parasitology*, **60**, 460–5.

Philipp, M. (1984), Acetylcholinesterase secreted by intestinal nematodes: a reinterpretation of its putative role of 'biochemical holdfast', *Transactions of the Royal Society of Tropical Medicine and Hygiene*, **78**, 138–9.

Philipp, M. and Rumjaneck, F.D. (1984), Antigenic and dynamic properties of helminth surface structures, *Molecular and Biochemical Parasitology*, **10**, 245–68.

Philipp, M., Parkhouse, R.M.E. and Ogilvie, B.M. (1980), Changing proteins in the surface of a parasitic nematode, *Nature*, **287**, 538–40.

Piessens, W.F., McGreevy, P.B., Piessens, P.W., McGreevy, M., Koiman, I., Saroso, J.S. and Dennis, D.T. (1980a), Immune responses in human infections with *Brugia malayi*, specific cellular unresponsiveness to filarial antigens, *Journal of Clinical Investigation*, **65**, 172–9.

Piessens, W.F., Ratiwayanto, S., Tuti, S., Palmieri, J.H., Piessens, P.W., Koiman, I. and Dennis, D.T. (1980b), Antigen-specific suppressor cells and suppressor factors in human filariasis with *Brugia malayi*, *New England Journal of Medicine*, **302**, 833–7.

Plaut, A.G. (1983), The IgA1 proteases of pathogenic bacteria, *Annual Reviews of Microbiology*, **37**, 603–22.

Ponnudurai, T., Denham, D.A., Nelson, G.S. and Rogers, R. (1974), Studies with *Brugia pahangi*. 4. Antibodies against adult and microfilarial stages, *Journal of Helminthology*, **48**, 107–11.

Pritchard, D.I. (1986), Antigens of gastrointestinal nematodes, *Transactions of the Royal Society of Tropical Medicine and Hygiene*, **80**, 728–34.

Rajasekariah, G.R. and Howell, M.J. (1977), *Fasciola hepatica* in rats: effect of age and infective dose, *International Journal of Parasitology*, **7**, 119–21.

Read, C.P. (1967), Longevity of the tapeworm, *Hymenolepis diminuta*, *Journal of Parasitology*, **53**, 1055–6.

Richie, T.L. (1988), Interactions between malaria parasites infecting the same vertebrate host, *Parasitology*, **96**, 607–39.

Ridley, D.S. and Hedge, E.C. (1977), Immunofluorescent reactions with microfilariae. 2. Bearing on host-parasite relations, *Transactions of the Royal Society of Tropical Medicine and Hygiene*, **71**, 522–5.

Riley, J. (1986), The biology of the pentastomids, *Advances in Parasitology*, **25**, 46–128.

Riley, J., James, J.L. and Banaja, A.A. (1979), The possible role of the frontal and sub-parietal gland systems of the pentastomid *Reighardia sternae* (Diesing, 1864) in the evasion of the host immune response, *Parasitology*, **78**, 53–66.

Roberts, J.M.D., Neumann, E., Gockel, C.W. and Highton, R.B. (1967), Onchocerciasis in Kenya, 9, 11 and 18 years after elimination of the vector, *Bulletin of the World Health Organisation*, **37**, 195–212.

Roberts, S.M., Aitken, R., Vojvodic, M., Wells, E. and Wilson, R.A. (1983), Identification of exposed components on the surface of adult *Schistosoma mansoni* by lactoperoxidase- catalysed iodination, *Molecular and Biochemical Parasitology*, **9**, 129–43.

Robinson, E.J. (1961), Survival of *Trichinella* in stressed hosts, *Journal of Parasitology*, **47**, Supplement, 16–17.

Ross, R. and Tomson, D. (1910), A case of sleeping sickness studied by precise enumerative methods: regular periodical increase of the parasites disclosed, *Proceedings of the Royal Society of London*, **B82**, 411–5.

Saunders, N., Wilson, R.A. and Coulson, P.S. (1987), The outer bilayer of the adult schistosome tegument surface has a low turnover rate *in vitro* and *in vivo*, *Molecular and Biochemical Parasitology*, **25**, 123–31.

Selby, G.R. and Wakelin, D. (1975), Suppression of the immune response to *Trichuris muris* in lactating mice, *Parasitology*, **71**, 77–85.

Selkirk, M.E., Yazdanbakhsh, M., Blaxter, M., Gregory, W. and Maizels, R.M. (1988), Organization, synthesis and structure of cuticular proteins of *Brugia* sp., *Transactions of the Royal Society of Tropical Medicine and Hygiene*, **82**, 820.

Sher, A., Hall, B.F. and Vadan, M.A. (1978), Acquisition of murine major histocompatibility complex gene products by schistosomula of *Schistosoma mansoni*, *Journal of Experimental Medicine*, **148**, 46–57.

Slater, A.F.G. (1988), The influence of dietary proteins on the experimental epidemiology of *Heligmosomoides polygyrus* (Nematoda) in the laboratory mouse, *Proceedings of the Royal Society of London*, **B234**, 239–54.

Slater, A.F.G. and Keymer, A.E. (1986), *Heligmosomoides polygyrus* (Nematoda): the influence of dietary proteins on the dynamics of repeated infection, *Proceedings of the Royal Society of London*, **B229**, 69–83.

Smith, N.C. and Bryant, C. (1986), The role of host generated free radicals in helminth infections: *Nippostrongylus brasiliensis* and *Nematospiroides dubius* compared, *International Journal for Parasitology*, **16**, 617–22.

Smithers, S.R. and Terry, R.J. (1969), Immunity in Schistosomiasis, *Annals of the New York Academy of Sciences*, **160**, 826–40.

Smithers, S.R. and Terry, R.J. (1976), The immunology of schistosomiasis, *Advances in Parasitology*, **14**, 399–422.

Smithers, S.R., Terry, R.J. and Hockley, D.J. (1969), Host antigens in schistosomiasis, *Proceedings of the Royal Society of London*, **B171**, 483–94.

Solomon, G.B. (1969), Host hormones in parasitic infections, *International Review of Tropical Medicine*, **3**, 101–58.

Sprent, J.F.A. (1962), Parasitism, immunity and evolution, in Leeper, G.W.

(Ed.), *The Evolution of Living Organisms*, pp. 144–65, Melbourne: Melbourne University Press.

Steinart, M. and Pays, E. (1986), Selective expression of surface antigen genes in African trypanosomes, *Parasitology Today*, **2**, 15–19.

Stephenson, L.S. (1987), *The impact of helminth infections on human nutrition, Schistosomes and soil-transmitted helminths*, London: Taylor & Francis.

Storey, N., Wakelin, D. and Behnke, J.M. (1985), The genetic control of host responses to *Dipetalonema viteae* (Filarioidea) infections in mice, *Parasite Immunology*, **7**, 349–58.

Strober, S. (1984), Natural suppressor (NS) cells, neonatal tolerance and total lymphoid irradiation, *Annual Reviews in Immunology*, **2**, 219–37.

Subrahmanyam, D., Mehta, K., Nelson, D.S., Rao, Y.V.B.G. and Rao, C.K. (1978), Immune reactions in human filariasis, *Journal of Clinical Microbiology*, **8**, 228–32.

Takayanagi, T., Takayanagi, M., Yabu, Y. and Kato, H. (1978), *Trypanosoma gambiense*: immune responses of neonatal rats receiving antibodies from the female, *Experimental Parasitology*, **44**, 82–91.

Tosato, G. (1987), The Epstein-Barr virus and the immune system, *Advances in Cancer Research*, **49**, 75–125.

Turner, J.H., Kates, K.C. and Wilson, G.I. (1962), The interaction of concurrent infections of the abomasal nematodes *Haemonchus contortus*, *Ostertagia circumcincta* and *Trichostrongylus axei* (Trichostrongylidae) in lambs, *Proceedings of the Helminthological Society of Washington*, **29**, 210–6.

Turner, M.J. (1984), Antigenic variation in parasites, *Parasitology*, **88**, 613–21.

Urquhart, G.M., Jarrett, W.F.H., Jenning, F.W. and McIntyre, W.W. (1966), Immunity to *Haemonchus contortus* infection: failure of X-irradiated larvae to immunize young lambs, *American Journal of Veterinary Research*, **27**, 1641–3.

Vetter, J.C.M. and Klaver-Wesseling, J.C. (1978), IgG antibody binding to the outer surface of infective larvae of *Ancylostoma caninum*, *Zeitschrift für Parasitenkunde*, **58**, 91–6.

Vickerman, K. and Barry, J.D (1982), African trypanosomiasis, in Cohen, S. and Warren, K.S. (Eds.), *Immunology of Parasitic Infections*, 2nd edition, pp. 204–60, Oxford: Blackwell Scientific Publications.

Viens, P., Targett, G.A.T., Wilson, V.C.L.C. and Edwards, C.I. (1972), The persistence of *Trypanosoma* (*Herpetosoma*) *musculi* in the kidneys of immune CBA mice, *Transactions of the Royal Society of Tropical Medicine and Hygiene*, **66**, 669–70.

Voller, A. and Rossan, R.N. (1969), Immunological studies with simian malaria. I. Antigenic variants of *Plasmodium cynomolgi bastianelli*, *Transactions of the Royal Society of Tropical Medicine and Hygiene*, **63**, 46–56.

Wade, A.W. and Szewczuk, M.R. (1984), Aging, idiotype repertoire shifts and compartmentalization of the mucosal-associated lymphoid system, *Advances in Immunology*, **36**, 143–88.

Wakelin, D. (1975), Genetic control of immune response to parasites: selection for responsiveness and non-responsiveness to *Trichuris muris* in random-bred mice, *Parasitology*, **71**, 377–84.

Wassom, D.L., Guess, V.M. and Grundmann, A.W. (1973), Host resistance in a natural host-parasite system. Resistance to *Hymenolepis citelli* by *Peromyscus maniculatus*, *Journal of Parasitology*, **59**, 117–21.

Wedderburn, N. (1974), Immunodepression produced by malarial infection in mice, in *Parasites in the Immunized Host: Mechanisms of Survival, Ciba Foundation Symposium*, **25**, 123–59.

Weiss, N. (1970), Parasitologische und immunbiologische Untersuchungen uber die durch *Dipetalonema viteae* erzeugte Nagetierfilariose, *Act Tropica*, **27**, 219–59.

Weiss, D. (1978), Studies on *Dipetalonema viteae* (Filarioidea), I Microfilaraemia in hamsters in relation to worm burden and humoral immune response, *Acta Tropica*, **35**, 137–50.

Wilkins, H.A., Goll, P.H., Marshall, T.F. and Moore, P.J. (1984), Dynamics of *Schistosoma haematobium* infection in a Gambian community. I. The pattern of human infection in the study area, *Transactions of the Royal Society of Tropical Medicine and Hygiene*, **78**, 216–21.

Wilson, R.A. and Barnes, P.E. (1977), The formation and turnover of the membranocalyx on the tegument of *Schistosoma mansoni*, *Parasitology*, **74**, 61–71.

Wing, E.J., Magee, D.M. and Barczynski, L.K. (1988), Acute starvation in mice reduces the number of T cells and suppresses the development of T-cell mediated immunity, *Immunology*, **63**, 677–82.

Wittenberger, J.F. (1981), *Animal Social Behaviour*, Boston: Duxbury Press.

14. Vaccination against parasites

F.E.G. Cox

14.1. INTRODUCTION

Vaccination or immunization against infections predates chemotherapy as a method for the control of infectious diseases. By the 5th century B.C., in Greece, it was well known that recovery from infection with smallpox or plague rendered individuals resistant to reinfection and that this resistance was specific. Nobody knows when the practice of immunization began, but over 2500 years ago the Chinese were known to inhale a powder made from smallpox scabs to protect them from this disease and by the 17th century, and probably earlier, the injection of material from active lesions as a prophylactic measure against smallpox was widespread and well documented in the Middle East. A similar kind of immunization against cutaneous leishmaniasis dates from the same period, and variations of this procedure persist to the present day. This empirical approach became scientifically acceptable with Jenner's famous experiments in which he used cowpox, vaccinia, material to protect individuals against smallpox; ever since, the word vaccination has been used to describe immunization procedures.

It was Pasteur, at the end of the 19th century, who established the basis of modern vaccination when he immunized chickens against chicken cholera using an attenuated strain and later applied the same technique to anthrax in sheep and rabies in humans. From such beginnings stemmed the development of the modern vaccines, based on attenuated bacteria and viruses, that have eradicated smallpox and reduced the prevalence of poliomyelitis and tuberculosis to insignificant levels in industrialized countries. Today over 50% of the world's children are being immunized against these diseases and also diphtheria, whooping cough, measles and tetanus. Vaccines against most of the major infectious diseases of man and domesticated animals are now available, and this apparent success has stimulated immunologists to develop similar vaccines against parasites. However the majority of available vaccines are not totally satisfactory in terms of affordability, safety or efficacy, and as parasitic diseases have proved to be particularly refactory to immunization, the search is now on for new generations of vaccines that do not have these disadvantages.

14.2. TYPES OF VACCINES

For convenience, vaccines can be grouped into five categories:

1. Live wild type
2. Attenuated
3. Inactivated/killed
4. Subunit
   (a) Extract or metabolic product
   (b) Synthetic
   (c) Recombinant
5. Anti-idiotype

The earliest known vaccines were of the live wild type, e.g. smallpox, and these were followed by attenuated vaccines, e.g. tuberculosis. Live vaccines are very effective but potentially dangerous in that they can revert to virulent forms or cause infections in immunocompromized or genetically susceptible individuals and are also difficult to manufacture and deliver. Inactivated or killed vaccines, e.g. whooping cough (Pertussis), overcome these difficulties but are less effective and involve the administration of extraneous antigens which could cause serious side effects. This can be avoided by the use of subunit vaccines, eg. diphtheria and tetanus toxoids, in which only the putatively protective antigens are used. Until recently, it was difficult to produce satisfactory subunit antigens, but with the techniques of modern molecular biology it became possible to identify potentially protective antigens with great precision.

The bulk of antigens identified in this way are peptides, and it is relatively easy to select the actual peptides required, usually exposed portions, and to synthesize molecules containing the whole sequence or relevant parts of it (Figure 14.1). A number of synthetic peptide vaccines have been produced in this way including foot and mouth disease, poliomyelitis, hepatitis B, influenza and Epstein Barr viruses, bacterial toxins and an experimental malaria vaccine (Steward and Howard, 1987). The advantages of synthetic vaccines are their purity and ease of preparation and the ability to select sequences that induce known immunological functions such as protective T-cell activation while avoiding adverse reactions such as suppressor T-cell stimulation.

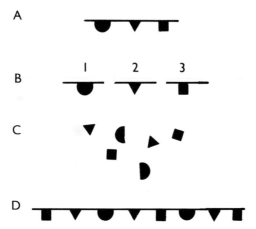

Figure 14.1. Preparation of a synthetic vaccine. The native antigen is identified (A) and its amino acid sequence determined (B). This information is then used as a guide for the assembly of amino acids (C) to produce a polypeptide with the same sequence as the native protein (D). The synthetic polypeptide can then be used as a vaccine. The techniques involved are standard biochemical ones.

An alternative method of producing molecules of immunological importance is the use of gene cloning technology. This is done by isolating the relevant mRNA, copying it to complementary DNA and inserting this, in a recombinant plasmid, into *Escherichia coli* or other bacterial, yeast or mammalian cells which then synthesize the appropriate peptide (Figure 14.2). Vaccines produced in this way are known as recombinant vaccines and include foot and mouth disease, hepatitis B, cholera toxin and an experimental malaria vaccine (Zanetti, *et al.*, 1987).

Synthetic or recombinant molecules can be used in place of more conventional vaccines but suffer from the same disadvantage that non-living preparations do not produce such good immunity as living ones. This problem can be overcome by inserting the DNA for the relevant peptide into a carrier such as the vaccinia virus which is then injected and multiplies to produce both native and the required vaccine antigens. The vaccinia virus is particularly useful as it can accommodate a large amount of foreign DNA. So far this system has been used experimentally for the Epstein-Barr, hepatitis B, influenza and rabies viruses (Zanetti *et al.*, 1987).

An additional advantage of the vaccinia carrier is that several different foreign genes can be inserted simultaneously, for example, genes coding for molecules such as the interleukins that control the immune response. In the field of parasitology, it should not be too difficult to insert genes coding for several antigens as is likely to be required for a malaria vaccine. One disadvantage of the vaccinia virus is that nobody is yet sure what would happen if given to people with antibodies to the virus. However this need

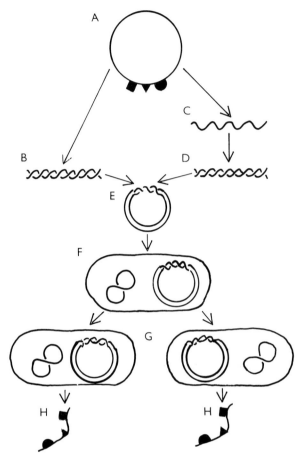

Figure 14.2. Preparation of a recombinant vaccine. Starting with the parasite (A), appropriate DNA segments (B) are extracted. Alternatively, messenger RNA (C) can be used as a template for the synthesis of double-stranded complementary DNA (D). In either case the DNA is usually inserted into a plasmid (E) which is then introduced into suitable cells (F) which are cloned (G) and the recombinant molecules produced (H) used as a vaccine. The techniques involved are standard molecular ones.

not be a problem because other carriers, such as *Salmonella*, are being developed.

The most recent kind of vaccine is based on the fact that when an antibody recognizes an antigen it forms an internal mirror image of the antigen (Kennedy *et al.*, 1986). An antibody against the first antibody would then form an internal image of the first internal image and thus mimic the original antigen (Figure 14.3). This kind of antigen is still at the developmental stage but has already been used for experimental studies on several viruses including hepatitis B and rabies, the bacteria *Streptococcus pnenumoniae* and *Listeria*

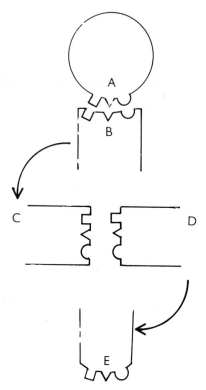

Figure 14.3. Preparation of an anti-idiotype vaccine. When an antigen (A) is recognized by an antibody it forms a mirror image of the original antigen (B). This antibody can then be used to elicit a second immune response in a different animal (C), and as the antibody elicited (D) is a mirror image of the first antibody it mimics the original antigen and can be used to replace it as a vaccine (E).

monocytogenes and the parasites *Trypanosoma b. rhodesiense, T. cruzi* and *Schistosoma mansoni* (Zanetti *et al.*, 1987).

At present there are very few anti-parasite vaccines available commercially. These are, in chickens, anti-coccidial vaccines based on mixtures of infective oocysts; in cattle, anti-babesial vaccines based on attenuated strains, anti-theilerial vaccines based on infection plus chemotherapy and anti-*Dictyocaulus* vaccines based on irradiation-attenuated worms. In man, the only vaccine currently in use is an anti-leishmanial one involving the use of living cultured parasites. All these vaccines are of the attenuated type and depend on the partial development of an infection and all the problems and potential dangers associated with such procedures.

14.3. PROBLEMS OF DEVELOPING ANTI-PARASITE VACCINES

Vaccines against viruses and bacteria depend on the fact that these micro-organisms are relatively simple and present a limited repertoire of antigens to the host. For example, the human immunodeficiency virus that causes AIDS has only two kinds of glycoprotein molecule in its coat. All parasites carry a myriad of surface proteins, glycoproteins and glycolipids (Chapter 2). In addition, parasites have complex life cycles and each stage may occupy a new site in the body and exhibit different antigens. Thus an infection with a parasite presents the host with a moving target to which the host must respond with a sequence of immunological counterattacks. Even this complexity is an oversimplification because parasites have also evolved sophisticated ways of evading the immune response thus compounding the problems faced by the host (See Chapter 13).

Immunity to parasitic infections is slow to develop and is usually incomplete, thus infections characteristically last for months or years. During this time the ineffective or misdirected response may cause major immunopathological changes. It is against this background that the development of vaccines must take place, and nobody believes that this will be a simple task.

14.4. ANIMAL MODELS

Animal models abound in parasitology. At one extreme the actual parasite of human or veterinary importance, such as *Trypanosoma brucei*, can be maintained in a convenient laboratory host, and at the other a parasite from a wild animal, such as *Plasmodium berghei*, maintained in a laboratory animal, can be regarded as an analogue of an important human pathogen. It is essential to bear in mind that such models can only provide a minimal amount of information about what happens in real life. Similarly, the requirements of a vaccine for human use are very demanding and are seldom fulfilled in laboratory experiments where the main aim is usually to obtain some protection, no matter how minimal, without particular regard to the potential requirements of the desired vaccine or any adverse reactions on the part of the host. Nevertheless there is virtually no parasitic infection in which some kind of protection has not been achieved in laboratory animals. Such findings are extremely useful if interpreted with caution but can give rise to false optimism if extrapolations from animals to man are too unrealistic.

Despite this animal experiments have led to a number of potential successes, for example the human malaria anti-sporozoite vaccine was based on initial experiments with *Plasmodium berghei* in mice, then with *P. knowlesi* in monkeys, the human parasite *P. falciparum* in *Aotus* monkeys and eventually this parasite in man. On the other hand the vaccine against leishmaniasis in man never passed through these stages and would probably have been rejected had it done so, and the vaccines against coccidiosis, *Dictyocaulus viviparus*, *Ancylostoma caninum*, babesiosis and theileriosis were developed directly in their natural animal hosts.

14.5. VACCINES AGAINST PARASITES

There are relatively few vaccines against parasites currently available or being developed and these will be considered under two headings, human and veterinary.

Vaccines against human parasites

Protozoa

LEISHMANIASIS

Vaccination against Old World cutaneous leishmaniasis, caused by *Leishmania tropica* and *L. major*, has been practised for centuries. These infections are usually self limiting and vaccination is usually carried out to avoid disfiguring or inconvenient lesions (Section 8.2). In its crudest form, the buttocks of infants are exposed to the bites of sandflies, but alternative methods include the inoculation of material from lesions or, since 1910, the injection of promastigote forms from cultures. The net result is not really protection but the induction of a controlled lesion which, after it has healed, confers long-term immunity.

There have been a number of clinical trials, mainly in the USSR, Israel and Iran, involving thousands of people and reported success exceeds 70% (Greenblatt, 1988). However such vaccines are difficult to produce and deliver and often have severe side effects including disseminated or non-healing vaccine-induced lesions, allergic reactions and psoriasis so are unlikely to be widely acceptable. In Brazil there has been some success in vaccinating people against New World cutaneous leishmaniasis using killed *L. mexicana* promastigotes (Antunes *et al.*, 1986; Mayrink *et al.*, 1986). Based on this success, the WHO is planning Phase I trials using crude killed *L. major* parasites. The most promising subunit vaccine for the future is a surface glycoprotein, designated gp63, present in most *Leishmania* species, which has been effective in laboratory studies (Bordier, 1987, Sections 2.2 and 7.3). The gene for gp63 is now being cloned and expressed in a variety of recombinant vectors.

MALARIA

The development of a vaccine against malaria is regarded as one of the major immunological challenges. This has proved to be extremely difficult, partly because so little is understood about immunity to malaria and partly because the different stages in the parasite life cycle—sporozoite, exoerythrocytic stage, erythrocytic stage and gametocytes—possess different repertoires of antigen making it virtually impossible to consider a vaccine based on a single simple antigen (Figure 14.4 and Section 2.2). At present there are several candidate antigens. The sporozoite that initiates the infection is an obvious target and fortunately possesses a single dominant antigen, the circumsporozoite (CS) protein. The genes for the CS proteins of *P. falciparum* have been cloned and the amino acid sequences determined, and it has been found that the immunogenicity resides in a short repeated sequence, asparagine-alanine-asparagine-proline (NANP).

Two vaccines based on this segment of the CS protein have been tested in volunteers. A synthetic vaccine, (NANP)₃ conjugated to tetanus toxoid, was

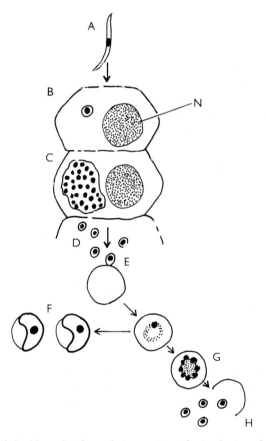

Figure 14.4. Part of the life cycle of a malaria parasite. Infection begins when the sporozoite (A) injected by a mosquito enters a liver cell (B) where multiplication occurs (C) resulting in the production of numerous merozoites (D) which invade red blood cells (E). Some develop into sexual stages or gametocytes (F) while others undergo further multiplication (G) to produce more merozoites (H) which invade further red blood cells. Possible targets for vaccine-induced attack are the sporozoites, liver stages, blood stages, merozoites and gametocytes. N is the host cell nucleus.

used to immunize volunteers who developed antibodies, and out of three challenged one was completely protected while two experienced slightly delayed infections (Herrington *et al.*, 1987). In another trial, a recombinant vaccine consisting of 32 repeats of four amino acids (NANP) produced similar results (Ballou *et al.*, 1987). These results indicated that an anti-sporozoite vaccine might be possible but that it is still a long way away.

Other candidate antigens for vaccines include a number of blood-stage proteins, many of which have elicited protective responses in laboratory animals. An antigen associated with ring-infected red blood cells (ring-infected erythrocyte surface antigen, RESA) has been identified and characterized (Section 2.2) and repetitive sequences used to immunize *Aotus* monkeys against *P. falciparum* with promising but not completely successful results (Collins *et al.*, 1986). However, a mixture of three synthetic peptides based on *P. falciparum* blood-stage antigens has been more successful in protecting *Aotus* monkeys, and two synthetic hybrid molecules based on these peptides have been tested in humans with encouraging results protecting one out of nine volunteers and partially protecting four others (Patarroyo *et al.*, 1988).

Other candidate antigens include those associated with the sexual stages that elicit development-inhibiting antibodies if immune serum is taken up together with gametocytes by mosquitoes (Section 2.2). Such antigens successfully block transmission experimentally but could not be used ethically on their own. A final class of antigens, those associated with the exoerythrocytic stages in the liver, is just beginning to receive attention, and the CS protein expressed in *Salmonella typhimurium* and given orally has been shown to induce an antibody-independent, local immune response effective against the early liver stages of *P. berghei* in rats (Sadoff *et al.*, 1988). This may have major implications for the future development of vaccines for human use.

Nobody now believes that a malaria vaccine will be based on a simple peptide as are many viral vaccines but will probably have to contain a cocktail of antigens against various stages in the life cycle together with epitopes that stimulate T cells plus immunomodulators that enhance the production of cytokines.

CHAGAS' DISEASE

There is no vaccine against Chagas' disease, caused by *Trypanosoma cruzi*, although, because of a lack of effective and safe drugs and difficulties in early diagnosis, there is a need for one. *Trypanosoma cruzi* can be maintained in a number of laboratory animals including mice and marmosets in which some, but only partial, protection has been achieved using surface antigens. Possible subunit vaccines include an 85 kDa surface glycoprotein and a synthetic peptide, arginine-glycine-asparagine-serine (RGNS), both of which are involved in the invasion of host cells but neither is completely protective. A major problem in the development of a vaccine against Chagas' disease is however, the real possibility of inducing auto-immune reactions to antigens shared between parasites and host cells (Hudson, 1985), and it has been argued that a conventional vaccine against this disease is unlikely (Brener, 1986).

SLEEPING SICKNESS

Sleeping sickness, caused by *Trypanosoma brucei rhodesiense* and *T.b. gambiense*, is characterized by waves of parasitaemia as the parasites continually switch their dominant surface glycoprotein coats (Section 2.2.) giving rise to a limitless series of antigenic variants which thus successfully evade otherwise effective immune responses (Section 13.3). The only hope for a vaccine would be one containing all the variant antigen types present in a particular locality and, although there is some evidence that *T.b. gambiense* has a limited repertoire, there is no possibility of such a vaccine for human use at present.

TOXOPLASMOSIS

Toxoplasma gondii is one of the most common and widespread human protozoan parasites. Recovery from, and immunity to, infection is the rule, but severe damage can be done to an unborn foetus or an immunocompromized host. Immunity involving vaccination with whole organisms or subunits is relatively easy to achieve in experimental animals. There is currently no progress towards a vaccine for human use largely because it is not at all clear if or how it should be used.

INTESTINAL PROTOZOA

The two most important intestinal protozoa are *Entamoeba histolytica*, causing amoebiasis, and *Giardia lamblia*, causing giardiasis. Some protection against both can be achieved using crude vaccines in laboratory animals, but because both infections can be easily drug cured there is no urgent need for the development of any vaccines against these parasites in man.

Helminths

SCHISTOSOMIASIS

Attempts to develop vaccines against human schistosomiasis caused by *Schistosoma mansoni*, *S. haematobium* and *S. japonicum* have been hindered by arguments as to which of the several laboratory models, mainly mice, rats or baboons, is most appropriate (See Capron and Capron, 1986) and whether, in fact, humans acquire any immunity at all. Current opinion now favours the concept of acquired immunity in man (Butterworth, 1987; Butterworth and Hagan, 1987). The acquisition of immunity is, however, not a simple process. Early in an infection, adult worms establish themselves and lay eggs. The host responds to egg antigens by producing antibody which unfortunately also binds to cross-reacting antigens on the surface of the larval schistosome, or schistosomulum, (Section 13.3). This antibody blocks antibody-dependent, cell-mediated cytotoxic (ADCC) killing of the parasites in the skin (Section

8.4). However, individuals eventually switch from an essentially blocking response to a protective one and become immune. The aim is to develop a vaccine that stimulates the protective response and not the blocking one and to give it to a child before it first experiences an infection.

Immunity to schistosomiasis is complex and involves an attack system consisting of antibodies and cells (Chapter 5) and an equally complex series of countermeasures on the part of the host (Figure 14.5; Chapter 13). The various antigens involved are discussed by Simpson and Smithers (1985), and the arguments for possible immunological intervention in *Acta Tropica*, supplement 12 (1987; see also Section 2.3). There is no immediate prospect of a vaccine although experimental evidence suggests that immunization can induce better immunity than that which occurs naturally. In natural infections adult worms evade the immune response by disguising themselves with host antigens (13.3), but larval stages are susceptible to immune attack and it is against these stages that a vaccine will have to be directed. Experiments with cattle vaccinated with irradiated larvae have provided encouraging results (Taylor *et al.*, 1986) but such a vaccine is not suitable for human use. Progress has been made in the development of non-living vaccines (James and Sher,

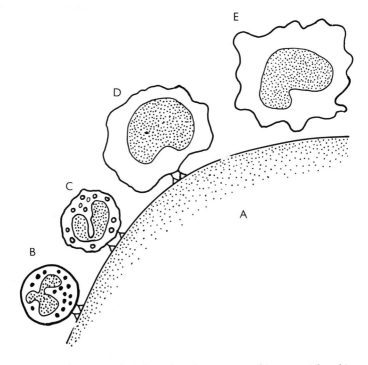

Figure 14.5. Targets for immunological attack against a young schistosome. The schistosomulum (A) has been shown to be attacked by neutrophils (B), eosinophils (C) and macrophages (D) via antibody bridges. Complement-mediated attack also occurs (not shown) as does antibody-independent attack by activated macrophages (E).

1986) and an anti-idiotype vaccine (Grzych *et al.*, 1985), but this is still at an early experimental stage. Nevertheless, there is general optimism about the prospects for an eventual vaccine against human schistosomiasis although what form it will take is still unclear.

FILARIASIS

There is very little real information available about immunity to filariasis in man (WHO, 1984), and most of what is known is derived from studies on a range of animal models. The evidence available suggests that protective immunity is directed against infective third-stage larvae. Irradiated larvae do induce immunity in birds, dogs, cats and monkeys and, although unacceptable as human vaccines, this success has stimulated a great deal of effort directed at the identification, characterization, cloning and expression of particular surface antigens (Philipp *et al.*, 1988). Any vaccine will have to be very carefully designed in order to produce a protective immune response and not to induce immunopathological changes; there is no immediate prospect of such a vaccine.

INTESTINAL NEMATODES

The most important nematodes for which a vaccine might be required are the hookworms, *Ancylostoma duodenale* and *Necator americanus*. It is debatable whether or not humans acquire immunity to these nematodes, but it would appear that antibody in the intestine has a slow adverse effect on adult worms leaving younger parasites unscathed (Behnke, 1987). Evidence from experimental models suggest that resistance can be induced following exposure to third-stage larvae, and a considerable amount of effort is now being devoted to characterizing the possible antigens involved (Almond and Parkhouse, 1985; Hotez *et al.*, 1987; Pritchard *et al.*, 1988). No vaccine is currently being considered.

Vaccines against parasites of veterinary importance

Protozoa

BABESIOSIS

Babesiosis embraces a number of tick-transmitted infections mainly in dogs (*Babesia canis*), horses (*B. equi*) and cattle (*B. bovis*, *B. bigemina* and *B. divergens*). An attenuated vaccine against *B. bovis* has been in use for nearly a century in Australia. In its present form, the parasite is blood-passaged in calves until it loses its virulence. The actual vaccine consists of parasites freed from red blood cells which initiate low-level infections in cattle. Fortunately the attenuation process also inhibits development in the tick so the vaccine, although crude, is effective (Purnell, 1980). Similar vaccines are in use in other parts of the world. Because of problems inherent in producing and delivering live vaccines, alternative strategies have been considered, including the use of irradiated infected blood, with limited success (Purnell, 1980). The use of subunit vaccines based on crude parasite antigens and supernatants from cultured blood parasites have been used experimentally, and current research is concerned with the identification, characterization and production of relevant antigens. A vaccine based on metabolic products released into culture, marketed under the name Pirodog, has been used successfully to vaccinate dogs against *B. canis* in France (Moreau, 1986).

THEILERIOSIS

Two species of *Theileria*, *T. parva* which causes east coast fever in Africa, and *T. annulata* causing tropical theileriosis, are major pathogens of cattle. Immunity to theileriosis has early humoral and late cell-mediated components (Figure 14.6), and vaccines are still at the experimental field trial level. The vaccines are live ones. The *T. annulala* vaccine consists of culture-attenuated, infected lymphoid cells and is widely used in Israel, Iran, India and other endemic areas (Hashemi-Fesharki, 1988). Attenuated cell lines are less effective against *T. parva*, and vaccination is achieved by injecting cryopreserved sporozoites followed by treatment with tetracycline, halofuginone or parva-quone (Dolan, 1987). These vaccines are effective but difficult and expensive to produce and deliver, and the possibility of using non-living vaccines or subunits is currently being investigated.

COCCIDIOSIS

Coccidiosis, caused by various species of *Eimeria*, is common in domesticated animals and is a major disease in poultry reared under intensive methods. Vaccination can be achieved by infecting young birds with low numbers of oocysts sufficient to stimulate immunity without causing disease (Rose and Long, 1980; Long and Jeffers, 1986). The use of irradiated oocysts or oocysts plus anti-coccidial drugs are no more effective than controlled infections or killed parasites and, as non-living preparations are ineffective, attention has switched to the possibility of using drug-sensitive, attenuated strains. Research is also being carried out on the identification and characterization of parasite antigens despite the comparative lack of success in immunizing birds with non-living material (Rose, 1985).

TRYPANOSOMIASIS

The most important trypanosomes of veterinary importance are *T.b. brucei*, *T. congolense* and *T. vivax*. There are no immediate prospects of developing vaccines for these because of the antigenic variation exhibited by trypanosomes, but immunity does develop after an infection has been cured and herd immunity can be induced by the selective drug treatment of infected cattle (Holmes, 1980). Although this immunity is specific to local strains of trypanosomes, it does indicate that trypanosomiasis might be susceptible to immunological intervention, and research is now being concentrated on

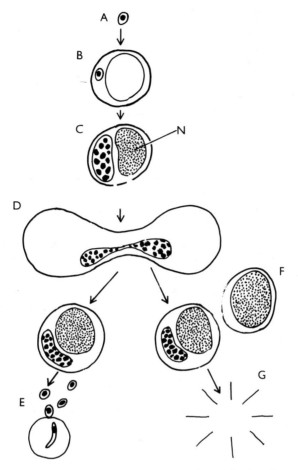

Figure 14.6. Part of the life cycle of *Theileria* spp. Infection begins when a sporozoite (A) injected by a tick enters a lymphocyte (B) where multiplication occurs (C) causing the infected cell to divide (D). Eventually merozoites are produced, and these enter red blood cells (E) to produce sexual stages. Immune attack against the sporozoites is antibody-mediated while cytotoxic T lymphocytes (F) are involved in the killing of parasites in lymphocytes causing infected cells to lyse (G).

alternative methods of turning the immune response in favour of the host (Nantulya, 1986).

Helminths

NEMATODES
The most successful and satisfactory anti-parasite vaccine is undoubtably that against the cattle lungworm *Dictyocaulus viviparus* (Peacock and Poynter, 1980). The vaccine, which was first introduced 30 years ago, is based on

irradiated third-stage larvae which survive until the fourth stage and then die but in the meantime elicit a protective immune response (Chapter 9). This vaccine relies on the normal immune response but prevents pathological damage caused by the adult worms.

The success of the *Dictyocaulus viviparus* vaccine has stimulated comparable studies on other nematodes with some success. An irradiated larva vaccine against *Dictyocaulus filaria*, the sheep lungworm, has been developed and is available commercially in several parts of the world including Eastern Europe and India (Sharma *et al.* 1988). The development of irradiated vaccines against *Trichostrongylus colubriformis* in lambs has been hindered by host genetic differences, and a vaccine against *Haemonchus contortus*, which is effective in sheep, is ineffective in lambs where it is required (Section 13.2). Similar restrictions apply to *Ostertagia ostertagi* in calves. On the other hand, foals can be successfully immunized against *Strongylus vulgaris* with irradiated larvae (Klei, 1986). The success of an irradiated larva vaccine does not necessarily ensure its commercial viability, and an extremely effective vaccine against *Ancylostoma caninum* in dogs was discontinued mainly because of its lack of acceptance by the American veterinary profession (Miller, 1978; Urquhart, 1980).

FLUKES

Cattle can be protected against *Schistosoma bovis* and sheep against *S. mattheei* by immunizing them with radiation-attenuated larvae; in field trials in the Sudan the *S. bovis* vaccine has been very effective in calves (Taylor, 1980). Similar vaccines, however, are ineffective against *Fasciola hepatica* in sheep (Hughes, 1985).

TAPEWORMS

The most important cestode diseases are those caused by larval stages of taeniid tapeworms, hydatidosis and cysticercosis. Immunity has been achieved following immunization of dogs with irradiated larvae of *Echinococcus granulosus*; however, some animals develop serious infections as a result of the vaccine. The metabolic products of larval cestodes grown in culture also protect dogs against *E. granulosus* and induce strong immunity against *Taenia ovis* in sheep and *T. saginata* in cattle (Urquhart, 1980; Rickard and Williams, 1982). Present research is being devoted towards exploiting established successes and the identification and characterization of the antigens involved (Harrison and Parkhouse, 1985). However, there has been a major breakthrough with the development of a recombinant vaccine, based on larval-stage antigens, that protects lambs against *T. ovis* (Johnson *et al.*, 1989), and, as this vaccine is based on sound immunological principles, there is no reason why it should not become commercially available with considerable implications for other diseases caused by larval tapeworms.

14.6. FUTURE VACCINES AND THEIR EVALUATION

This summary of what advances have been made in developing vaccines against parasites can be viewed from two completely different standpoints, depression or optimism. It is true that very little has actually been achieved, but this must be seen in the context of a much wider realization that the majority of currently available, anti-microbial vaccines are not completely acceptable and that new generations of vaccines will have to be developed. There are now three possible approaches, the development of synthetic vaccines, recombinant vaccines and internal image or anti-idiotype vaccines. All hold out the promise of producing better immunity than that which occurs naturally, and no doubt there will be further development as carbohydrate and lipid antigens become better understood.

The success of a vaccine does not just depend on its ability to prevent infection, and commercial development parallels that of chemotherapy where the majority of drugs fall by the wayside for reasons other than efficacy. Vaccines for human use have to fulfil certain WHO guidelines and pass through five phases.

Phase 0	Assessment of all the experimental evidence in order to justify a trial in humans.
Phase I	Trials in healthy volunteers to establish the degree of immune responsiveness and that the vaccine can be tolerated.
Phase IIA	Trials to establish that protective immunity to experimental challenge does occur.
Phase IIB	Trials to establish efficacy against natural challenge.
Phase III	Trials in selected areas where the infection occurs naturally.
Phase IV	Large-scale evaluation of the vaccine under natural conditions.

Each of these phases involves the development and use of several immunological and epidemiological techniques so the whole process can take several years. The general principles outlined above can be illustrated by reference to malaria vaccines (Perlmann, 1986), and it is obvious that the time span before there is any vaccine against human parasites is going to be very long; for most parasitic diseases there is much basic work to be done before even the first trials begin.

14.7. VACCINATION IN ERADICATION AND CONTROL

The possibility of developing a successful vaccine should not be allowed to overshadow considerations of the real need for that vaccine and the uses to which it is to be put. Vaccines can be used in three ways: for eradication or containment of a disease or for the selective treatment of individuals at particular risk. Although these overlap, they are quite distinct, and each has particular requirements.

Eradication of major diseases is the ultimate aim spurred on by the success of the smallpox vaccine. Ideally a vaccine should be cheap, stable, given orally at birth and capable of protecting 100% of those vaccinated for the rest of their lives. Obviously such a vaccine is unlikely, but progress in the right direction can be made. The actual design of the vaccine must be the responsibility of immunologists, but epidemiological considerations must be taken into account in deciding if eradication is to be attempted and, if so, who to vaccinate and when. The theory of vaccination has been elaborated by Anderson (1982) and Anderson and May (1982, 1985). It is not possible to go into the mathematical considerations here but essentially they revolve around the basic reproductive rate, R, which is a measure of the number of other individuals infected from one infected individual, and the age at which exposure to infection usually occurs. If $R=1$ the infection in a population will remain stable; if $R>1$, it will increase; and if $R<1$ it will decline.

In eradication programmes, the aim is to achieve an $R<1$ situation. Vaccination can only be successful if carried out before the average age of infection (A). Thus if R is high and A is low, eradication by vaccination would be difficult to achieve. To give some ideas of what these figures mean, for mumps, rubella and polio $R=4-7$ and for measles and whooping cough $R=13-17$. Children in industrialized countries become infected with measles at four to six years and for eradication over 90% need to be vaccinated before this time. In industrialized countries the percentage of children vaccinated is about 70% and in Africa about 25%. This points to the difficulties in achieving mass vaccination against parasitic diseases where R tends to be high and A low (see also Chapter 12).

Containment should not be regarded as a second-best alternative to eradication and should be considered wherever practicable to prevent epidemics or the spread of infection into susceptible populations.

The use of vaccines for eradication or containment has many limitations. Firstly, there is the problem of vaccinating children early enough, and the belief that very young children should not be vaccinated. However, although there is some risk with live vaccines, BCG and oral polio vaccines are now given routinely to newborn infants. Measles vaccines cannot be given at birth because of interference by maternally acquired antibodies, a problem that is almost certain to occur in parasitic infections. Vaccines are seldom 100% effective and repeated vaccinations are likely to be necessary, the number of booster vaccinations depending on the duration of the efficacy of a particular vaccine. There are additional problems here in that certain individuals are genetically low-responders in whom even the most effective vaccines will not take (see Chapter 6), and the immunodepression that accompanies many infectious diseases (Section 13.3) frequently renders vaccination ineffective.

Selective vaccination is a promising area because individuals at particular risk, such as travellers, medical and service personnel, can be carefully selected and monitored and cost and length of efficacy are not major considerations.

The use of vaccines for eradication, control and selective immunization present separate problems, but these are no different for parasitologists than for microbiologists; the lessons learned in one field can easily be transferred to others. A comprehensive account of vaccination against tropical diseases, including parasites, is given in Liew (1989).

14.8. CONCLUSIONS

Vaccination against parasitic diseases has always been a dream of parasitologists but has been brought sharply into focus by the realization that other control methods are being hindered by a lack of drugs, resistance to those that are in use and widespread vector resistance to pesticides. This, fortunately, has coincided with an explosion of activity in basic immunology and molecular biology and the possibilities of novel and better vaccines. The development of vaccines against parasites is no longer constrained by considerations of a bewildering array of antigens because those of protective importance can now be selected and used in synthetic or recombinant forms, and delivery of complex antigens is now possible using vectors such as the vaccina virus. Developments in the production of vaccines against viruses and bacteria are well under way, and vaccines against parasites are likely to follow. The major problems are whether or not these vaccines can be commercially produced economically and, if so, how effective they will be in eradicating parasitic diseases. Epidemiological, logistic, financial, political and ethical problems will all have to be overcome before anti-parasitic vaccines are widely available. In the meantime safer and more efficient drugs will also become available, and the future roles of vaccination and chemotherapy are as yet uncertain.

REFERENCES

Acta Tropica (1987), Prospects for immunological intervention in human schistosomiasis, *Acta Tropica*, **44, Supplement 12**, 1–117.

Almond, N.M. and Parkhouse, R.M.E. (1985), Nematode antigens, in Parkhouse, R.M.E. (Ed.), *Parasite Antigens in Protection, Diagnosis and Escape. Current Topics in Microbiology and Immunology*, **Vol. 120**, pp. 173–203. Berlin: Springer Verlag.

Anderson, R.M. (1982), Epidemiology, in Cox, F.E.G. (Ed.), *Modern Parasitology*, pp. 204–51, Oxford: Blackwell Scientific Publications.

Anderson, R.M. and May, R.M. (1982), Directly transmitted infectious diseases: Control by vaccination. *Science*, **215**, 1053–60.

Anderson, R.M. and May, R.M. (1985), Helminth infections of humans: Mathematical models, population dynamics, and control, *Advances in Parasitology*, **24**, 1–101.

Antunes, C.M.F., Mayrink, W. and Magalhaes, P.A. (1986), Controlled field trials of a vaccine against New World cutaneous leishmaniasis, *International Journal of Epidemiology*, **15**, 572–80.

Ballou, W.R., Hoffman, S.L., Sherwood, J.A., Hollingdale, M.R., Neva, F.A., Hockmeyer, W.T., Gordon, D.M., Schneider, I., Wirtz, R.A., Young, J.F., Wasserman, G.F., Reeve, P., Diggs, C.L. and Chulay, J.D. (1987), Safety and efficacy of a recombinant DNA *Plasmodium falciparum* sporozoite vaccine, *Lancet*, 1, 1277–81.

Behnke, J.M. (1987), Do hookworms elicit protective immunity in man? *Parasitology Today*, 3, 200–6.

Bordier, C. (1987), The promastigote surface protease of *Leishmania*, *Parasitology Today*, 3, 151–3.

Brener, Z. (1986), Why vaccines do not work in Chagas' disease, *Parasitology Today*, 2, 196–7.

Butterworth, A.E. (1987), Immunity in human schistosomiasis, *Acta Tropica*, 44 Supplement 12, 31–40.

Butterworth, A.E. and Hagan, P. (1987), Immunity in human schistosomiasis, *Parasitology Today*, 3, 11–16.

Capron, M. and Capron, A. (1986), Rats, mice and men—models for immune effector mechanisms against schistosomiasis, *Parasitology Today*, 2, 69–75.

Collins, W.E., Anders, R.F. Pappaioanou, M., Campbell, G.H., Brown, G.V., Kemp, D.J., Coppel, R.L., Skinner, J.C., Andrysiak, P.M., Favaloro, J.M., Corcoran, L.M., Broderson, J.R., Mitchell, G.F. and Campbell, C.C. (1986), Immunization of *Aotus* monkeys with recombinant proteins of a surface antigen of *Plasmodium falciparum*, *Nature*, 323, 259–62.

Dolan, T.T. (1987), Control of East Coast Fever, *Parasitology Today*, 3, 4–6.

Greenblatt, C.L. (1988), Cutaneous leishmaniasis: the prospects for a killed vaccine, *Parasitology Today*, 4, 53–4.

Grzych, J.M., Capron, M., Lambert, P.H., Dissous, C., Torres, S. and Capron, A. (1985), An anti-idiotype vaccine against experimental schistosomiasis, *Nature*, 316, 74–6.

Harrison, L.J.S. and Parkhouse, R.M.E. (1985), Antigens of taeniid cestodes in protection, diagnosis and escape. in Parkhouse, R.M.E. (Ed.), *Parasite Antigens in Protection Diagnosis and Escape. Current Topics in Microbiology and Immunology*, Vol. 120, pp. 159–72. Berlin: Springer-Verlag.

Hashemi-Fesharki, R. (1988), Control of *Theileria annulata* in Iran, *Parasitology Today*, 4, 36–40.

Herrington, D.A., Clyde, D.F., Losonsky, G., Cortesia, M., Murphy, J.R., Davis, J., Baqard, S., Felix, A.M., Heimer, E.P., Gillessen, D., Nardin, E., Nussenzweig, R.S., Nussenzweig, V., Hollingdale, M.R. and Levine, M.M. (1987), Safety and Immunogenicity in man of a synthetic peptide malaria vaccine against *Plasmodium falciparum* sporozoites, *Nature*, 328, 257–9.

Holmes, P.H. (1980), Vaccination against trypanosomes, in Taylor, A.E.R. and Muller, R. (Eds.), *Vaccines against Parasites*, pp. 75–105, Oxford: Blackwell Scientific Publications.

Hotez, P.J., Le Trang, N. and Cerami, A. (1987), Hookworm antigens: the potential for vaccination, *Parasitology Today*, 3, 247–9.

Hudson, L. (1985), Autoimmune phenomena in chronic chagasic cardiopathy, *Parasitology Today*, 1, 6–7.

Hughes, H.P.A. (1985), Toxoplasmosis: the need for improved diagnostic techniques and accurate risk assessment, in Parkhouse, R.M.E. (Ed.), *Parasite Antigens in Protection, Diagnosis and Escape*, Vol. 120, pp. 105–39, Berlin Springer-Verlag.

James, S.L. and Sher, A. (1986), Prospects for a nonliving vaccine against schistosomiasis, *Parasitology Today*, 2, 134–7.

Johnson, K.S., Harrison, G.B.L., Lightowlers, M.W., O'Hoy, K.L., Cougle, W.G., Dempster, R.P., Lawrence, S.B., Vinton, J.G., Heath, D.D. and Rickard, M.D.

(1989), Vaccination against ovine cysticerosis using a defined recombinant antigen, *Nature*, **338**, 585–7.

Kennedy, R.G., Melnick, J.L. and Dreesman, G.R. (1986), Anti-idiotypes and immunity, *Scientific American*, **255**, 40–8.

Klei, T.R. (1986), Development of vaccines against equine helminths, *Parasitology Today*, **2**, 80–2.

Liew, F.Y. (Ed.), (1989), *Vaccination Strategies for Tropical Disease*, Boca Raton: CRC Press Inc.

Long, P.L. and Jeffers, T.K. (1986), Control of chicken coccidiosis, *Parasitology Today*, **2**, 236–40.

Mayrink, W., Antunes, C.M.F., Costa, C.A. da, Melo, M.N., Dias, M., Michalick, M.S., Magalhaes, P.A., de Oliveira, L.A. and Williams, P. (1986), Further trials of a vaccine against American cutaneous leishmaniasis, *Transactions of the Royal Society of Tropical Medicine and Hygiene*, **80**, 1001.

Miller, T.A. (1978), Industrial development and field use of the canine hookworm vaccine, *Advances in Parasitology*, **16**, 333–42.

Moreau, Y. (1986), Immunologie parasitaire: réalités et perspectives, *Point Veterinaire*, **18**, 467–73.

Nantulya, V.M. (1986), Immunological approaches to the control of animal trypanosomiasis, *Parasitology Today*, **2**, 168–73.

Patarroyo, M.E., Amador, R., Clavijo, P., Moreno, A., Guzman, F., Romero, P., Tascon, R., Franco, A., Murillo, L.A., Ponton, G. and Trujillo, G. (1988), A synthetic vaccine protects humans against challenge with asexual blood stages of *Plasmodium falciparum* malaria, *Nature*, **332**, 158–61.

Peacock, R. and Poynter, D. (1980), Field experience with a bovine lungworm vaccine, in Taylor, A.E.R. and Muller, R. (Eds.), *Vaccines Against Parasites*, pp. 141–8. Oxford: Blackwell Scientific Publications.

Perlmann, P. (1986), Immunogenicity assays for clinical trials of malaria vaccines, *Parasitology Today*, **2**, 127–30.

Philipp, M., Davis, T.B., Storey, N. and Carlow, C.K.S. (1988), Immunity in filariasis: perspectives for vaccine development, *Annual Reviews of Microbiology*, **42**, 685–716.

Pritchard, D.I., McKean, P.G. and Rogan, M.T. (1988), Cuticular collagens-a concealed target for immune attack in hookworms, *Parasitology Today*, **4**, 239–41.

Purnell, R., (1980), Vaccines against piroplasms, in Taylor, A.E.R. and Muller, R. (Eds.), *Vaccines Against Parasites*, pp. 25–55, Oxford: Blackwell Scientific Publications.

Rickard, M.D. and Williams, J.F. (1982), Hydatosis cysticercosis: immune mechanisms and immunization against infection, *Advances in Parasitology*, **21**, 229–96.

Rose, M.E. (1985), The *Eimeria*, in Parkhouse, R.M.E. (Ed.), *Parasite Antigens in Protection, Diagnosis and Escape. Current Topics in Microbiology and Immunology*, **Vol. 120**, pp. 7–17, Berlin: Springer-Verlag.

Rose, M.E. and Long, P.L. (1980), Vaccination against coccidiosis in chickens, in Taylor, A.E.R. and Muller, R. (Eds.), *Vaccines Against Parasites*, pp. 57–74, Oxford: Blackwell Scientific Publications.

Sadoff, J.C., Ballou, W.R., Baron, L.S., Majarian, W.R., Brey, R.N., Hockmeyer, W.T., Young, J.F., Cryz, S.J., Ou, J., Lowell, G.H., and Chulay, J.D. (1988), Oral *Salmonella typhimurium* vaccine expressing circumsporozoite protein protects against malaria, *Science*, **240**, 336–8.

Sharma, R.L., Bhat, T.K. and Dhar, D.N. (1988), Control of sheep lungworm in India, *Parasitology Today*, **4**, 33–6.

Simpson, A.J.G. and Smithers, S.R. (1985), Schistosomes: surface, egg and circulating antigens, in Parkhouse, R.M.E. (Ed.), *Parasite Antigens in Protection, Diagnosis and Escape. Current Topics in Microbiology and Immunology*, **Vol. 120**, pp. 205–39, Berlin: Springer-Verlag.

Steward, M.W. and Howard, C.R. (1987), Synthetic peptides: a next generation of vaccines? *Immunology Today*, **8**, 51–8.

Taylor, M.G. (1980), Vaccination against trematodes, in Taylor, A.E.R. and Muller, R. (Eds.), *Vaccines against Parasites*, pp. 115–40, Oxford: Blackwell Scientific Publications.

Taylor, M.G., Bickle, Q.D., James, S.L. and Sher, A. (1986), Irradiated schistosome vaccines, *Parasitology Today*, **2**, 132–4.

Urquhart, G.M. (1980), Immunity to cestodes, in Taylor, A.E.R. and Muller, R. (Eds.), *Vaccines Against Parasites*, pp. 107–14, Oxford: Blackwell Scientific Publications.

WHO, (1984), Lymphatic Filariasis, *WHO Technical Report Series*, **No. 702**.

Zanetti, M., Sercarz, E. and Salk, J. (1987), The immunology of new generation vaccines, *Immunology Today*, **8**, 18–25.

Index